Interactive Multiobjective Decision Making Under Uncertainty

Interactive Multiobjective Decision Making Under Uncertainty

Hitoshi Yano
Graduate School of Humanities and Social Sciences
Nagoya City University
Nagoya, Japan

CRC Press
Taylor & Francis Group
Boca Raton London New York

CRC Press is an imprint of the
Taylor & Francis Group, an **informa** business
A SCIENCE PUBLISHERS BOOK

CRC Press
Taylor & Francis Group
6000 Broken Sound Parkway NW, Suite 300
Boca Raton, FL 33487-2742

Fisrt issued in hardback 2019

ISBN-13: 978-1-4987-6354-7 (hbk)

Visit the Taylor & Francis Web site at
http://www.taylorandfrancis.com

and the CRC Press Web site at
http://www.crcpress.com

To my families

Preface

In this book, we present multiobjective programming problems involving uncertainty and obtain satisfactory solutions to these problems via a single decision maker or multiple decision makers. Although many books on multiobjective programming problems have been published, very few have been published in a unified way, i.e., with a focus on interactive decision making for multiobjective programming problems involving random variable coefficients, fuzzy coefficients, and/or fuzzy random variable coefficients with a single decision maker or multiple decision makers in a hierarchical decision structure.

Since multiple objective functions are often incommensurable and conflict with one another, we introduce Pareto optimal solutions for multiobjective programming problems instead of optimal solutions for single objective programming problems. In general, Pareto optimal solutions are plural. As a result, the decision maker must find a satisfactory solution according to his or her preference structure. Since it is very difficult for the decision maker to find such a satisfactory solution directly, various interactive methods for multiobjective programming problems have been proposed; in such interactive methods, candidates for satisfactory solutions are iteratively obtained via multiple interactions with the decision maker.

Unfortunately, many kinds of uncertainties are inherented in such multiobjective programming problems. Decision makers often describe their preferences rather ambiguously. Further, experts who formulate decision problems as multiobjective programming problems often express parameters of the objective functions and constraints imprecisely. In addition, some parameters of the objective functions and constraints may not be identified as fixed values, for example water resource allocation problems may have random future inflows. To handle such fuzziness and randomness, multiobjective fuzzy programming approaches and multiobjective stochastic programming approaches have been proposed. Fusing these two types of approaches, multiobjective fuzzy stochastic programming approaches have also been proposed. Recently, using the concept of fuzzy random variables, which are essentially a hybrid concept combining fuzzy sets and random variables, multiobjective fuzzy random programming approaches have been proposed.

Throughout this book, we focus on the latest results with respect to interactive decision making for multiobjective programming problems involving uncertainty; our overarching goal is to obtain satisfactory solutions via a decision maker or multiple decision makers in a hierarchical decision structure. Here, multiobjective programming problems involving uncertainty are usually transformed into multiobjective programming problems via a variety of mathematical techniques. Thus, transformed multiobjective programming problems involve many parameters, almost all of which are subjectively specified by decision makers. In practice, it is rather difficult for decision makers to specify such parameters appropriately. Therefore, in this book, some parameters are automatically set, thus reducing the processing load of preference judgment on the decision maker.

In addition to the introductory section, we have organized this book as follows. In Chapter 2, we formulated multiobjective stochastic programming problems (MOSPs) and propose fuzzy approaches and interactive algorithms corresponding to two types of MOSPs. In Chapter 3, we formulated multiobjective fuzzy random programming problems (MOFRPs) and proposed several interactive algorithms corresponding to two key types of MOFRPs through chance constraint methods. MOFRPs with simple recourse, are also formulated, and we proposed an interactive algorithm for this. Next, in Chapter 4, we formulated hierarchical multiobjective programming problems (HMOPs) involving uncertainty with multiple decision makers in a hierarchical decision structure. Further, we proposed interactive algorithms for hierarchical multiobjective stochastic programming problems (HMOSPs) and hierarchical multiobjective fuzzy random programming problems (HMOFRPs). In Chapter 5, we formulated multiobjective two-person zero-sum games as multiobjective programming problems under the assumption that each player (i.e., decision maker) makes the worst of his or her expected payoffs (i.e., objective functions); we proposed interactive algorithms for this. In Chapter 6, we formulated generalized multiobjective programming problems (GMOPs) and propose interactive algorithms in which fuzzy coefficients, random variable coefficients, and fuzzy random variable coefficients comprise the objective functions. In Chapter 7, we described applications of farm planning and solve corresponding problems by obtaining satisfactory solutions using a hypothetical decision maker that implements the interactive algorithms proposed in the previous chapters. Note that all numerical examples and applications can be easily solved via EXCEL solver and EXCEL VBA Macro except for those presented in Sections 3.4 and 7.3.

Target readers of this book include graduate students and researchers in management science, operations research, industrial engineering, systems science, and computer science.

We express our sincere appreciation to emeritus professor Masatoshi Sakawa of Hiroshima University. We also thank professor Ichiro Nishizaki of Hiroshima University. Finally, we thank Science Publishers and CRC Press for their assistance in publication. This work was supported by JSPS KAKENHI Grant Number JP26350432.

Nagoya City University, Japan Hitoshi Yano

Contents

Chapter 1

Introduction

CONTENTS

1.1 Multiobjective decision making problems (MOPs)

Given the rapid increase in the diversity and complexity of human societies, we often encounter real-world decision making problems in which single or plural decision makers must find a solution from a feasible set of alternatives using their own preferences and evaluation criteria. To handle such real-world decision making problems, a variety of multicriteria decision making problem models have been formulated, corresponding solution concepts defined, and methods to obtain such solutions developed [9, 20, 25, 44, 59, 68, 71, 77, 87, 88, 89, 92, 95, 137].

In general, the modeling process defined from real-world decision making problems to corresponding conclusions via various multicriteria decision making methods requires three steps [64]. The first step is to formulate mathematical models of the multicriteria decision making problems that adequately reflect real-world decision situations. For this step, we classify numerous types of multicriteria decision making problems, because we are often faced with various categories of decision making problems that are different from one another in their mathematical structure. The second step is to solve the mathematical models and obtain solutions in which the preferences of the decision makers are somehow reflected. Corresponding to the types of multicriteria decision making problems, various multicriteria decision making techniques have been developed. The third step is to find a real solution for the decision maker in real-world decision making by using a solution of a mathematical model, obtained via one or more multicriteria decision making techniques.

Roughly speaking, multicriteria decision making problems can be classified into the following two kinds of problems [59].

- Multiattribute decision problems [20, 25, 44, 68, 89, 142]
- Multiobjective programming problems [1, 9, 71, 77, 87, 88, 95]

Multiattribute decision problems are formulated to find the preferred alternative of the decision maker or to rank all possibilities from a set of alternatives based on the decision maker's evaluation criteria (i.e., attributes). Multiattribute decision problems include multiattribute utility theory [20, 44, 89], multiattribute value theory [20], the analytic hierarchy process [68], and outranking methods [25].

Conversely, when we formulate decision problems as mathematical programming problems with decision variables, constraints, and multiple objective functions, we are solving multiobjective programming problems. In such multiobjective programming problems, decision variables are controlled by the decision maker; further, the decision maker identifies the preferred alternative under the given constraints that reflects his or her preferred multiple objective functions. Since the objective functions conflict with one another, the decision maker should consider the balance between multiple objective functions according to his or her preferences.

Throughout this book, we focus on multiobjective programming problems that are formulated as follows.

MOP 1.1

$$\min_{\mathbf{x}\in X} (f_1(\mathbf{x}),\cdots,f_k(\mathbf{x}))$$

where $\mathbf{x}$ is a vector of the decision variables, X is a constraint set, and $f_i(\mathbf{x}), i = 1,\cdots,k$ are multiple objective functions that are incommensurable and conflict with one another. MOP1.1 is a natural extension of a single objective programming problem; however, obtaining an optimal solution for a single objective programming problem cannot be directly applied to MOP1.1, because the multiple objective functions conflict with one another, i.e., an improvement to one objective function can only be achieved at the expense of another. Therefore, instead of obtaining the usual optimal solution, we define a Pareto optimal solution as follows.

Definition 1.1 $\mathbf{x}^* \in X$ is a Pareto optimal solution to MOP1.1 if and only if there does not exist another $\mathbf{x} \in X$ such that $f_i(\mathbf{x}) \leq f_i(\mathbf{x}^*), i = 1, \cdots, k$, with strict inequality holding for at least one i.

One approach to solving MOP1.1 is to find the preference function of the decision maker. If we can identify such a preference function for the objective functions of MOP1.1, then, MOP1.1 is reduced to the following single objective programming problem, where $U(\cdot)$ is a preference function of the decision maker.

MOP 1.2

$$\max_{\mathbf{x} \in X} U(f_1(\mathbf{x}), \cdots, f_k(\mathbf{x}))$$

Unfortunately, since it seems to be very difficult to explicitly identify such a preference function $U(\cdot)$, we cannot directly solve MOP1.2 to obtain the preferred solution of the decision maker. Therefore, the following two types of approaches have been developed to obtain a satisfactory solution of the decision maker for MOP1.1 [71].

- Goal programming approaches [10, 29, 30, 50]
- Interactive programming approaches [9, 22, 45, 69, 80, 107, 141]

In goal programming approaches, it is assumed that the decision maker can set the goals of the objective functions, and a solution that is close to his or her goals in some sense is obtained. Conversely, in interactive programming approaches, candidates of satisfactory solutions are updated through interactions with the decision maker under the assumption that the decision maker cannot provide global information regarding his or her preference for multiple objective functions, but rather give local information iteratively. As a hybrid between these two types of approaches, interactive goal programming approaches have also been proposed [19, 58].

1.2 Multiobjective decision making problems (MOPs) under uncertainty

When formulating real-world decision making problems as multiobjective programming problems, we are often faced with the difficulty of handling uncertainty. For example, an expert may not be able to specify some parameters of the objective functions and constraints as fixed values, because the decision problem is too complicated to identify such parameters, but he or she may be able to estimate such values subjectively. Here, some parameters may be represented as random variables, because such parameters are essentially uncertain and the corresponding probability density functions can be obtained through past information. To handle such uncertainty in the decision making process, the following two types of approaches to multiobjective programming problems involving uncertainty have been investigated [77, 87].

- Fuzzy mathematical programming approaches [48, 66, 67, 70, 71, 81, 139]
- Stochastic programming approaches [14, 42, 43, 72, 73, 93, 97]

Zimmermann [139] first formulated a multiobjective fuzzy linear programming problem to obtain a satisfactory solution based on a fuzzy decision [6, 71, 140] under the assumption that the decision maker has fuzzy goals for the objective functions and such fuzzy goals can be specified by corresponding membership functions. Next, Sakawa et al. [71, 81] proposed interactive fuzzy decision making methods for multiobjective linear, linear fractional, and nonlinear programming problems, all focused on obtaining a satisfactory solution for the decision maker by simultaneously incorporating interactive methods and fuzzy mathematical programming techniques. Sakawa and Yano [82, 83, 84, 85] and Yano and Sakawa [130] also formulated multiobjective linear, linear fractional, and nonlinear programming problems with fuzzy parameters and proposed interactive fuzzy decision making methods in which a new Pareto optimal solution called α-Pareto optimal solution based on the α-level set [18] was introduced.

In contrast, Contini [14] first formulated and solved multiobjective stochastic programming problems in which parameters of the objective functions and constraints were defined as random variables by applying goal programming approaches [29, 30] and stochastic programming techniques [7, 33]. An overview of multiobjective stochastic programming techniques through 1990 was published by Stancu-Minasian [94]. By incorporating fuzzy mathematical approaches into multiobjective stochastic programming problems, Hulsurkar et al. [28] proposed a method to obtain a satisfactory solution based on the fuzzy decision and the product operator [6, 71, 140].

Using four chance constrained programming techniques, i.e., the expectation model (E-model), variance model (V-model), probability model (P-model), and fractile model (F-model), Sakawa et al. [56, 72, 73, 74] formulated fuzzy multiobjective stochastic programming problems, defined the corresponding Pareto optimal solution, and proposed interactive algorithms to obtain satisfactory solutions from a Pareto optimal solution set.

As a hybrid of random variables and fuzzy numbers [18], Kwakernaak [46] first defined the concept of a fuzzy random variable, while Puri and Ralescu [65] investigated the mathematical properties of fuzzy random variables. Recently, using a possibility measure [18], Katagiri et al. [38, 39] proposed two types of interactive methods for multiobjective fuzzy random programming problems using a probability model and a fractile model.

In any of the interactive methods for multiobjective programming problems discussed above, we assume that a single decision maker subjectively finds a satisfactory solution that reflects his or her preferences; however, in real-world decision making situations, a decision making system may consist of many divisions, with each division controlled by separate decision makers. To handle such multilevel programming problems involving multiple decision makers who are noncooperative with one another, Stackelberg games [2, 105] were formulated and corresponding algorithms proposed to obtain what is referred to as Stackelberg solutions. In contrast to Stackelberg games in which decision makers are noncooperative with one another, Lai [47] and Shih et al. [91] formulated multilevel programming problems and proposed interactive methods for obtaining satisfactory solutions from multiple decision makers under the assumption that multiple decision makers cooperate with one another via a hierarchical decision structure. In their approaches, through decision powers [47, 49] specified by multiple decision makers, a hierarchical decision structure between multiple decision makers is reflected in the candidates for a satisfactory solution. To overcome various difficulties in using their methods, Sakawa et al. [57] proposed interactive fuzzy decision making methods for multiobjective two-level programming problems. As a natural extension, Sakawa et al. [77, 78, 79] also proposed interactive fuzzy decision making methods for multiobjective two-level programming problems with fuzzy parameters and random variables.

1.3 Content organization

As discussed above, to formulate multiobjective programming problems involving uncertainty, it is necessary to apply various types of techniques, fuzzy programming approaches, stochastic programming approaches, and/or a possibility measure [18]. The corresponding formulated problems thereby become complicated and involve many parameters specified by a decision maker. As a result, if some parameters specified by a decision maker are not appropriate, the corresponding solution will not reflect the decision maker's preference. In this book, we systematically explain the latest research regarding interactive methods for multiobjective programming problems involving uncertainty. Since almost all the interactive decision making methods discussed in this book are based on a linear or convex programming technique, except for those in Sections 3.4 and 7.3, readers can easily solve the numerical examples in each chapter or applications to farm planning in Chapter 7 by using EXCEL solver and EXCEL VBA Macro (Microsoft Corp.). Throughout each chapter, we introduce the corresponding Pareto optimal solutions to handle multiobjective programming problems involving uncertainty; further, we develop interactive algorithms to obtain satisfactory solutions from a Pareto optimal solution set.

In Chapter 2, we focus on multiobjective stochastic programming problems (MOSPs) in which random variable coefficients exist in the objective functions and constraints. We also describe four methods applicable to MOSPs through probability maximization and fractile optimization models. In our proposed methods, by using the fuzzy decision [6, 71, 140], we automatically compute such parameters as permissible objective levels and permissible probability levels. This idea was first proposed by Inuiguchi and Sakawa [31] for single objective stochastic programming problems. As a result, it is possible for decision makers to reduce their processing loads during their decision making processes. In Section 2.1, we propose a fuzzy approach for MOSPs with a special structure [111] based on the fuzzy decision. Similarly, in Section 2.2, we propose a fuzzy approach for MOSPs with variance covariance matrices [121] based on the fuzzy decision. In Section 2.3, we develop an interactive linear programming algorithm for MOSPs with a special structure [117] for obtaining satisfactory solutions. In Section 2.4, we present an interactive convex programming algorithm for MOSPs with variance covariance matrices [116] for obtaining satisfactory solutions.

In Chapter 3, we focus on multiobjective fuzzy random programming problems (MOFRPs) in which fuzzy random variable coefficients exist in the objective functions or constraints [123, 124, 125]. In Section 3.1, using a possibility measure [18], we formulate MOFRPs with a special structure through probability maximization and fractile optimization models [131, 132, 133, 134]. We propose an interactive linear programming algorithm for obtaining satisfactory solutions. In the proposed methods, permissible possibility and probability levels are automatically computed through the fuzzy decision. Similarly, in Section 3.2, using a possibility measure [18], we formulate MOFRPs with variance covariance matrices through probability maximization and fractile optimization models [114, 119, 120]. Next, we propose an interactive convex programming algorithm for obtaining satisfactory solutions. In Section 3.3, we formulate MOFRPs through expectation and variance models [127, 128]. Here, we propose an interactive convex programming algorithm for obtaining satisfactory solutions. In Section 3.4, we formulate MOFRPs with simple recourse [135]; we develop an interactive method for MOFRPs with simple recourse for obtaining satisfactory solutions.

In Chapter 4, we focus on hierarchical multiobjective programming problems (HMOPs) involving uncertainty in which we have multiple decision makers in a hierarchical decision structure [109]. In HMOPs, multiple decision makers must find a satisfactory solution that reflects not only their own preference structure for their own objective functions but also their overall hierarchical decision structure. In Sections 4.1 and 4.2, we formulate hierarchical multiobjective stochastic programming problems (HMOSPs) through a probability maximization model and a fractile optimization model, respectively [110, 112]. Here, we also develop interactive algorithms for obtaining satisfactory solutions of multiple decision makers. In Section 4.3, we develop an interactive decision making method for

HMOSPs [113, 115, 118] based on the fuzzy decision in which appropriate values of permissible objective levels or permissible probability levels are automatically computed. In Section 4.4, we formulate hierarchical multiobjective fuzzy random programming problems (HMOFRPs) [126] based on the fuzzy decision and develop an interactive algorithm for obtaining satisfactory solutions of multiple decision makers in which a hierarchical decision structure between multiple decision makers is reflected through their decision powers [47, 49, 91].

In Chapter 5, we discuss multiobjective two-person zero-sum games [129]. Essentially, in such games, rather than a Pareto optimal solution, we define an equilibrium solution; however, given the assumption that each player (i.e., decision maker) makes the worst of his or her expected payoffs (i.e., objective functions), multiobjective two-person zero-sum games are transformed into multiobjective programming problems with the introduction of a pessimistic Pareto optimal solution for each player. In Section 5.1, we formulate two-person zero-sum games with vector payoffs as multiobjective programming problems and develop an interactive algorithm for obtaining a pessimistic compromise solution from a pessimistic Pareto optimal solution set. Similarly, in Section 5.2, we formulate two-person zero-sum games with vector fuzzy payoffs as multiobjective programming problems using a possibility measure [18] and develop an interactive algorithm for obtaining a pessimistic compromise solution from a pessimistic Pareto optimal solution set.

In Chapter 6, we focus on generalized multiobjective programming problems (GMOPs) in which several uncertainty parameters, including fuzzy coefficients, random variable coefficients, and fuzzy random variable coefficients, exist in the objective functions. In Section 6.1, we formulate GMOPs with a special structure [122] and develop an interactive algorithm based on a linear programming technique to obtain satisfactory solutions from a generalized Pareto optimal solution set. Similarly, in Section 6.2, we formulate GMOPs with variance covariance matrices and develop an interactive algorithm based on convex programming techniques to obtain satisfactory solutions from a generalized Pareto optimal solution set.

Finally, in Chapter 7, we formulate applications in farm planning and solve such problems to show the efficiency of the interactive algorithms that we propose in the previous chapters using the hypothetical decision maker. In Section 7.1, a farmer must decide the planting ratio for ten crops to maximize his or her total income, maximize total leisure hours for one year, and maximize minimum leisure hours for one period. To handle the uncertainty of the profit coefficients of the ten crops, we formulate the corresponding farm planning problem as a GMOP and apply Algorithm 6.3 to obtain a satisfactory solution from a generalized Pareto optimal solution set. In Section 7.2, a farmer must decide the planting ratio for seven crops to maximize his or her total income for one year and minimize total work hours for one year. Similar to the previous section, to handle the uncertainty of the profit coefficients of the seven crops, the corresponding farm planning problem is formulated as a GMOP and Algorithm 6.3 is applied to a second crop planning problem to obtain a satisfactory solution from a generalized Pareto optimal solution set. In Section 7.3, we formulate a farm planning problem with simple recourse [135] as a two objective fuzzy random simple recourse programming problem in which the parameter representing the water availability constraint is defined as a fuzzy random variable. We apply Algorithm 3.7 to this farm planning problem with simple recourse to obtain a satisfactory solution from a γ-Pareto optimal solution set. In Section 7.4, we formulate a vegetable shipment problem [129] as a two objective two-person zero-sum game in which Player 1 is a farmer, Player 2 is Nature, and two payoff matrices consist of the price coefficients of two kinds of vegetables. A farmer (i.e., Player 1) must decide his or her mixed strategy corresponding to each month and each vegetable. We apply Algorithm 5.1 to the two objective two-person zero-sum game to obtain a pessimistic compromise solution from a pessimistic Pareto optimal solution set.

Chapter 2

Multiobjective Stochastic Programming Problems (MOSPs)

CONTENTS

In this chapter, we focus on multiobjective stochastic programming problems (MOSPs) in which random variable coefficients exist in the objective functions and constraints. To properly handle MOSPs, the following two types of approaches have been developed:

- Chance constraint approaches [11, 13, 41]
- Two-stage programming approaches [3, 16, 101, 106]

In chance constraint approaches, four models were proposed (expectation model, variance model, probability maximization model, and fractile optimization model). In the probability maximization and fractile optimization models, the decision maker is asked to specify permissible levels in advance and he or she can control the probabilities or the objective function values under the constraints such that the specified permissible levels are satisfied. Conversely, in two-stage programming approaches, the expected values of the difference between the left-hand and right-hand-sides in the equality constraints with random variable coefficients are expressed as penalties against the objective functions. Such a penalty is called a recourse. As a special case of two-stage programming problems, simple recourse programming problems have been investigated and applied to many real-world decision making problems, including water resource programming problems [96, 104].

In the sections that follow, we adopt probability maximization and fractile optimization models to handle MOSPs. Note that a significant feature of our proposed methods is that appropriate values of permissible objective levels or permissible probability levels are automatically computed. This idea was first proposed by Inuiguchi and Sakawa [31] for single objective stochastic programming problems. In Section 2.1, we formulate MOSPs with a special structure [111], obtaining a satisfactory solution based on the fuzzy decision [6, 71, 140] by using linear programming techniques. In Section 2.2, we formulate MOSPs with variance covariance matrices [121], obtaining a satisfactory solution based on the fuzzy decision by using convex programming techniques. In Section 2.3, we formulate MOSPs with a special structure [117] and propose interactive linear programming algorithms to obtain a satisfactory solution from an MP-Pareto optimal solution set or an MF-Pareto optimal solution set. Similarly, in Section 2.4, we formulate MOSPs with variance covariance matrices [116] and propose interactive convex programming algorithms to obtain a satisfactory solution from an MP-Pareto optimal solution set or an MF-Pareto optimal solution set.

2.1 Fuzzy approaches for MOSPs with a special structure

In this section, we formulate MOSPs with a special structure and propose fuzzy approaches based on the fuzzy decision [6, 71, 140] to obtain a satisfactory solution under the assumption that each objective function involves only one random variable coefficient [111]. In Section 2.1.1, we formulate two multiobjective programming problems by using probability maximization and fractile optimization models, then define the corresponding Pareto optimal solutions, i.e., P-Pareto and F-Pareto optimal solutions, respectively. We investigate the relationships between them, and show that a P-Pareto optimal solution set is equivalent to an F-Pareto optimal solution set. In Section 2.1.2, with the assumption that the decision maker has fuzzy goals not only for permissible objective levels but also for corresponding distribution functions of the probability maximization model, we formulate the maxmin problem to obtain a satisfactory solution based on the fuzzy decision. Likewise, with the assumption that the decision maker has fuzzy goals not only for permissible probability levels but also for corresponding objective functions of the fractile optimization model, we formulate the maxmin problem to obtain a satisfactory solution based on the fuzzy decision. Here, the two maxmin problems can be easily solved using linear programming techniques. In Section 2.1.3, we apply the proposed method in Section 2.1.2 to a numerical example and compare the results to a probability maximization model to show the efficiency of our proposed method.

2.1.1 Formulations using probability and fractile models

In this subsection, we focus on multiobjective stochastic programming problems (MOSPs) formulated as follows.

MOP 2.1

$$\min \bar{\mathbf{z}}(\mathbf{x}) = (\bar{z}_1(\mathbf{x}), \cdots, \bar{z}_k(\mathbf{x})) \tag{2.1}$$

subject to

$$A\mathbf{x} \leq \bar{\mathbf{b}},\ \mathbf{x} \geq \mathbf{0} \tag{2.2}$$

where $\mathbf{x} = (x_1, x_2, \cdots, x_n)^T$ is an n-dimensional decision variable column vector,

$$\bar{z}_i(\mathbf{x}) = \bar{\mathbf{c}}_i\mathbf{x} + \bar{\alpha}_i, i = 1, \cdots, k, \tag{2.3}$$

are multiobjective functions involving random variable coefficients, $\bar{\mathbf{c}}_i$ is a n-dimensional random variable row vector expressed by

$$\bar{\mathbf{c}}_i = \mathbf{c}_i^1 + \bar{t}_i\mathbf{c}_i^2, \tag{2.4}$$

where $\bar{t}_i$ is a random variable, and $\bar{\alpha}_i$ is a random variable row vector expressed by

$$\bar{\alpha}_i = \alpha_i^1 + \bar{t}_i\alpha_i^2. \tag{2.5}$$

Further, A is an $(m \times n)$-dimensional coefficient matrix, $\bar{\mathbf{b}} = (\bar{b}_1, \cdots, \bar{b}_m)^T$ is an m-dimensional column vector, whose elements $\bar{b}_j, j = 1, \cdots, m$ are random variables independent of one another. We denote distribution functions of random variables are random variables $\bar{t}_i$ and $\bar{b}_j$ as $T_i(\cdot)$ and $F_j(\cdot)$, respectively, which are strictly increasing and continuous.

Since MOP2.1 contains random variable coefficients in the objective functions and the right-hand-side of the constraints, mathematical programming techniques cannot be directly applied. Instead, to handle MOP2.1, we interpret constraints (2.2) as chance constrained conditions [11], meaning that the j-th constraint of (2.2) must be satisfied with a certain probability β_j (the constraint probability level), i.e.,

$$X(\beta) = \{\mathbf{x} \geq \mathbf{0} \mid \Pr(\omega \mid \mathbf{a}_j\mathbf{x} \leq b_j(\omega)) \geq \beta_j, j = 1, \cdots, m\}, \tag{2.6}$$

where $\mathbf{a}_j$ is the j-th row vector of A and $\beta = (\beta_1, \cdots, \beta_m)$.

Given chance constrained conditions (2.6), if we adopt a probability maximization model for objective functions $\bar{z}_i(\mathbf{x}), i = 1, \cdots, k$ in MOP2.1, we can substitute the minimization of the objective functions for the maximization of the probability that each objective function $\bar{z}_i(\mathbf{x})$ is less than or equal to a certain permissible level $\hat{f}_i$ called a permissible objective level. Such a probability function can be defined as

$$p_i(\mathbf{x}, \hat{f}_i) \stackrel{\text{def}}{=} \Pr(\omega \mid z_i(\mathbf{x}, \omega) \leq \hat{f}_i), i = 1, \cdots, k, \tag{2.7}$$

where

$$\hat{\mathbf{f}} = (\hat{f}_1, \cdots, \hat{f}_k) \tag{2.8}$$

is a k-dimensional vector of permissible objective levels specified subjectively by the decision maker. Then, MOP2.1 can be converted into the following form, called a probability maximization model.

MOP 2.2 $(\hat{\mathbf{f}}, \beta)$

$$\max_{\mathbf{x} \in X(\beta)} (p_1(\mathbf{x}, \hat{f}_1), \cdots, p_k(\mathbf{x}, \hat{f}_k)) \tag{2.9}$$

In MOP2.2$(\hat{\mathbf{f}}, \beta)$, assuming that $\mathbf{c}_i^2\mathbf{x} + \alpha_i^2 > 0, i = 1, \cdots, k$ for any $\mathbf{x} \in X(\beta)$, we can rewrite the objective function $p_i(\mathbf{x}, \hat{f}_i)$ as

$$\begin{aligned} p_i(\mathbf{x}, \hat{f}_i) &= \Pr(\omega \mid z_i(\mathbf{x}, \omega) \leq \hat{f}_i) \\ &= \Pr(\omega \mid \mathbf{c}_i(\omega)\mathbf{x} + \alpha_i(\omega) \leq \hat{f}_i) \\ &= \Pr\left(\omega \mid t_i(\omega) \leq \frac{\hat{f}_i - (\mathbf{c}_i^1\mathbf{x} + \alpha_i^1)}{\mathbf{c}_i^2\mathbf{x} + \alpha_i^2}\right) \\ &= T_i\left(\frac{\hat{f}_i - (\mathbf{c}_i^1\mathbf{x} + \alpha_i^1)}{\mathbf{c}_i^2\mathbf{x} + \alpha_i^2}\right). \end{aligned} \tag{2.10}$$

To handle MOP2.2$(\hat{\mathbf{f}}, \beta)$, we introduce a P-Pareto optimal solution.

Definition 2.1 $\mathbf{x}^* \in X(\beta)$ is a P-Pareto optimal solution to MOP2.2$(\hat{\mathbf{f}},\beta)$ if and only if there does not exist another $\mathbf{x} \in X(\beta)$ such that $p_i(\mathbf{x},\hat{f}_i) \geq p_i(\mathbf{x}^*,\hat{f}_i)$, $i = 1,\cdots,k$, with strict inequality holding for at least one i.

If we instead adopt a fractile optimization model for the objective functions in MOP2.2$(\hat{\mathbf{f}},\beta)$ under chance constrained conditions (2.6), we can convert MOP2.2$(\hat{\mathbf{f}},\beta)$ into the multiobjective programming problem shown below, where

$$\hat{\mathbf{p}} = (\hat{p}_1,\cdots,\hat{p}_k) \tag{2.11}$$

is a k-dimensional vector of permissible probability levels specified subjectively by the decision maker.

MOP 2.3 $(\hat{\mathbf{p}},\beta)$

$$\min_{\mathbf{x}\in X(\beta), f_i\in R^1, i=1,\cdots,k} (f_1,\cdots,f_k) \tag{2.12}$$

subject to

$$p_i(\mathbf{x},f_i) \geq \hat{p}_i, i = 1,\cdots,k \tag{2.13}$$

In MOP2.3$(\hat{\mathbf{p}},\beta)$, since a distribution function $T_i(\cdot)$ is continuous and strictly increasing and continuous, constraints (2.13) can be transformed into the following form.

$$\begin{aligned} \hat{p}_i &\leq p_i(\mathbf{x},f_i) = T_i\left(\frac{f_i - (\mathbf{c}_i^1\mathbf{x} + \alpha_i^1)}{\mathbf{c}_i^2\mathbf{x} + \alpha_i^2}\right) \\ \Leftrightarrow f_i &\geq T_i^{-1}(\hat{p}_i)\cdot(\mathbf{c}_i^2\mathbf{x} + \alpha_i^2) + (\mathbf{c}_i^1\mathbf{x} + \alpha_i^1). \end{aligned} \tag{2.14}$$

Let us define the right-hand-side of inequality (2.14) as

$$f_i(\mathbf{x},\hat{p}_i) \stackrel{\text{def}}{=} T_i^{-1}(\hat{p}_i)\cdot(\mathbf{c}_i^2\mathbf{x} + \alpha_i^2) + (\mathbf{c}_i^1\mathbf{x} + \alpha_i^1). \tag{2.15}$$

Then, MOP2.3$(\hat{\mathbf{p}},\beta)$ can be equivalently reduced to the following simple form, where $f_i, i = 1,\cdots,k$ have disappeared.

MOP 2.4 $(\hat{\mathbf{p}},\beta)$

$$\min_{\mathbf{x}\in X(\beta)} (f_1(\mathbf{x},\hat{p}_1),\cdots,f_k(\mathbf{x},\hat{p}_k)). \tag{2.16}$$

To handle MOP2.4$(\hat{\mathbf{p}},\beta)$, we introduce an F-Pareto optimal solution as follows.

Definition 2.2 $\mathbf{x}^* \in X(\beta)$ is an F-Pareto optimal solution to MOP2.4$(\hat{\mathbf{p}},\beta)$ if and only if there does not exist another $\mathbf{x} \in X(\beta)$ such that $f_i(\mathbf{x},\hat{p}_i) \leq f_i(\mathbf{x}^*,\hat{p}_i)$, $i = 1,\cdots,k$, with strict inequality holding for at least one i.

The relationship between a P-Pareto optimal solution to MOP2.2($\hat{\mathbf{f}}, \beta$) and an F-Pareto optimal solution to MOP2.4($\hat{\mathbf{p}}, \beta$) can be characterized by the following theorem.

Theorem 2.1
(1) If $\mathbf{x}^* \in X(\beta)$ is a P-Pareto optimal solution to MOP2.2($\hat{\mathbf{f}}, \beta$) for some given permissible objective levels $\hat{\mathbf{f}}$, then, there exist some permissible probability levels $\hat{\mathbf{p}}$ such that $\mathbf{x}^* \in X(\beta)$ is an F-Pareto optimal solution to MOP2.4($\hat{\mathbf{p}}, \beta$).

(2) If $\mathbf{x}^* \in X(\beta)$ is an F-Pareto optimal solution to MOP2.4($\hat{\mathbf{p}}, \beta$) for some given permissible probability levels $\hat{\mathbf{p}}$, then, there exist some permissible objective levels $\hat{\mathbf{f}}$ such that $\mathbf{x}^* \in X(\beta)$ is a P-Pareto optimal solution to MOP2.2($\hat{\mathbf{f}}, \beta$).

(Proof)
(1) Assume that $\mathbf{x}^* \in X(\beta)$ is not an F-Pareto optimal solution to MOP2.4($\hat{\mathbf{p}}, \beta$) for any permissible probability levels $\hat{\mathbf{p}}$. Then, from Definition 2.1, there exists $\mathbf{x} \in X(\beta)$ such that

$$\begin{aligned} & f_i(\mathbf{x}, \hat{p}_i) \leq f_i(\mathbf{x}^*, \hat{p}_i) \\ \Leftrightarrow\ & T_i^{-1}(\hat{p}_i) \cdot (\mathbf{c}_i^2\mathbf{x} + \alpha_i^2) + (\mathbf{c}_i^1\mathbf{x} + \alpha_i^1) \\ & \leq T_i^{-1}(\hat{p}_i) \cdot (\mathbf{c}_i^2\mathbf{x}^* + \alpha_i^2) + (\mathbf{c}_i^1\mathbf{x}^* + \alpha_i^1), i = 1, \cdots, k, \end{aligned} \tag{2.17}$$

with strict inequality holding for at least one i. If we define $\hat{f}_i$ as

$$\hat{f}_i \stackrel{\text{def}}{=} T_i^{-1}(\hat{p}_i) \cdot (\mathbf{c}_i^2\mathbf{x}^* + \alpha_i^2) + (\mathbf{c}_i^1\mathbf{x}^* + \alpha_i^1), i = 1, \cdots, k, \tag{2.18}$$

then, because of $\mathbf{c}_i^2\mathbf{x} + \alpha_i^2 > 0$ and the property of $T_i(\cdot)$, inequalities (2.17) can be equivalently transformed into

$$\hat{p}_i \leq T_i\left(\frac{\hat{f}_i - (\mathbf{c}_i^1\mathbf{x} + \alpha_i^1)}{\mathbf{c}_i^2\mathbf{x} + \alpha_i^2}\right). \tag{2.19}$$

On the other hand, from (2.18), it holds that

$$\hat{p}_i = T_i\left(\frac{\hat{f}_i - (\mathbf{c}_i^1\mathbf{x}^* + \alpha_i^1)}{\mathbf{c}_i^2\mathbf{x}^* + \alpha_i^2}\right). \tag{2.20}$$

As a result, inequalities (2.17) become

$$\begin{aligned} & T_i\left(\frac{\hat{f}_i - (\mathbf{c}_i^1\mathbf{x} + \alpha_i^1)}{\mathbf{c}_i^2\mathbf{x} + \alpha_i^2}\right) \geq T_i\left(\frac{\hat{f}_i - (\mathbf{c}_i^1\mathbf{x}^* + \alpha_i^1)}{\mathbf{c}_i^2\mathbf{x}^* + \alpha_i^2}\right) \\ \Leftrightarrow\ & p_i(\mathbf{x}, \hat{f}_i) \geq p_i(\mathbf{x}^*, \hat{f}_i), i = 1, \cdots, k. \end{aligned} \tag{2.21}$$

This means that $\mathbf{x}^* \in X(\beta)$ is not a P-Pareto optimal solution to MOP2.2($\hat{\mathbf{f}}, \beta$) for any permissible objective levels $\hat{f}_i, i = 1, \cdots, k$ defined by (2.18).

(2) Assume that $\mathbf{x}^* \in X(\beta)$ is not a P-Pareto optimal solution to MOP2.2($\hat{\mathbf{p}}, \beta$) for any permissible probability levels $\hat{\mathbf{p}}$. Then, from Definition 2.2, there exists $\mathbf{x} \in X(\beta)$ such that

$$\begin{aligned} & p_i(\mathbf{x}, \hat{f}_i) \geq p_i(\mathbf{x}^*, \hat{f}_i) \\ \Leftrightarrow \quad & T_i\left(\frac{\hat{f}_i - (\mathbf{c}_i^1\mathbf{x} + \alpha_i^1)}{\mathbf{c}_i^2\mathbf{x} + \alpha_i^2}\right) \geq T_i\left(\frac{\hat{f}_i - (\mathbf{c}_i^1\mathbf{x}^* + \alpha_i^1)}{\mathbf{c}_i^2\mathbf{x}^* + \alpha_i^2}\right), i = 1, \cdots, k, \end{aligned} \tag{2.22}$$

with strict inequality holding for at least one i. If we define $\hat{p}_i$ as

$$\hat{p}_i \stackrel{\text{def}}{=} T_i\left(\frac{\hat{f}_i - (\mathbf{c}_i^1\mathbf{x}^* + \alpha_i^1)}{\mathbf{c}_i^2\mathbf{x}^* + \alpha_i^2}\right), i = 1, \cdots, k, \tag{2.23}$$

then, because of $\mathbf{c}_i^2\mathbf{x} + \alpha_i^2 > 0$ and the property of $T_i(\cdot)$, inequalities (2.22) can be equivalently transformed into

$$\begin{aligned} & T_i\left(\frac{\hat{f}_i - (\mathbf{c}_i^1\mathbf{x} + \alpha_i^1)}{\mathbf{c}_i^2\mathbf{x} + \alpha_i^2}\right) \geq \hat{p}_i \\ \Leftrightarrow \quad & \hat{f}_i - (\mathbf{c}_i^1\mathbf{x} + \alpha_i^1) \geq T_i^{-1}(\hat{p}_i) \cdot (\mathbf{c}_i^2\mathbf{x} + \alpha_i^2) \\ \Leftrightarrow \quad & \hat{f}_i \geq f_i(\mathbf{x}, \hat{p}_i), i = 1, \cdots, k. \end{aligned} \tag{2.24}$$

On the other hand, from (2.23), it holds that

$$\begin{aligned} \hat{f}_i &= T_i^{-1}(\hat{p}_i) \cdot (\mathbf{c}_i^2\mathbf{x}^* + \alpha_i^2) + (\mathbf{c}_i^1\mathbf{x}^* + \alpha_i^1) \\ &= f_i(\mathbf{x}^*, \hat{p}_i). \end{aligned} \tag{2.25}$$

As a result, inequalities (2.22) become

$$f_i(\mathbf{x}, \hat{p}_i) \leq f_i(\mathbf{x}^*, \hat{p}_i), i = 1, \cdots, k. \tag{2.26}$$

This means that $\mathbf{x}^* \in X(\beta)$ is not an F-Pareto optimal solution to MOP2.4($\hat{\mathbf{p}}, \beta$) for any permissible probability levels $\hat{p}_i, i = 1, \cdots, k$ defined by (2.23). □

Sakawa et al. [73] formulated MOSPs through a probability maximization model. In their approach, assuming that the decision maker has a fuzzy goal for each distribution function $p_i(\mathbf{x}, \hat{f}_i)$, a satisfactory solution is derived from an M-Pareto optimal solution set by updating reference membership levels; however, in the probability maximization model, permissible objective levels $\hat{f}_i, i = 1, \cdots, k$ for original objective functions $\bar{z}_i(\mathbf{x}), i = 1, \cdots, k$ must be specified in advance by the decision maker. The fewer the values for permissible objective levels $\hat{f}_i, i = 1, \cdots, k$ in MOP2.2($\hat{\mathbf{f}}, \beta$), the fewer the values for corresponding distribution functions $p_i(\mathbf{x}, \hat{f}_i), i = 1, \cdots, k$. As a result, if the decision maker specifies fewer values for permissible objective levels $\hat{f}_i, i = 1, \cdots, k$, he or she may not be able to accept any satisfactory solution candidate, because the corresponding values of the distribution functions $p_i(\mathbf{x}, \hat{f}_i), i = 1, \cdots, k$ will approach zero. Even if we adopt a fractile optimization model for MOSPs, similar problems arise.

2.1.2 Fuzzy approaches based on linear programming

In this subsection, we propose two fuzzy approaches for MOSPs considering both permissible objective levels in a probability maximization model and permissible probability levels in a fractile optimization model.

First, we consider multiobjective stochastic programming problems MOP2.2($\hat{\mathbf{f}}, \beta$) using a probability maximization model. To handle MOP2.2($\hat{\mathbf{f}}, \beta$), the decision maker must specify his or her permissible objective levels $\hat{\mathbf{f}}$ in advance. Considering the imprecise nature of the decision maker's judgment, it is natural to assume that the decision maker has fuzzy goals not only for each objective function in MOP2.2($\hat{\mathbf{f}}, \beta$) but also for permissible objective levels $\hat{\mathbf{f}}$. We assume that such fuzzy goals can be quantified by eliciting the corresponding membership functions. Here, we denote the membership function of $p_i(\mathbf{x}, \hat{f}_i)$ as $\mu_{p_i}(p_i(\mathbf{x}, \hat{f}_i))$, and the membership function of permissible objective level $\hat{f}_i$ as $\mu_{\hat{f}_i}(\hat{f}_i)$ respectively. Throughout this subsection, we make the following assumptions for membership functions $\mu_{\hat{f}_i}(\hat{f}_i)$, $\mu_{p_i}(p_i(\mathbf{x}, \hat{f}_i))$, $i = 1, \cdots, k$.

Assumption 2.1
$\mu_{\hat{f}_i}(\hat{f}_i), i = 1, \cdots, k$ are strictly decreasing and continuous with respect to $\hat{f}_i \in [\hat{f}_{i\min}, \hat{f}_{i\max}] \in R^1$, where $f_{i\min}$ and $f_{i\max}$ are a sufficiently satisfactory maximum value and an acceptable maximum value, respectively, with $\mu_{\hat{f}_i}(\hat{f}_{i\min}) = 1, \mu_{\hat{f}_i}(\hat{f}_{i\max}) = 0, i = 1, \cdots, k$.

Assumption 2.2
$\mu_{p_i}(p_i(\mathbf{x}, \hat{f}_i)), i = 1, \cdots, k$ are strictly increasing and continuous with respect to $p_i(\mathbf{x}, \hat{f}_i) \in [p_{i\min}, p_{i\max}] \in R^1$, where $p_{i\min}(> 0)$ and $p_{i\max}(< 1)$ are an acceptable maximum value and a sufficiently satisfactory minimum value, respectively, with $\mu_{p_i}(p_{i\min}) = 0, \mu_{p_i}(p_{i\max}) = 1, i = 1, \cdots, k$.

For example, we can define membership functions $\mu_{\hat{f}_i}(\hat{f}_i)$, $\mu_{p_i}(p_i(\mathbf{x}, \hat{f}_i)), i = 1, \cdots, k$ as follows. After the decision maker specifies a sufficiently satisfactory maximum value $f_{i\min}$ and an acceptable maximum value $f_{i\max}$, he or she defines membership functions $\mu_{\hat{f}_i}(\hat{f}_i), i = 1, \cdots, k$ on interval $[f_{i\min}, f_{i\max}]$ subjectively according to Assumption 2.1. Corresponding to interval $[f_{i\min}, f_{i\max}]$, $p_{i\max}$ can be obtained by solving the following problem.

$$p_{i\max} = \max_{\mathbf{x} \in X(\beta)} p_i(\mathbf{x}, \hat{f}_{i\max}), i = 1, \cdots, k \tag{2.27}$$

Note that the above problem is equivalent to the following linear fractional programming problem [12] because distribution functions $T_i(\cdot), i = 1, \cdots, k$ are strictly increasing and continuous.

$$\max_{\mathbf{x} \in X(\beta)} \left(\frac{f_{i\max} - (\mathbf{c}_i^1 \mathbf{x} + \alpha_i^1)}{\mathbf{c}_i^2 \mathbf{x} + \alpha_i^2} \right), i = 1, \cdots, k$$

Conversely, to obtain $p_{i\min}$, we first solve

$$\max_{\mathbf{x}\in X(\beta)} p_i(\mathbf{x}, \hat{f}_{i\min}), i = 1, \cdots, k, \tag{2.28}$$

and denote the corresponding optimal solution as $\mathbf{x}_i, i = 1, \cdots, k$. Using optimal solutions $\mathbf{x}_i, i = 1, \cdots, k$, $p_{i\min}$ can be obtained as [139]

$$p_{i\min} = \min_{\ell=1,\cdots,k,\ell\neq i} p_i(\mathbf{x}_\ell, \hat{f}_{i\min}), i = 1, \cdots, k. \tag{2.29}$$

According to Assumption 2.2, the decision maker can define his or her membership function $\mu_{p_i}(p_i(\mathbf{x}, \hat{f}_i))$ on interval $[p_{i\min}, p_{i\max}]$.

Then, if the decision maker adopts the fuzzy decision [6, 71, 140] as an aggregation operator, a satisfactory solution is obtained by solving the following maxmin problem.

MAXMIN 2.1

$$\max_{\mathbf{x}\in X(\beta), \hat{f}_i\in F_i, i=1,\cdots,k, \lambda\in[0,1]} \lambda \tag{2.30}$$

subject to

$$\mu_{p_i}(p_i(\mathbf{x}, \hat{f}_i)) \geq \lambda, i = 1, \cdots, k \tag{2.31}$$

$$\mu_{\hat{f}_i}(\hat{f}_i) \geq \lambda, i = 1, \cdots, k \tag{2.32}$$

where

$$F_i \stackrel{\text{def}}{=} [f_{i\min}, f_{i\max}], i = 1, \cdots, k.$$

The relationships between optimal solution $\mathbf{x}^* \in X, \hat{f}_i^* \in F_i, i = 1, \cdots, k, \lambda^* \in [0, 1]$ of MAXMIN2.1, P-Pareto optimal solutions to MOP2.2$(\hat{\mathbf{f}}, \beta)$, and F-Pareto optimal solutions to MOP2.4$(\hat{\mathbf{p}}, \beta)$ can be characterized by the following theorem.

Theorem 2.2
Assume that $\mathbf{x}^* \in X, \hat{f}_i^* \in F_i, i = 1, \cdots, k, \lambda^* \in [0, 1]$ is a unique optimal solution of MAXMIN2.1. Then,

(1) $\mathbf{x}^* \in X(\beta)$ is a P-Pareto optimal solution to MOP2.2$(\hat{\mathbf{f}}^*, \beta)$, where $\hat{\mathbf{f}}^* \stackrel{\text{def}}{=} (f_1^*, \cdots, f_k^*)$

(2) $\mathbf{x}^* \in X(\beta)$ is an F-Pareto optimal solution to MOP2.4$(\hat{\mathbf{p}}^*, \beta)$, where $\hat{p}_i^* \stackrel{\text{def}}{=} p_i(\mathbf{x}^*, \hat{f}_i^*), i = 1, \cdots, k$, $\hat{\mathbf{p}}^* \stackrel{\text{def}}{=} (\hat{p}_1^*, \cdots, \hat{p}_k^*)$.

(Proof)
(1) If $\mathbf{x}^* \in X(\beta)$ is not a P-Pareto optimal solution to MOP2.2($\hat{\mathbf{f}}^*, \beta$), because of Assumption 2.2, there exists some $\mathbf{x} \in X(\beta)$ such that

$$\begin{aligned} & p_i(\mathbf{x}, \hat{f}_i^*) \geq p_i(\mathbf{x}^*, \hat{f}_i^*) \\ \Leftrightarrow \quad & \mu_{p_i}(p_i(\mathbf{x}, \hat{f}_i^*)) \geq \mu_{p_i}(p_i(\mathbf{x}^*, \hat{f}_i^*)), \end{aligned} \tag{2.33}$$

with strict inequality holding for at least one i. This contradicts the fact that $\mathbf{x}^* \in X, \hat{f}_i^* \in F_i, i = 1, \cdots, k, \lambda^* \in [0,1]$ is a unique optimal solution to MAXMIN2.1.

(2) If $\mathbf{x}^* \in X(\beta)$ is not an F-Pareto optimal solution to MOP2.4($\hat{\mathbf{p}}^*, \beta$), because of (2.15) and Assumption 2.2, there exists some $\mathbf{x} \in X(\beta)$ such that

$$\begin{aligned} & f_i(\mathbf{x}, \hat{p}_i^*) \leq f_i(\mathbf{x}^*, \hat{p}_i^*) \\ \Leftrightarrow \quad & T_i^{-1}(\hat{p}_i^*) \cdot (\mathbf{c}_i^2\mathbf{x} + \alpha_i^2) + (\mathbf{c}_i^1\mathbf{x} + \alpha_i^1) \\ & \leq T_i^{-1}(\hat{p}_i^*) \cdot (\mathbf{c}_i^2\mathbf{x}^* + \alpha_i^2) + (\mathbf{c}_i^1\mathbf{x}^* + \alpha_i^1) \stackrel{\text{def}}{=} \hat{f}_i^* \\ \Leftrightarrow \quad & \hat{f}_i^* - (\mathbf{c}_i^1\mathbf{x} + \alpha_i^1) \geq T_i^{-1}(\hat{p}_i^*) \cdot (\mathbf{c}_i^2\mathbf{x} + \alpha_i^2) \\ \Leftrightarrow \quad & T_i\left(\frac{\hat{f}_i^* - (\mathbf{c}_i^1\mathbf{x} + \alpha_i^1)}{\mathbf{c}_i^2\mathbf{x} + \alpha_i^2}\right) \geq \hat{p}_i^* \\ \Leftrightarrow \quad & p_i(\mathbf{x}, \hat{f}_i^*) \geq p_i(\mathbf{x}^*, \hat{f}_i^*) \\ \Leftrightarrow \quad & \mu_{p_i}(p_i(\mathbf{x}, \hat{f}_i^*)) \geq \mu_{p_i}(p_i(\mathbf{x}^*, \hat{f}_i^*)), \end{aligned} \tag{2.34}$$

with strict inequality holding for at least one i. This contradicts the fact that $\mathbf{x}^* \in X(\beta)$ is a unique optimal solution to MAXMIN2.1. □

In inequality (2.31) of MAXMIN2.1, the following relations hold because of Assumption 2.2 and $\mathbf{c}_i^2\mathbf{x} + \alpha_i^2 > 0$, $i = 1, \cdots, k$.

$$\begin{aligned} & \mu_{p_i}(p_i(\mathbf{x}, \hat{f}_i)) \geq \lambda \\ \Leftrightarrow \quad & p_i(\mathbf{x}, \hat{f}_i) \geq \mu_{p_i}^{-1}(\lambda) \\ \Leftrightarrow \quad & T_i\left(\frac{\hat{f}_i - (\mathbf{c}_i^1\mathbf{x} + \alpha_i^1)}{\mathbf{c}_i^2\mathbf{x} + \alpha_i^2}\right) \geq \mu_{p_i}^{-1}(\lambda) \\ \Leftrightarrow \quad & \hat{f}_i - (\mathbf{c}_i^1\mathbf{x} + \alpha_i^1) \geq T_i^{-1}(\mu_{p_i}^{-1}(\lambda)) \cdot (\mathbf{c}_i^2\mathbf{x} + \alpha_i^2), \\ & i = 1, \cdots, k \end{aligned} \tag{2.35}$$

Conversely, inequality (2.35) can be transformed into the following form because $\hat{f}_i \leq \mu_{\hat{f}_i}^{-1}(\lambda)$.

$$\mu_{\hat{f}_i}^{-1}(\lambda) - (\mathbf{c}_i^1\mathbf{x} + \alpha_i^1) \geq T_i^{-1}(\mu_{p_i}^{-1}(\lambda)) \cdot (\mathbf{c}_i^2\mathbf{x} + \alpha_i^2), i = 1, \cdots, k \tag{2.36}$$

Therefore, MAXMIN2.1 is equivalently transformed into MAXMIN2.2, and permissible objective levels $\bar{f}_i, i = 1, \cdots, k$ have disappeared.

MAXMIN 2.2

$$\max_{\mathbf{x}\in X(\beta),\lambda\in[0,1]} \lambda \tag{2.37}$$

subject to

$$\mu_{\hat{f}_i}^{-1}(\lambda)-(\mathbf{c}_i^1\mathbf{x}+\alpha_i^1)\geq T_i^{-1}(\mu_{p_i}^{-1}(\lambda))\cdot(\mathbf{c}_i^2\mathbf{x}+\alpha_i^2), i=1,\cdots,k \tag{2.38}$$

Note that constraints (2.38) of MAXMIN2.2 can be reduced to a set of linear inequalities for some fixed value $\lambda\in[0,1]$. This means that an optimal solution $(\mathbf{x}^*,\lambda^*)$ of MAXMIN2.2 is obtained by the combined use of the bisection method with respect to $\lambda\in[0,1]$ and the first phase of the two-phase simplex method of linear programming.

Similar to the formulation using the probability maximization model, we propose a fuzzy approach to MOP2.4($\hat{\mathbf{p}},\beta$) using a fractile optimization model. To handle MOP2.4($\hat{\mathbf{p}},\beta$), the decision maker must specify permissible probability levels $\hat{\mathbf{p}}$ in advance.

Considering the imprecise nature of the decision maker's judgment, we assume that the decision maker has fuzzy goals for each objective function $f_i(\mathbf{x},\hat{p}_i)$ in MOP2.4($\hat{\mathbf{p}},\beta$). Such fuzzy goals can be quantified by eliciting the corresponding membership functions. Here, we denote a membership function of objective function $f_i(\mathbf{x},\hat{p}_i)$ as $\mu_{f_i}(f_i(\mathbf{x},\hat{p}_i))$, and a membership function of permissible probability level $\hat{p}_i$ as $\mu_{\hat{p}_i}(\hat{p}_i)$ respectively. Throughout this subsection, we make the following assumptions for $\mu_{\hat{p}_i}(\hat{p}_i)$, $\mu_{f_i}(f_i(\mathbf{x},\hat{p}_i))$, $i=1,\cdots,k$.

Assumption 2.3
$\mu_{\hat{p}_i}(\hat{p}_i), i=1,\cdots,k$ are strictly increasing and continuous with respect to $\hat{p}_i\in[p_{i\min},p_{i\max}]$, where $p_{i\min}(>0)$ is an acceptable minimum value and $p_{i\max}(<1)$ is a sufficiently satisfactory minimum value, with $\mu_{\hat{p}_i}(\hat{p}_{i\min})=0, \mu_{\hat{p}_i}(\hat{p}_{i\max})=1, i=1,\cdots,k$.

Assumption 2.4
$\mu_{f_i}(f_i(\mathbf{x},\hat{p}_i)), i=1,\cdots,k$ are strictly decreasing and continuous with respect to $f_i(\mathbf{x},\hat{p}_i)\in[f_{i\min},f_{i\max}]$, where $f_{i\min}$ is a sufficiently satisfactory maximum value and $p_{i\max}$ is an acceptable maximum value, and $\mu_{f_i}(f_{i\min})=1, \mu_{f_i}(f_{i\max})=0, i=1,\cdots,k$.

For example, we can define membership functions $\mu_{\hat{p}_i}(\hat{p}_i)$ and $\mu_{f_i}(f_i(\mathbf{x},\hat{p}_i))$ as follows. First, the decision maker specifies intervals $[p_{i\min},p_{i\max}], i=1,\cdots,k$ subjectively, defining membership function $\mu_{\hat{p}_i}(\hat{p}_i)$ according to Assumption 2.3. Corresponding to interval $[p_{i\min},p_{i\max}]$, we can compute interval $[f_{i\min},f_{i\max}]$ as follows.

$$f_{i\min}=\min_{\mathbf{x}\in X(\beta)} f_i(\mathbf{x},\hat{p}_{i\min}), i=1,\cdots,k \tag{2.39}$$

To obtain $f_{i\max}$, we first solve the following k linear programming problems.

$$\min_{\mathbf{x}\in X(\beta)} f_i(\mathbf{x},\hat{p}_{i\max}), i=1,\cdots,k \tag{2.40}$$

Let $\mathbf{x}_i, i = 1, \cdots, k$ be the above optimal solution. Using optimal solutions $\mathbf{x}_i, i = 1, \cdots, k$, $f_{i\max}$ can be obtained as follows [139].

$$f_{i\max} = \max_{\ell=1,\cdots,k,\ell\neq i} f_i(\mathbf{x}_\ell, \hat{p}_{i\max}), i = 1, \cdots, k \tag{2.41}$$

According to Assumption 2.4, the decision maker can define his or her membership function $\mu_{f_i}(f_i(\mathbf{x}, \hat{p}_i))$ on interval $[f_{i\min}, f_{i\max}]$.

Then, if the decision maker adopts the fuzzy decision [6, 71, 140] as an aggregation operator, a satisfactory solution is obtained by solving the following maxmin problem.

MAXMIN 2.3

$$\max_{\mathbf{x}\in X(\beta), \hat{p}_i \in P_i, i=1,\cdots,k, \lambda\in[0,1]} \lambda \tag{2.42}$$

subject to

$$\mu_{f_i}(f_i(\mathbf{x}, \hat{p}_i)) \geq \lambda, i = 1, \cdots, k \tag{2.43}$$

$$\mu_{\hat{p}_i}(\hat{p}_i) \geq \lambda, i = 1, \cdots, k \tag{2.44}$$

where

$$P_i \stackrel{\text{def}}{=} [p_{i\min}, p_{i\max}], i = 1, \cdots, k.$$

Similar to Theorem 2.2, the relationships between optimal solution $(\mathbf{x}^*, \lambda^*)$ of MAXMIN2.3, P-Pareto optimal solutions to MOP2.2($\mathbf{f}^*, \beta$), and F-Pareto optimal solutions to MOP2.4($\hat{\mathbf{p}}^*, \beta$) can be characterized by the following theorem.

Theorem 2.3

Assume that $\mathbf{x}^* \in X(\beta), \hat{p}_i^* \in P_i, i = 1, \cdots, k, \lambda^* \in [0, 1]$ is a unique optimal solution of MAXMIN2.3. Then,

(1) $\mathbf{x}^* \in X(\beta)$ is a P-Pareto optimal solution to MOP2.2($\mathbf{f}^*, \beta$), where $f_i^* \stackrel{\text{def}}{=} f_i(\mathbf{x}^*, \hat{p}_i^*), i = 1, \cdots, k$, $\mathbf{f}^* \stackrel{\text{def}}{=} (f_1^*, \cdots, f_k^*)$.

(2) $\mathbf{x}^* \in X(\beta)$ is an F-Pareto optimal solution to MOP2.4($\hat{\mathbf{p}}^*, \beta$).

Because of Assumption 2.4 and $\mathbf{c}_i^2\mathbf{x} + \alpha_i^2 > 0$, $i = 1, \cdots, k$, constraints (2.43) of MAXMIN2.3 can be transformed as follows.

$$\begin{aligned}
& \mu_{f_i}(f_i(\mathbf{x}, \hat{p}_i)) \geq \lambda \\
\Leftrightarrow\ & f_i(\mathbf{x}, \hat{p}_i) \leq \mu_{f_i}^{-1}(\lambda) \\
\Leftrightarrow\ & T_i^{-1}(\hat{p}_i) \cdot (\mathbf{c}_i^2\mathbf{x} + \alpha_i^2) + (\mathbf{c}_i^1\mathbf{x} + \alpha_i^1) \leq \mu_{f_i}^{-1}(\lambda) \\
\Leftrightarrow\ & \hat{p}_i \leq T_i\left(\frac{\mu_{f_i}^{-1}(\lambda) - (\mathbf{c}_i^1\mathbf{x} + \alpha_i^1)}{\mathbf{c}_i^2\mathbf{x} + \alpha_i^2}\right)
\end{aligned} \tag{2.45}$$

From constraints (2.44) and Assumption 2.3, $\hat{p}_i \geq \mu_{\tilde{p}_i}^{-1}(\lambda)$ holds. Therefore, constraint (2.45) can be reduced to the following inequality in which permissible probability levels $\hat{p}_i, i = 1, \cdots, k$ have disappeared.

$$\mu_{\tilde{p}_i}^{-1}(\lambda) \leq T_i\left(\frac{\mu_{\tilde{f}_i}^{-1}(\lambda) - (\mathbf{c}_i^1\mathbf{x} + \alpha_i^1)}{\mathbf{c}_i^2\mathbf{x} + \alpha_i^2}\right)$$

$$\Leftrightarrow \quad \mu_{\tilde{f}_i}^{-1}(\lambda) - (\mathbf{c}_i^1\mathbf{x} + \alpha_i^1) \geq T_i^{-1}(\mu_{\tilde{p}_i}^{-1}(\lambda)) \cdot (\mathbf{c}_i^2\mathbf{x} + \alpha_i^2) \tag{2.46}$$

Then, MAXMIN2.3 is equivalently transformed into the following problem.

MAXMIN 2.4

$$\max_{\mathbf{x} \in X(\beta), \lambda \in [0,1]} \lambda \tag{2.47}$$

subject to

$$\mu_{\tilde{f}_i}^{-1}(\lambda) - (\mathbf{c}_i^1\mathbf{x} + \alpha_i^1) \geq T_i^{-1}(\mu_{\tilde{p}_i}^{-1}(\lambda)) \cdot (\mathbf{c}_i^2\mathbf{x} + \alpha_i^2), i = 1, \cdots, k \tag{2.48}$$

Note that constraints (2.48) can be reduced to a set of linear inequalities for some fixed value $\lambda \in [0,1]$. This means that an optimal solution $(\mathbf{x}^*, \lambda^*)$ of MAXMIN2.4 is obtained by the combined use of the bisection method with respect to $\lambda \in [0,1]$ and the first phase of the two-phase simplex method of linear programming.

2.1.3 A numerical example

In this subsection, we consider the following three objective programming problem, formulated by Sakawa et al. [73], to demonstrate the feasibility of our proposed fuzzy approach from Section 2.1.2.

MOP 2.5

$$\begin{array}{ll} \min & (\mathbf{c}_1^1 + \bar{t}_1\mathbf{c}_1^2)\mathbf{x} + (\alpha_1^1 + \bar{t}_1\alpha_1^2) \\ \min & (\mathbf{c}_2^1 + \bar{t}_2\mathbf{c}_2^2)\mathbf{x} + (\alpha_2^1 + \bar{t}_2\alpha_2^2) \\ \min & (\mathbf{c}_3^1 + \bar{t}_3\mathbf{c}_3^2)\mathbf{x} + (\alpha_3^1 + \bar{t}_3\alpha_3^2) \end{array}$$

subject to

$$\mathbf{x} \in X \stackrel{\text{def}}{=} \{\mathbf{x} \in \mathrm{R}^{10} \mid \mathbf{a}_i\mathbf{x} \leq \bar{b}_i, i = 1, \cdots, 7, \mathbf{x} \geq \mathbf{0}\}$$

where $\mathbf{a}_i, i = 1, \cdots, 7$, $\mathbf{c}_1^1, \mathbf{c}_1^2$, $\mathbf{c}_2^1, \mathbf{c}_2^2$, $\mathbf{c}_3^1, \mathbf{c}_3^2$ are ten-dimensional coefficient row vectors (as defined in Table 2.1), and $\alpha_1^1 = -18, \alpha_1^2 = 5, \alpha_2^1 = -27, \alpha_2^2 = 6, \alpha_3^1 = -10, \alpha_3^2 = 4$. Further, $\bar{t}_i, i = 1, \cdots, 3$, $\bar{b}_j, j = 1, \cdots, 7$ are Gaussian random variables defined as follows.

Table 2.1: Parameters of the objective functions and constraints in MOP2.5

$\mathbf{x}$	x_1	x_2	x_3	x_4	x_5	x_6	x_7	x_8	x_9	x_{10}
$\mathbf{c}_1^1$	19	48	21	10	18	35	46	11	24	33
$\mathbf{c}_1^2$	3	2	2	1	4	3	1	2	4	2
$\mathbf{c}_2^1$	12	–46	–23	–38	–33	–48	12	8	19	20
$\mathbf{c}_2^2$	1	2	4	2	2	1	2	1	2	1
$\mathbf{c}_3^1$	–18	–26	–22	–28	–15	–29	–10	–19	–17	–28
$\mathbf{c}_3^2$	2	1	3	2	1	2	3	3	2	1
$\mathbf{a}_1$	12	–2	4	–7	13	–1	–6	6	11	–8
$\mathbf{a}_2$	–2	5	3	16	6	–12	12	4	–7	–10
$\mathbf{a}_3$	3	–16	–4	–8	–8	2	–12	–12	4	–3
$\mathbf{a}_4$	–11	6	–5	9	–1	8	–4	6	–9	6
$\mathbf{a}_5$	–4	7	–6	–5	13	6	–2	–5	14	–6
$\mathbf{a}_6$	5	–3	14	–3	–9	–7	4	–4	–5	9
$\mathbf{a}_7$	–3	–4	–6	9	6	18	11	–9	–4	7

$$\begin{array}{lll} \bar{t}_1 \sim \mathrm{N}(4,2^2), & \bar{t}_2 \sim \mathrm{N}(3,3^2), & \bar{t}_3 \sim \mathrm{N}(3,2^2) \\ \bar{b}_1 \sim \mathrm{N}(164,30^2), & \bar{b}_2 \sim \mathrm{N}(-190,20^2), & \bar{b}_3 \sim \mathrm{N}(-184,15^2) \\ \bar{b}_4 \sim \mathrm{N}(99,22^2), & \bar{b}_5 \sim \mathrm{N}(-150,17^2), & \bar{b}_6 \sim \mathrm{N}(154,35^2) \\ \bar{b}_7 \sim \mathrm{N}(142,42^2) & & \end{array}$$

Let us assume that the hypothetical decision maker sets constraint probability levels to $\beta = (\beta_1, \cdots, \beta_7) = (0.85, 0.95, 0.8, 0.9, 0.85, 0.8, 0.9)$ for MOP2.5, which are the same values as the numerical example [73]. In the numerical example [73], a probability maximization model is adopted to formulate MOP2.5. To compare our proposed fuzzy approach with the results of [73], we apply MAXMIN2.2 to MOP2.5.

First, we set membership functions $\mu_{p_i}(p_i(\mathbf{x}, \hat{f}_i)), i = 1,2,3$, as follows, which are the same as those of [73].

$$\begin{aligned} \mu_{p_1}(p_1(\mathbf{x}, \hat{f}_1)) &= \frac{p_1(\mathbf{x}, \hat{f}_1) - 0.502}{(0.88 - 0.502)} \\ \mu_{p_2}(p_2(\mathbf{x}, \hat{f}_2)) &= \frac{p_2(\mathbf{x}, \hat{f}_2) - 0.06}{(0.783 - 0.06)} \\ \mu_{p_3}(p_3(\mathbf{x}, \hat{f}_3)) &= \frac{p_3(\mathbf{x}, \hat{f}_3) - 0.446}{(0.664 - 0.446)} \end{aligned}$$

In the numerical example [73],

$$\hat{\mathbf{f}} = (\hat{f}_1, \hat{f}_2, \hat{f}_3) = (2150, 450, -950)$$

Table 2.2: Comparison of our proposed method with a probability model in MOP2.5

	our proposed method	probability model
$\hat{f}_1$	2297	2150
$\hat{f}_2$	545.3	450
$\hat{f}_3$	–941.3	–950
$p_1(\mathbf{x}, \hat{f}_1)$	0.8145	0.719
$p_2(\mathbf{x}, \hat{f}_2)$	0.6577	0.474
$p_3(\mathbf{x}, \hat{f}_3)$	0.6262	0.571

are adopted as permissible objective levels for the probability maximization model. Corresponding to these values, we assume that the hypothetical decision maker sets his or her membership functions $\mu_{\hat{f}_i}(\hat{f}_i)$ for permissible objective levels $\hat{f}_i$, $i = 1,2,3$ as follows.

$$\mu_{\hat{f}_1}(\hat{f}_1) = \frac{3000 - \hat{f}_1}{(3000 - 2150)}$$

$$\mu_{\hat{f}_2}(\hat{f}_2) = \frac{1000 - \hat{f}_2}{(1000 - 450)}$$

$$\mu_{\hat{f}_3}(\hat{f}_3) = \frac{-900 - \hat{f}_3}{((-900) - (-950))}$$

For these membership functions, we solve MAXMIN2.2 by the combined use of the bisection method with respect to $\lambda \in [0,1]$ and the first phase of the two-phase simplex method of linear programming. Optimal solution $\lambda^* = 0.8267$ is obtained and the corresponding membership function values are obtained as $\lambda^* = \mu_{p_i}(\cdot) = \mu_{\hat{f}_i}(\cdot) = 0.8267, i = 1,2,3$. Both the optimal solution of MAXMIN2.2 and the probability maximization model with permissible objective levels $(\hat{f}_1, \hat{f}_2, \hat{f}_3) = (2150, 450, -950)$ are shown in Table 2.2.

As shown in Table 2.2, at the optimal solution of MAXMIN2.2, $p_i(\mathbf{x}^*, \hat{f}_i^*), i = 1,2,3$ are improved at the expense of the permissible objective levels in comparison with the results of the probability maximization model [73]. From these results, we expect that the proper balance between $p_i(\mathbf{x}, \hat{f}_i)$ and permissible objective level $\hat{f}_i$ in the probability maximization model will be attained at the optimal solution of MAXMIN2.2. Similarly, we expect that the proper balance between objective function $f_i(\mathbf{x}, \hat{p}_i)$ and permissible probability level $\hat{p}_i$ in the fractile optimization model will be attained at the optimal solution of MAXMIN2.4.

2.2 Fuzzy approaches for MOSPs with variance covariance matrices

In MOSPs, such as the crop planning problem at the farm level [24, 26, 32, 98], we often must consider the influence of variance covariance matrices between random variable coefficients in the objective functions. To handle such situations, we formulate MOSPs with variance covariance matrices and propose fuzzy approaches [121] to obtain satisfactory solutions based on the fuzzy decision [6, 71, 140].

In Section 2.2.1, we formulate two multiobjective programming problems using probability maximization and fractile optimization models; we also define the corresponding Pareto optimal solutions (i.e., P-Pareto and F-Pareto optimal solutions) and investigate the relationships between them. In Section 2.2.2, given the assumption that the decision maker has fuzzy goals not only for permissible objective levels but also for corresponding distribution functions of the probability maximization model, we formulate the maxmin problem to obtain a satisfactory solution based on the fuzzy decision. Similarly, given the assumption that the decision maker has fuzzy goals not only for permissible probability levels but also for corresponding objective functions of a fractile optimization model, we formulate the maxmin problem to obtain a satisfactory solution based on the fuzzy decision. These two maxmin problems can be solved using convex programming techniques. In Section 2.2.3, we apply our proposed method (from Section 2.2.2) to a numerical example and compare the results with a fractile optimization model to show the efficiency of our proposed method.

2.2.1 Formulations using probability and fractile models

In this subsection, we focus on the following multiobjective stochastic programming problem, in which random variable coefficients of the objective functions are correlated with one another.

MOP 2.6

$$\min \bar{\mathbf{C}}\mathbf{x} = (\bar{\mathbf{c}}_1\mathbf{x}, \cdots, \bar{\mathbf{c}}_k\mathbf{x}) \tag{2.49}$$

subject to

$$A\mathbf{x} \leq \bar{\mathbf{b}},\ \mathbf{x} \geq \mathbf{0}, \tag{2.50}$$

where $\mathbf{x} = (x_1, \cdots, x_n)^T$ is an n-dimensional decision variable column vector, $\bar{\mathbf{c}}_i = (\bar{c}_{i1}, \cdots, \bar{c}_{in})$ is an n-dimensional random variable row vector, corresponding elements $\bar{c}_{i\ell}, i = 1, \cdots, k, \ell = 1, \cdots, n$ are Gaussian random variables

$$\bar{c}_{i\ell} \sim \mathrm{N}(E[\bar{c}_{i\ell}], \sigma_{i\ell\ell}), i = 1, \cdots, k, \ell = 1, \cdots, n, \tag{2.51}$$

and the $(n \times n)$-dimensional variance covariance matrices $V_i, i = 1, \cdots, k$ between Gaussian random variables $\bar{c}_{i\ell}, \ell = 1, \cdots, n$ are given as follows.

$$V_i = \begin{pmatrix} \sigma_{i11} & \sigma_{i12} & \cdots & \sigma_{i1n} \\ \sigma_{i21} & \sigma_{i22} & \cdots & \sigma_{i2n} \\ \vdots & \vdots & \ddots & \vdots \\ \sigma_{in1} & \sigma_{in2} & \cdots & \sigma_{inn} \end{pmatrix}, i = 1, \cdots, k \tag{2.52}$$

We denote vectors of the expectation for random variable row vector $\bar{\mathbf{c}}_i, i = 1, \cdots, k$ as

$$\mathbf{E}[\bar{\mathbf{c}}_i] = (E[\bar{c}_{i1}], \cdots, E[\bar{c}_{in}]), i = 1, \cdots, k. \tag{2.53}$$

Then, using variance covariance matrices V_i, we formally express vector $\bar{\mathbf{c}}_i$ as

$$\bar{\mathbf{c}}_i = (\bar{c}_{i1}, \cdots, \bar{c}_{in}) \sim \mathrm{N}(\mathbf{E}[\bar{\mathbf{c}}_i], V_i), i = 1, \cdots, k. \tag{2.54}$$

From the properties of Gaussian random variables, objective function $\bar{\mathbf{c}}_i\mathbf{x}$ becomes Gaussian random variable.

$$\bar{\mathbf{c}}_i\mathbf{x} \sim \mathrm{N}(\mathbf{E}[\bar{\mathbf{c}}_i]\mathbf{x}, \mathbf{x}^T V_i \mathbf{x}), i = 1, \cdots, k. \tag{2.55}$$

Here, A is an $(m \times n)$-dimensional coefficient matrix, $\bar{\mathbf{b}} = (\bar{b}_1, \cdots, \bar{b}_m)^T$ is an m-dimensional column vector, whose elements $\bar{b}_j, j = 1, \cdots, m$ are random variables independent of one another, and $F_j(\cdot), j = 1, \cdots, m$ are distribution functions of $\bar{b}_j, j = 1, \cdots, m$.

To handle MOP2.6, we interpret constraints (2.50) as chance constrained conditions [11], which means that the j-th constraint of (2.50) must be satisfied with a certain probability β_j (called a constraint probability level), i.e.,

$$X(\beta) \stackrel{\text{def}}{=} \{\mathbf{x} \geq \mathbf{0} \mid \Pr(\omega \mid \mathbf{a}_j\mathbf{x} \leq b_j(\omega)) \geq \beta_j, j = 1, \cdots, m\}, \tag{2.56}$$

where $\mathbf{a}_j$ is the j-th row vector of A and $\beta \stackrel{\text{def}}{=} (\beta_1, \cdots, \beta_m)$. Under chance constrained conditions (2.56), if we adopt a probability maximization model for MOP2.6, we can substitute the minimization of objective functions $\bar{\mathbf{c}}_i\mathbf{x}, i = 1, \cdots, k$ for the maximization of the probability that each objective function $\bar{\mathbf{c}}_i\mathbf{x}$ is less than or equal to a certain permissible level $\hat{f}_i$ (called the permissible objective level). Such a probability function can be defined as

$$p_i(\mathbf{x}, \hat{f}_i) \stackrel{\text{def}}{=} \Pr(\omega \mid \mathbf{c}_i(\omega)\mathbf{x} \leq \hat{f}_i), i = 1, \cdots, k, \tag{2.57}$$

where

$$\hat{\mathbf{f}} = (\hat{f}_1, \cdots, \hat{f}_k) \tag{2.58}$$

is a k-dimensional vector of permissible objective levels specified subjectively by the decision maker. Then, MOP2.6 can be converted into the following form.

MOP 2.7 $(\hat{\mathbf{f}}, \beta)$

$$\max_{\mathbf{x} \in X(\beta)} (p_1(\mathbf{x}, \hat{f}_1), \cdots, p_k(\mathbf{x}, \hat{f}_k)) \tag{2.59}$$

Because of (2.55), we can rewrite objective function $p_i(\mathbf{x}, \hat{f}_i)$ as

$$\begin{aligned} p_i(\mathbf{x}, \hat{f}_i) &= \Pr(\omega \mid \mathbf{c}_i(\omega)\mathbf{x} \leq \hat{f}_i) \\ &= \Pr\left(\omega \mid \frac{\mathbf{c}_i(\omega)\mathbf{x} - \mathbf{E}[\bar{\mathbf{c}}_i]\mathbf{x}}{\sqrt{\mathbf{x}^T V_i \mathbf{x}}} \leq \frac{\hat{f}_i - \mathbf{E}[\bar{\mathbf{c}}_i]\mathbf{x}}{\sqrt{\mathbf{x}^T V_i \mathbf{x}}}\right) \\ &= \Phi\left(\frac{\hat{f}_i - \mathbf{E}[\bar{\mathbf{c}}_i]\mathbf{x}}{\sqrt{\mathbf{x}^T V_i \mathbf{x}}}\right). \end{aligned} \tag{2.60}$$

Here, $\Phi(\cdot)$ is a distribution function of the standard Gaussian random variable denoted as $\mathrm{N}(0,1)$.

To handle MOP2.7$(\hat{\mathbf{f}}, \beta)$, we introduce a P-Pareto optimal solution as follows.

Definition 2.3 $\mathbf{x}^* \in X(\beta)$ is a P-Pareto optimal solution to MOP2.7$(\hat{\mathbf{f}}, \beta)$ if and only if there does not exist another $\mathbf{x} \in X(\beta)$ such that $p_i(\mathbf{x}, \hat{f}_i) \geq p_i(\mathbf{x}^*, \hat{f}_i)$, $i = 1, \cdots, k$ with strict inequality holding for at least one i.

Conversely, if we adopt a fractile optimization model for objective functions $\bar{\mathbf{c}}_i\mathbf{x}, i = 1, \cdots, k$ in MOP2.6, we can convert MOP2.6 to the following multiobjective programming problem.

MOP 2.8 $(\hat{\mathbf{p}}, \beta)$

$$\min_{\mathbf{x} \in X(\beta), f_i \in R^1, i=1,\cdots,k} (f_1, \cdots, f_k) \tag{2.61}$$

subject to

$$p_i(\mathbf{x}, f_i) \geq \hat{p}_i, i = 1, \cdots, k \tag{2.62}$$

where

$$\hat{\mathbf{p}} = (\hat{p}_1, \cdots, \hat{p}_k) \tag{2.63}$$

is a k-dimensional vector of permissible probability levels specified subjectively by the decision maker.

In MOP2.8$(\hat{\mathbf{p}}, \beta)$, the decision maker minimizes target values $(f_1, \cdots, f_k)$ after he or she specifies certain permissible probability levels $\hat{\mathbf{p}} = (\hat{p}_1, \cdots, \hat{p}_k)$ such that each of the objective functions is less than or equal to the target values. Constraints (2.62) can be transformed into the following form.

$$\begin{aligned} \hat{p}_i &\leq p_i(\mathbf{x}, f_i) = \Phi\left(\frac{\hat{f}_i - \mathbf{E}[\bar{\mathbf{c}}_i]\mathbf{x}}{\sqrt{\mathbf{x}^T V_i \mathbf{x}}}\right) \\ \Leftrightarrow f_i &\geq \Phi^{-1}(\hat{p}_i) \cdot \sqrt{\mathbf{x}^T V_i \mathbf{x}} + \mathbf{E}[\bar{\mathbf{c}}_i]\mathbf{x} \end{aligned} \tag{2.64}$$

We define the right-hand-side of inequality (2.64) as

$$f_i(\mathbf{x},\hat{p}_i) \stackrel{\text{def}}{=} \Phi^{-1}(\hat{p}_i)\cdot\sqrt{\mathbf{x}^T V_i \mathbf{x}} + \mathbf{E}[\bar{\mathbf{c}}_i]\mathbf{x}. \tag{2.65}$$

Then, MOP2.8($\hat{\mathbf{p}},\beta$) can be reduced to the following simple form, where the target values $f_i, i = 1,\cdots,k$ have disappeared.

MOP 2.9 $(\hat{\mathbf{p}},\beta)$

$$\min_{\mathbf{x}\in X(\beta)} (f_1(\mathbf{x},\hat{p}_1),\cdots,f_k(\mathbf{x},\hat{p}_k)) \tag{2.66}$$

To handle MOP2.9($\hat{\mathbf{p}},\beta$), we introduce the F-Pareto optimal solution as follows.

Definition 2.4 $\mathbf{x}^* \in X(\beta)$ is an F-Pareto optimal solution to MOP2.9($\hat{\mathbf{p}},\beta$) if and only if there does not exist another $\mathbf{x} \in X(\beta)$ such that $f_i(\mathbf{x},\hat{p}_i) \le f_i(\mathbf{x}^*,\hat{p}_i)$, $i = 1,\cdots,k$, with strict inequality holding for at least one i.

The relationships between a P-Pareto optimal solution to MOP2.7($\hat{\mathbf{f}},\beta$) and an F-Pareto optimal solution to MOP2.9($\hat{\mathbf{p}},\beta$) can be characterized by the following theorem.

Theorem 2.4
(1) If $\mathbf{x}^* \in X(\beta)$ is a P-Pareto optimal solution to MOP2.7($\hat{\mathbf{f}},\beta$) for some given permissible objective levels $\hat{\mathbf{f}}$, then, there exist some permissible probability levels $\hat{\mathbf{p}}$ such that $\mathbf{x}^* \in X(\beta)$ is an F-Pareto optimal solution to MOP2.9($\hat{\mathbf{p}},\beta$).

(2) If $\mathbf{x}^* \in X(\beta)$ is an F-Pareto optimal solution to MOP2.9($\hat{\mathbf{p}},\beta$) for some given permissible probability levels $\hat{\mathbf{p}}$, then, there exist some permissible objective levels $\hat{\mathbf{f}}$ such that $\mathbf{x}^* \in X(\beta)$ is a P-Pareto optimal solution to MOP2.7($\hat{\mathbf{f}},\beta$).

(Proof)
(1) Assume that $\mathbf{x}^* \in X(\beta)$ is not an F-Pareto optimal solution to MOP2.9($\hat{\mathbf{p}},\beta$) for any permissible probability levels $\hat{\mathbf{p}}$. Then, there exists $\mathbf{x} \in X(\beta)$ such that

$$\begin{aligned} & f_i(\mathbf{x},\hat{p}_i) \le f_i(\mathbf{x}^*,\hat{p}_i) \\ \Leftrightarrow\ & \Phi^{-1}(\hat{p}_i)\cdot\sqrt{\mathbf{x}^T V_i \mathbf{x}} + \mathbf{E}[\bar{\mathbf{c}}_i]\mathbf{x} \le \Phi^{-1}(\hat{p}_i)\cdot\sqrt{\mathbf{x}^{*T} V_i \mathbf{x}^*} + \mathbf{E}[\bar{\mathbf{c}}_i]\mathbf{x}^*, \\ & i = 1,\cdots,k, \end{aligned} \tag{2.67}$$

with strict inequality holding for at least one i. On the right-hand-side of (2.67), if we set

$$\hat{f}_i \stackrel{\text{def}}{=} \Phi^{-1}(\hat{p}_i)\cdot\sqrt{\mathbf{x}^{*T} V_i \mathbf{x}^*} + \mathbf{E}[\bar{\mathbf{c}}_i]\mathbf{x}^*, i = 1,\cdots,k, \tag{2.68}$$

then, inequalities (2.67) can be equivalently transformed into the following form, because $\Phi(\cdot)$ is continuous and strictly increasing.

$$\hat{p}_i \leq \Phi\left(\frac{\hat{f}_i - \mathbf{E}[\bar{\mathbf{c}}_i]\mathbf{x}}{\sqrt{\mathbf{x}^T V_i \mathbf{x}}}\right), i = 1, \cdots, k, \tag{2.69}$$

with strict inequality holding for at least one i. Conversely, from relationship (2.68), it immediately holds that

$$\hat{p}_i = \Phi\left(\frac{\hat{f}_i - \mathbf{E}[\bar{\mathbf{c}}_i]\mathbf{x}^*}{\sqrt{\mathbf{x}^{*T} V_i \mathbf{x}^*}}\right), i = 1, \cdots, k. \tag{2.70}$$

Consequently, inequalities (2.67) are equivalent to

$$\begin{aligned} & \Phi\left(\frac{\hat{f}_i - \mathbf{E}[\bar{\mathbf{c}}_i]\mathbf{x}}{\sqrt{\mathbf{x}^T V_i \mathbf{x}}}\right) \geq \Phi\left(\frac{\hat{f}_i - \mathbf{E}[\bar{\mathbf{c}}_i]\mathbf{x}^*}{\sqrt{\mathbf{x}^{*T} V_i \mathbf{x}^*}}\right) \\ \Leftrightarrow \quad & p_i(\mathbf{x}, \hat{f}_i) \geq p_i(\mathbf{x}^*, \hat{f}_i), i = 1, \cdots, k, \end{aligned} \tag{2.71}$$

with strict inequality holding for at least one i. This means that there exists $\mathbf{x} \in X(\beta)$ such that inequalities (2.71) are satisfied for any permissible probability levels $\hat{\mathbf{p}}$. Therefore, from (2.68), $\mathbf{x}^* \in X(\beta)$ is not a P-Pareto optimal solution to MOP2.7($\hat{\mathbf{f}}, \beta$) for any permissible objective levels $\hat{\mathbf{f}}$.

(2) Assume that $\mathbf{x}^* \in X(\beta)$ is not a P-Pareto optimal solution to MOP2.7($\hat{\mathbf{f}}, \beta$) for any permissible objective levels $\hat{\mathbf{f}}$. Then, there exists $\mathbf{x} \in X(\beta)$ such that

$$\begin{aligned} & p_i(\mathbf{x}, \hat{f}_i) \geq p_i(\mathbf{x}^*, \hat{f}_i), \\ \Leftrightarrow \quad & \Phi\left(\frac{\hat{f}_i - \mathbf{E}[\bar{\mathbf{c}}_i]\mathbf{x}}{\sqrt{\mathbf{x}^T V_i \mathbf{x}}}\right) \geq \Phi\left(\frac{\hat{f}_i - \mathbf{E}[\bar{\mathbf{c}}_i]\mathbf{x}^*}{\sqrt{\mathbf{x}^{*T} V_i \mathbf{x}^*}}\right), i = 1, \cdots, k \end{aligned} \tag{2.72}$$

with strict inequality holding for at least one i. On the right-hand-side of (2.72), if we set

$$\hat{p}_i \stackrel{\text{def}}{=} \Phi\left(\frac{\hat{f}_i - \mathbf{E}[\bar{\mathbf{c}}_i]\mathbf{x}^*}{\sqrt{\mathbf{x}^{*T} V_i \mathbf{x}^*}}\right), i = 1, \cdots, k, \tag{2.73}$$

then, inequalities (2.72) can be equivalently transformed into the following form, because $\Phi(\cdot)$ is continuous and strictly increasing.

$$\begin{aligned} & \Phi\left(\frac{\hat{f}_i - \mathbf{E}[\bar{\mathbf{c}}_i]\mathbf{x}}{\sqrt{\mathbf{x}^T V_i \mathbf{x}}}\right) \geq \hat{p}_i \\ \Leftrightarrow \quad & \hat{f}_i - \mathbf{E}[\bar{\mathbf{c}}_i]\mathbf{x} \geq \Phi^{-1}(\hat{p}_i) \cdot \sqrt{\mathbf{x}^T V_i \mathbf{x}} \\ \Leftrightarrow \quad & \hat{f}_i \geq f_i(\mathbf{x}, \hat{p}_i), i = 1, \cdots, k, \end{aligned} \tag{2.74}$$

with strict inequality holding for at least one i. Conversely, from relationship (2.73), it immediately holds that

$$\hat{f}_i = \Phi^{-1}(\hat{p}_i) \cdot \sqrt{\mathbf{x}^{*T} V_i \mathbf{x}^*} + \mathbf{E}[\bar{\mathbf{c}}_i]\mathbf{x}^* = f_i(\mathbf{x}^*, \hat{p}_i). \tag{2.75}$$

As a result, inequalities (2.72) are equivalent to

$$f_i(\mathbf{x}, \hat{p}_i) \leq f_i(\mathbf{x}^*, \hat{p}_i), i = 1, \cdots, k, \tag{2.76}$$

with strict inequality holding for at least one i. This means that there exists $\mathbf{x} \in X(\beta)$ such that inequalities (2.76) are satisfied for any permissible objective levels $\hat{\mathbf{f}}$. Therefore, from (2.73), $\mathbf{x}^* \in X(\beta)$ is not an F-Pareto optimal solution to MOP2.9$(\hat{\mathbf{p}}, \beta)$ for any permissible probability levels $\hat{\mathbf{p}}$. □

In a probability maximization model MOP2.7$(\hat{\mathbf{f}}, \beta)$, permissible objective levels $\hat{f}_i, i = 1, \cdots, k$ must be specified by the decision maker in advance. Fewer values for permissible objective levels $\hat{f}_i, i = 1, \cdots, k$ will yield fewer values for corresponding distribution functions $p_i(\mathbf{x}, \hat{f}_i), i = 1, \cdots, k$. Conversely, in a fractile optimization model MOP2.9$(\hat{\mathbf{p}}, \beta)$, permissible probability levels $\hat{p}_i, i = 1, \cdots, k$ must also be specified by the decision maker in advance. Here, larger values for permissible probability levels $\hat{p}_i, i = 1, \cdots, k$ will yield larger values for objective functions $f_i(\mathbf{x}, \hat{p}_i), i = 1, \cdots, k$ for minimization problems. From this perspective, it is necessary for the decision maker to attain the appropriate balance between $\hat{f}_i$ and $p_i(\mathbf{x}, \hat{f}_i)$ in MOP2.7$(\hat{\mathbf{f}}, \beta)$, or $\hat{p}_i$ and $f_i(\mathbf{x}, \hat{p}_i)$ in MOP2.9$(\hat{\mathbf{p}}, \beta)$.

2.2.2 Fuzzy approaches based on convex programming

In this subsection, we first propose a fuzzy decision making method for MOP2.7$(\hat{\mathbf{f}}, \beta)$. To handle MOP2.7$(\hat{\mathbf{f}}, \beta)$, the decision maker must specify his or her permissible objective levels $\hat{\mathbf{f}}$ in advance; however, in general, the decision maker seems to prefer not only larger values of distribution function $p_i(\mathbf{x}, \hat{f}_i)$ but also smaller value of permissible objective level $\hat{f}_i$. Considering the imprecise nature of the decision maker's judgment, it is natural to assume that the decision maker has fuzzy goals not only for $p_i(\mathbf{x}, \hat{f}_i)$ but also for permissible objective level $\hat{f}_i$. We assume that such fuzzy goals can be quantified via corresponding membership functions. Let us denote a membership function of distribution function $p_i(\mathbf{x}, \hat{f}_i)$ as $\mu_{p_i}(p_i(\mathbf{x}, \hat{f}_i))$, and a membership function of permissible objective level $\hat{f}_i$ as $\mu_{\hat{f}_i}(\hat{f}_i)$ respectively. Throughout this subsection, we make the following assumptions.

Assumption 2.5
$\mu_{\hat{f}_i}(\hat{f}_i), i = 1, \cdots, k$ are strictly decreasing and continuous with respect to $\hat{f}_i \in [\hat{f}_{i\min}, \hat{f}_{i\max}]$, where $\hat{f}_{i\min}$ is a sufficiently satisfactory maximum value and $\hat{f}_{i\max}$ is an acceptable maximum value, i.e., $\mu_{\hat{f}_i}(\hat{f}_{i\min}) = 1, \mu_{\hat{f}_i}(\hat{f}_{i\max}) = 0, i = 1, \cdots, k$.

Assumption 2.6
$\mu_{p_i}(p_i(\mathbf{x},\hat{f}_i)), i=1,\cdots,k$ are strictly increasing and continuous with respect to $p_i(\mathbf{x},\hat{f}_i) \in [p_{i\min}, p_{i\max}]$, where $p_{i\min}(>0.5)$ is an acceptable minimum value and $p_{i\max}(<1)$ is a sufficiently satisfactory minimum value, i.e., $\mu_{p_i}(p_{i\min}) = 0, \mu_{p_i}(p_{i\max}) = 1, i=1,\cdots,k$.
It seems to be natural for the decision maker to accept Assumption 2.6 in real-world situations, as distribution function values in a probability maximization model should be greater than 0.5.

If the decision maker adopts the fuzzy decision [6, 71, 140] as an aggregation operator, a satisfactory solution is obtained by solving the following maxmin problem.

MAXMIN 2.5

$$\max_{\mathbf{x}\in X(\beta),\hat{f}_i\in F_i,i=1,\cdots,k,\lambda\in[0,1]} \lambda \tag{2.77}$$

subject to

$$\mu_{p_i}(p_i(\mathbf{x},\hat{f}_i)) \geq \lambda, i=1,\cdots,k \tag{2.78}$$
$$\mu_{\hat{f}_i}(\hat{f}_i) \geq \lambda, i=1,\cdots,k \tag{2.79}$$

where

$$F_i \stackrel{\text{def}}{=} [\hat{f}_{i\min}, \hat{f}_{i\max}], i=1,\cdots,k.$$

The relationships between optimal solution $\mathbf{x}^* \in X(\beta), \hat{f}_i^* \in F_i, i = 1,\cdots,k, \lambda^* \in [0,1]$ of MAXMIN2.5, P-Pareto optimal solutions to MOP2.7$(\hat{\mathbf{f}},\beta)$, and F-Pareto optimal solutions to MOP2.9$(\hat{\mathbf{p}},\beta)$ can be characterized by the following theorem.

Theorem 2.5
Assume that $\mathbf{x}^* \in X(\beta), \hat{f}_i^* \in F_i, i=1,\cdots,k, \lambda^* \in [0,1]$ is a unique optimal solution of MAXMIN2.5.
(1) $\mathbf{x}^* \in X(\beta)$ is a P-Pareto optimal solution to MOP2.7$(\hat{\mathbf{f}}^*,\beta)$.

(2) $\mathbf{x}^* \in X(\beta)$ is an F-Pareto optimal solution to MOP2.9$(\hat{\mathbf{p}}^*,\beta)$, where $\hat{p}_i^* \stackrel{\text{def}}{=} p_i(\mathbf{x}^*,\hat{f}_i^*), i=1,\cdots,k$, $\hat{\mathbf{p}}^* \stackrel{\text{def}}{=} (\hat{p}_1^*,\cdots,\hat{p}_k^*)$.

(Proof)
(1) Assume that $\mathbf{x}^* \in X(\beta)$ is not a P-Pareto optimal solution to MOP2.7$(\hat{\mathbf{f}}^*,\beta)$. Then, from Assumption 2.6, there exists $\mathbf{x} \in X(\beta)$ such that

$$\begin{aligned} & p_i(\mathbf{x},\hat{f}_i^*) \geq p_i(\mathbf{x}^*,\hat{f}_i^*) \\ \Leftrightarrow \quad & \mu_{\tilde{p}_i}(p_i(\mathbf{x},\hat{f}_i^*)) \geq \mu_{\tilde{p}_i}(p_i(\mathbf{x}^*,\hat{f}_i^*)), i=1,\cdots,k, \end{aligned}$$

with strict inequality holding for at least one i. This contradicts the fact that

$\mathbf{x}^* \in X(\beta), \hat{f}_i^* \in F_i, i = 1, \cdots, k, \lambda^* \in [0,1]$ is a unique optimal solution of MAXMIN2.5.

(2) Assume that $\mathbf{x}^* \in X(\beta)$ is not an F-Pareto optimal solution to MOP2.9 $(\hat{\mathbf{p}}^*, \beta)$, where $\hat{p}_i^* \stackrel{\text{def}}{=} p_i(\mathbf{x}^*, \hat{f}_i^*), i = 1, \cdots, k$, $\hat{\mathbf{p}}^* \stackrel{\text{def}}{=} (\hat{p}_1^*, \cdots, \hat{p}_k^*)$. Then, from (2.65) and Assumption 2.6, there exists $\mathbf{x} \in X(\beta)$ such that

$$
\begin{aligned}
& f_i(\mathbf{x}, \hat{p}_i^*) \leq f_i(\mathbf{x}^*, \hat{p}_i^*) \\
\Leftrightarrow\ & \Phi^{-1}(\hat{p}_i^*) \cdot \sqrt{\mathbf{x}^T V_i \mathbf{x}} + \mathbf{E}[\bar{\mathbf{c}}_i]\mathbf{x} \leq \Phi^{-1}(\hat{p}_i^*) \cdot \sqrt{\mathbf{x}^{*T} V_i \mathbf{x}^*} + \mathbf{E}[\bar{\mathbf{c}}_i]\mathbf{x}^* = \hat{f}_i^* \\
\Leftrightarrow\ & \hat{f}_i^* - \mathbf{E}[\bar{\mathbf{c}}_i]\mathbf{x} \geq \Phi^{-1}(\hat{p}_i^*) \cdot \sqrt{\mathbf{x}^T V_i \mathbf{x}} \\
\Leftrightarrow\ & \Phi\left(\frac{\hat{f}_i^* - \mathbf{E}[\bar{\mathbf{c}}_i]\mathbf{x}}{\sqrt{\mathbf{x}^T V_i \mathbf{x}}} \right) \geq \hat{p}_i^* \\
\Leftrightarrow\ & p_i(\mathbf{x}, \hat{f}_i^*) \geq p_i(\mathbf{x}^*, \hat{f}_i^*) \\
\Leftrightarrow\ & \mu_{\tilde{p}_i}(p_i(\mathbf{x}, \hat{f}_i^*)) \geq \mu_{\tilde{p}_i}(p_i(\mathbf{x}^*, \hat{f}_i^*)), i = 1, \cdots, k,
\end{aligned}
$$

with strict inequality holding for at least one i. This contradicts the fact that $\mathbf{x}^* \in X(\beta), \hat{f}_i^* \in F_i, i = 1, \cdots, k, \lambda^* \in [0,1]$ is a unique optimal solution of MAXMIN2.5. □

Due to Assumption 2.6, constraints (2.78) can be transformed into

$$
\begin{aligned}
& \mu_{p_i}(p_i(\mathbf{x}, \hat{f}_i)) \geq \lambda \\
\Leftrightarrow\ & p_i(\mathbf{x}, \hat{f}_i) \geq \mu_{p_i}^{-1}(\lambda) \\
\Leftrightarrow\ & \Phi\left(\frac{\hat{f}_i - \mathbf{E}[\bar{\mathbf{c}}_i]\mathbf{x}}{\sqrt{\mathbf{x}^T V_i \mathbf{x}}} \right) \geq \mu_{p_i}^{-1}(\lambda) \\
\Leftrightarrow\ & \hat{f}_i - \mathbf{E}[\bar{\mathbf{c}}_i]\mathbf{x} \geq \Phi^{-1}(\mu_{p_i}^{-1}(\lambda)) \cdot \sqrt{\mathbf{x}^T V_i \mathbf{x}}, i = 1, \cdots, k,
\end{aligned}
\tag{2.80}
$$

where $\mu_{p_i}^{-1}(\cdot)$ is an inverse function of $\mu_{p_i}(\cdot)$. From constraints (2.79) and Assumption 2.5, it holds that $\hat{f}_i \leq \mu_{\hat{f}_i}^{-1}(\lambda)$. Therefore, constraints (2.80) can be reduced to the following inequalities.

$$
\mu_{\hat{f}_i}^{-1}(\lambda) - \mathbf{E}[\bar{\mathbf{c}}_i]\mathbf{x} \geq \Phi^{-1}(\mu_{p_i}^{-1}(\lambda)) \cdot \sqrt{\mathbf{x}^T V_i \mathbf{x}}, i = 1, \cdots, k \tag{2.81}
$$

Next, MAXMIN2.5 is equivalently transformed into the following problem, where permissible objective levels $\hat{f}_i, i = 1, \cdots, k$ have disappeared.

MAXMIN 2.6

$$
\max_{\mathbf{x} \in X(\beta), \lambda \in [0,1]} \lambda \tag{2.82}
$$

subject to

$$
\mu_{\hat{f}_i}^{-1}(\lambda) - \mathbf{E}[\bar{\mathbf{c}}_i]\mathbf{x} \geq \Phi^{-1}(\mu_{p_i}^{-1}(\lambda)) \cdot \sqrt{\mathbf{x}^T V_i \mathbf{x}}, i = 1, \cdots, k \tag{2.83}
$$

To solve MAXMIN2.6, which is a nonlinear programming problem, we first define the following function for constraints (2.83).

$$g_i(\mathbf{x},\lambda) \stackrel{\text{def}}{=} \mu_{\hat{f}_i}^{-1}(\lambda) - \mathbf{E}[\bar{\mathbf{c}}_i]\mathbf{x} - \Phi^{-1}(\mu_{p_i}^{-1}(\lambda)) \cdot \sqrt{\mathbf{x}^T V_i \mathbf{x}}, i = 1,\cdots,k \quad (2.84)$$

Note that $g_i(\mathbf{x},\lambda), i = 1,\cdots,k$ are concave functions for any fixed $\lambda \in [0,1]$ because $\Phi^{-1}(\mu_{p_i}^{-1}(\lambda)) > 0$ under Assumption 2.6. Therefore, constraint set $G(\lambda)$ is a convex set for any $\lambda \in [0,1]$ and is defined as follows.

$$G(\lambda) \stackrel{\text{def}}{=} \{\mathbf{x} \in X(\beta) \mid g_i(\mathbf{x},\lambda) \geq 0, i = 1,\cdots,k\}. \quad (2.85)$$

Constraint set $G(\lambda)$ satisfies the following property for $\lambda \in [0,1]$.

Property 2.1
If $0 \leq \lambda_1 < \lambda_2 \leq 1$, then it holds that $G(\lambda_1) \supset G(\lambda_2)$.

(Proof)
From Assumption 2.5 and Assumption 2.6, it holds that $\mu_{\hat{f}_i}^{-1}(\lambda_1) > \mu_{\hat{f}_i}^{-1}(\lambda_2)$, $\Phi^{-1}(\mu_{p_i}^{-1}(\lambda_1)) < \Phi^{-1}(\mu_{p_i}^{-1}(\lambda_2)), i = 1,\cdots,k$. This means that $g_i(\mathbf{x},\lambda_1) > g_i(\mathbf{x},\lambda_2), i = 1,\cdots,k$ for any $\mathbf{x} \in X(\beta)$. Therefore, $G(\lambda_1) \supset G(\lambda_2)$ for any $0 \leq \lambda_1 < \lambda_2 \leq 1$. □

From Property 2.1, we can easily obtain optimal solution $\mathbf{x}^* \in X(\beta), \lambda^* \in [0,1]$ of MAXMIN2.6 by using the bisection method for $\lambda \in [0,1]$, where we assume $G(0) \neq \phi, G(1) = \phi$.

Algorithm 2.1

Step 1: *Set $\lambda_0 = 0, \lambda_1 = 1$ and $\lambda \leftarrow (\lambda_0 + \lambda_1)/2$.*
Step 2: *Solve the following convex programming problem for $\lambda \in [0,1]$. Let us denote the optimal solution as $\mathbf{x}^* \in X(\beta)$.*

$$\max_{\mathbf{x} \in X(\beta)} g_j(\mathbf{x},\lambda)$$

subject to

$$g_i(\mathbf{x},\lambda) \geq 0, i = 1,\cdots,k, i \neq j$$

Step 3: *If $\mid \lambda_1 - \lambda_0 \mid < \delta$ then go to Step 4, where δ is a sufficiently small positive constant. If $g_j(\mathbf{x}^*,\lambda) < 0$, or there exists the index $i \neq j$ such that $g_i(\mathbf{x}^*,\lambda) < 0$, then set $\lambda_1 \leftarrow \lambda, \lambda \leftarrow (\lambda_0 + \lambda_1)/2$, and go to Step 2. Otherwise, if $g_j(\mathbf{x}^*,\lambda) \geq 0$ and $g_i(\mathbf{x}^*,\lambda) \geq 0$ for any $i = 1,\cdots,k, i \neq j$, then set $\lambda_0 \leftarrow \lambda, \lambda \leftarrow (\lambda_0 + \lambda_1)/2$, and go to Step 2.*
Step 4: *Set $\lambda^* \leftarrow \lambda$, and optimal solution $\mathbf{x}^* \in X(\beta), \lambda^* \in [0,1]$ of MAXMIN2.6 is obtained.*

Similarly, we consider a fuzzy approach to MOP2.9($\hat{\mathbf{p}},\beta$), which is formulated by applying a fractile optimization model to MOP2.6. To handle

MOP2.9($\hat{\mathbf{p}}, \beta$), the decision maker must specify permissible probability levels $\hat{\mathbf{p}} = (\hat{p}_1, \cdots, \hat{p}_k)$ in advance.

Considering the imprecise nature of the decision maker's judgment, we assume the decision maker has a fuzzy goal for each objective function in MOP2.9($\hat{\mathbf{p}}, \beta$). Such fuzzy goals can be quantified via corresponding membership functions. Let us denote a membership function of objective function $f_i(\mathbf{x}, \hat{p}_i)$ as $\mu_{f_i}(f_i(\mathbf{x}, \hat{p}_i))$, and a membership function of permissible probability level $\hat{p}_i$ as $\mu_{\hat{p}_i}(\hat{p}_i)$. Throughout this subsection, we make the following assumptions for $\mu_{f_i}(f_i(\mathbf{x}, \hat{p}_i))$, $\mu_{\hat{p}_i}(\hat{p}_i), i = 1, \cdots, k$.

Assumption 2.7
$\mu_{\hat{p}_i}(\hat{p}_i), i = 1, \cdots, k$ are strictly increasing and continuous with respect to $\hat{p}_i \in [\hat{p}_{i\min}, \hat{p}_{i\max}]$, where $\hat{p}_{i\min}(> 0.5)$ is an acceptable minimum value and $\hat{p}_{i\max}(< 1)$ is a sufficiently satisfactory minimum value, i.e., $\mu_{\hat{p}_i}(\hat{p}_{i\min}) = 0, \mu_{\hat{p}_i}(\hat{p}_{i\max}) = 1, i = 1, \cdots, k$.

Assumption 2.8
$\mu_{f_i}(f_i(\mathbf{x}, \hat{p}_i)), i = 1, \cdots, k$ are strictly decreasing and continuous with respect to $f_i(\mathbf{x}, \hat{p}_i) \in [f_{i\min}, f_{i\max}]$, where $\hat{f}_{i\min}$ is a sufficiently satisfactory maximum value and $\hat{f}_{i\max}$ is an acceptable maximum value, i.e., $\mu_{f_i}(f_{i\min}) = 1, \mu_{f_i}(f_{i\max}) = 0, i = 1, \cdots, k$.

For example, we can define membership functions $\mu_{\hat{p}_i}(\hat{p}_i)$ and $\mu_{f_i}(f_i(\mathbf{x}, \hat{p}_i))$, $i = 1, \cdots, k$ as follows. According to Assumption 2.7, the decision maker first subjectively specifies permissible probability intervals $[\hat{p}_{i\min}, \hat{p}_{i\max}], i = 1, \cdots, k$, then defines membership function $\mu_{\hat{p}_i}(\hat{p}_i)$ on interval $[\hat{p}_{i\min}, \hat{p}_{i\max}]$. Corresponding to interval $[\hat{p}_{i\min}, \hat{p}_{i\max}]$, $f_{i\min}$ can be obtained by solving the following problem.

$$f_{i\min} = \min_{\mathbf{x} \in X(\beta)} f_i(\mathbf{x}, \hat{p}_{i\min}) \tag{2.86}$$

The above problem becomes a convex programming problem, because $\Phi^{-1}(\cdot) > 0$ under Assumption 2.7. To obtain $f_{i\max}$, we first solve the following k convex programming problems.

$$\min_{\mathbf{x} \in X(\beta)} f_i(\mathbf{x}, \hat{p}_{i\max}) \tag{2.87}$$

Let $\mathbf{x}_i, i = 1, \cdots, k$ be the above optimal solution. Using optimal solutions $\mathbf{x}_i, i = 1, \cdots, k$, $f_{i\max}$ can be obtained as follows [139].

$$f_{i\max} = \max_{\ell = 1, \cdots, k, \ell \neq i} f_i(\mathbf{x}_\ell, \hat{p}_{i\max}) \tag{2.88}$$

According to Assumption 2.8, the decision maker can define his or her membership function $\mu_{f_i}(f_i(\mathbf{x}, \hat{p}_i))$ on interval $[f_{i\min}, f_{i\max}]$.

Next, if the decision maker adopts the fuzzy decision [6, 71, 140] as an aggregation operator, a satisfactory solution is obtained by solving the following maxmin problem.

MAXMIN 2.7

$$\max_{\mathbf{x}\in X(\beta),\hat{p}_i\in P_i,i=1,\cdots,k,\lambda\in[0,1]} \lambda \tag{2.89}$$

subject to

$$\mu_{f_i}(f_i(\mathbf{x},\hat{p}_i)) \geq \lambda, i=1,\cdots,k \tag{2.90}$$

$$\mu_{\hat{p}_i}(\hat{p}_i) \geq \lambda, i=1,\cdots,k \tag{2.91}$$

Similar to Theorem 2.5, the following relationships between the optimal solutions of MAXMIN2.7, P-Pareto optimal solutions to MOP2.7($\hat{\mathbf{f}}^*,\beta$), and F-Pareto optimal solutions to MOP2.9($\hat{\mathbf{p}}^*,\beta$) can be characterized by the following theorem.

Theorem 2.6

Assume that $\mathbf{x}^* \in X(\beta), \hat{p}_i^* \in P_i, i=1,\cdots,k, \lambda^* \in [0,1]$ *is a unique optimal solution of MAXMIN2.7.*

(1) $\mathbf{x}^* \in X(\beta)$ *is an F-Pareto optimal solution to MOP2.9(*$\hat{\mathbf{p}}^*,\beta$*), where* $\hat{\mathbf{p}}^* \overset{\text{def}}{=} (\hat{p}_1^*,\cdots,\hat{p}_k^*)$.

(2) $\mathbf{x}^* \in X(\beta)$ *is a P-Pareto optimal solution to MOP2.7(*$\hat{\mathbf{f}}^*,\beta$*), where* $\hat{f}_i^* \overset{\text{def}}{=} f_i(\mathbf{x}^*,\hat{p}_i^*), i=1,\cdots,k$, $\hat{\mathbf{f}}^* \overset{\text{def}}{=} (\hat{f}_1^*,\cdots,\hat{f}_k^*)$.

Due to Assumption 2.8, the constraints in (2.90) can be transformed into

$$\begin{aligned} & \mu_{f_i}(f_i(\mathbf{x},\hat{p}_i)) \geq \lambda \\ \Leftrightarrow\quad & f_i(\mathbf{x},\hat{p}_i) \leq \mu_{f_i}^{-1}(\lambda) \\ \Leftrightarrow\quad & \Phi^{-1}(\hat{p}_i)\cdot\sqrt{\mathbf{x}^T V_i \mathbf{x}} + \mathbf{E}[\bar{\mathbf{c}}_i]\mathbf{x} \leq \mu_{f_i}^{-1}(\lambda) \\ \Leftrightarrow\quad & \hat{p}_i \leq \Phi\left(\frac{\mu_{f_i}^{-1}(\lambda) - \mathbf{E}[\bar{\mathbf{c}}_i]\mathbf{x}}{\sqrt{\mathbf{x}^T V_i \mathbf{x}}}\right), i=1,\cdots,k. \end{aligned} \tag{2.92}$$

From the constraints in (2.91) and Assumption 2.7, it holds that $\hat{p}_i \geq \mu_{\hat{p}_i}^{-1}(\lambda)$. Therefore, constraint (2.92) can be reduced to the following inequality in which permissible probability levels $\hat{p}_i, i=1,\cdots,k$ have disappeared.

$$\begin{aligned} & \mu_{\hat{p}_i}^{-1}(\lambda) \leq \Phi\left(\frac{\mu_{f_i}^{-1}(\lambda) - \mathbf{E}[\bar{\mathbf{c}}_i]\mathbf{x}}{\sqrt{\mathbf{x}^T V_i \mathbf{x}}}\right) \\ \Leftrightarrow\quad & \mu_{f_i}^{-1}(\lambda) - \mathbf{E}[\bar{\mathbf{c}}_i]\mathbf{x} \geq \Phi^{-1}(\mu_{\hat{p}_i}^{-1}(\lambda))\cdot\sqrt{\mathbf{x}^T V_i \mathbf{x}}, \\ & i=1,\cdots,k \end{aligned} \tag{2.93}$$

Then, MAXMIN2.7 is equivalently transformed into the following problem.

MAXMIN 2.8

$$\max_{\mathbf{x}\in X(\beta),\lambda\in[0,1]} \lambda \tag{2.94}$$

subject to

$$\mu_{f_i}^{-1}(\lambda)-\mathbf{E}[\bar{\mathbf{c}}_i]\mathbf{x}\geq\Phi^{-1}(\mu_{\hat{p}_i}^{-1}(\lambda))\cdot\sqrt{\mathbf{x}^T V_i\mathbf{x}}, i=1,\cdots,k \tag{2.95}$$

Note that MAXMIN2.8 is equivalent to MAXMIN2.6. Therefore, we can easily solve MAXMIN2.8 by applying Algorithm 2.1.

2.2.3 A numerical example

To demonstrate the efficiency of our proposed method for MOP2.6, we consider the following three objective stochastic programming problem [77].

MOP 2.10

$$\min \ (\bar{\mathbf{c}}_1\mathbf{x},\bar{\mathbf{c}}_2\mathbf{x},\bar{\mathbf{c}}_3\mathbf{x})$$

subject to

$$\mathbf{a}_j\mathbf{x}\leq\bar{b}_j, j=1,2,\mathbf{x}\geq\mathbf{0}$$

where $\mathbf{x}=(x_1,x_2,x_3,x_4)^T$ is a four-dimensional decision column vector, $\mathbf{a}_j, j=1,2$ are four-dimensional coefficient row vectors defined as $\mathbf{a}_1=(7,3,4,6), \mathbf{a}_2=(-5,-6,-7,-9)$, $(\bar{b}_1,\bar{b}_2)$ are Gaussian random variables defined as $b_1\sim \mathrm{N}(27,6^2)$, $b_2\sim\mathrm{N}(-15,7^2)$, $\bar{\mathbf{c}}_i, i=1,2,3$ are four-dimensional Gaussian random variable vectors with mean vectors

$$\mathbf{E}[\bar{\mathbf{c}}_1]=(2,3,2,4),\ \mathbf{E}[\bar{\mathbf{c}}_2]=(10,-7,1,-2),\ \mathbf{E}[\bar{\mathbf{c}}_3]=(-8,-5,-7,-14),$$

and corresponding variance covariance matrices $V_i, i=1,2,3$ are set as

$$V_1 = \begin{pmatrix} 25 & -1 & 0.8 & -2 \\ -1 & 4 & -1.2 & 1.2 \\ 0.8 & -1.2 & 4 & 2 \\ -2 & 1.2 & 2 & 9 \end{pmatrix}$$

$$V_2 = \begin{pmatrix} 16 & 1.4 & -1.2 & 1.4 \\ 1.4 & 1 & 1.5 & -0.8 \\ -1.2 & 1.5 & 25 & -0.6 \\ 1.4 & -0.8 & -0.6 & 4 \end{pmatrix}$$

$$V_3 = \begin{pmatrix} 4 & -1.9 & 1.5 & 1.8 \\ -1.9 & 25 & 0.8 & -0.4 \\ 1.5 & 0.8 & 9 & 2.5 \\ 1.8 & -0.4 & 2.5 & 36 \end{pmatrix}.$$

After setting permissible probability levels $\hat{p}_i, i=1,2,3$ through a fractile optimization model, MOP2.6 can be transformed into the following multiobjective programming problem.

MOP 2.11 $(\hat{\mathbf{p}}, \beta)$

$$\begin{array}{ll} \min & f_1(\mathbf{x}, \hat{p}_1) = \Phi^{-1}(\hat{p}_1) \cdot \sqrt{\mathbf{x}^T V_1 \mathbf{x}} + \mathbf{E}[\bar{\mathbf{c}}_1]\mathbf{x} \\ \min & f_2(\mathbf{x}, \hat{p}_2) = \Phi^{-1}(\hat{p}_2) \cdot \sqrt{\mathbf{x}^T V_2 \mathbf{x}} + \mathbf{E}[\bar{\mathbf{c}}_2]\mathbf{x} \\ \min & f_3(\mathbf{x}, \hat{p}_3) = \Phi^{-1}(\hat{p}_3) \cdot \sqrt{\mathbf{x}^T V_3 \mathbf{x}} + \mathbf{E}[\bar{\mathbf{c}}_3]\mathbf{x} \end{array}$$

subject to

$$\mathbf{x} \in X(\beta) = \{\mathbf{x} \geq \mathbf{0} \mid \Pr(\omega \mid \mathbf{a}_j \mathbf{x} \leq b_j(\omega)) \geq \beta_j, j = 1, 2\}$$

In MOP2.11$(\hat{\mathbf{p}}, \beta)$, we assume that the hypothetical decision maker first sets his or her constraint probability levels as $\beta = (\beta_1, \beta_2) = (0.7, 0.7)$, then subjectively specifies intervals $[\hat{p}_{i\min}, \hat{p}_{i\max}], i = 1, 2, 3$ for permissible probability levels $\hat{p}_i, i = 1, 2, 3$ according to Assumption 2.7 as follows.

$$\begin{array}{rcl} (\hat{p}_{1\min}, \hat{p}_{1\max}) & = & (0.6, 0.9) \\ (\hat{p}_{2\min}, \hat{p}_{2\max}) & = & (0.65, 0.95) \\ (\hat{p}_{3\min}, \hat{p}_{3\max}) & = & (0.6, 0.85) \end{array}$$

Suppose that the hypothetical decision maker defines the following linear membership functions $\mu_{\hat{p}_i}(\hat{p}_i), i = 1, 2, 3$ according to Assumption 2.7.

$$\mu_{\hat{p}_i}(\hat{p}_i) = \frac{\hat{p}_{i\min} - \hat{p}_i}{\hat{p}_{i\min} - \hat{p}_{i\max}}, i = 1, 2, 3 \tag{2.96}$$

Using $\hat{p}_{i\min}, \hat{p}_{i\max}, i = 1, 2, 3$, $\hat{f}_{i\min}$ and $\hat{f}_{i\max}, i = 1, 2, 3$ can be obtained as follows by solving convex programming problems (2.86), (2.87) and (2.88), i.e.,

$$\begin{array}{rcl} (f_{1\min}, f_{1\max}) & = & (4.05698, 55.9025) \\ (f_{2\min}, f_{2\max}) & = & (-66.4696, 13.6200) \\ (f_{3\min}, f_{3\max}) & = & (-62.7041, 1.83056). \end{array}$$

Considering these values, suppose that the hypothetical decision maker sets the following linear membership functions $\mu_{f_i}(f_i(\mathbf{x}, \hat{p}_i)), i = 1, 2, 3$ according to Assumption 2.8.

$$\mu_{\widetilde{f}_i}(f_i(\mathbf{x}, \hat{p}_i)) = \frac{f_i(\mathbf{x}, \hat{p}_i) - f_{i\max}}{f_{i\min} - f_{i\max}}, i = 1, 2, 3 \tag{2.97}$$

Then, using membership functions $\mu_{\hat{p}_i}(\hat{p}_i), \mu_{f_i}(f_i(\mathbf{x}, \hat{p}_i)), i = 1, 2, 3$, MAXMIN2.8 is formulated as

MAXMIN 2.9

$$\max_{\mathbf{x}\in X(\beta),\lambda\in[0,1]} \lambda$$

subject to

$$\mu_{f_i}^{-1}(\lambda) - \mathbf{E}[\bar{\mathbf{c}}_i]\mathbf{x} \geq \Phi^{-1}(\mu_{\hat{p}_i}^{-1}(\lambda)) \cdot \sqrt{\mathbf{x}^T V_i \mathbf{x}}, i = 1,2,3.$$

By applying Algorithm 2.1 to MAXMIN2.9, a satisfactory solution based on the fuzzy decision is easily obtained as

$$\begin{aligned} \lambda^* = \mu_{\hat{p}_i}(\hat{p}_i^*) = \mu_{f_i}(f_i(\mathbf{x}^*, \hat{p}_i^*)) &= 0.536605, i = 1,2,3 \\ (\hat{p}_1^*, \hat{p}_2^*, \hat{p}_3^*) &= (0.760982, 0.810982, 0.734151) \\ (f_1(\mathbf{x}^*, \hat{p}_1^*), f_2(\mathbf{x}^*, \hat{p}_2^*), f_3(\mathbf{x}^*, \hat{p}_3^*)) &= (28.0819, -29.3565, -32.7991). \end{aligned}$$

To show the efficiency of our proposed method, we compare our proposed method with previous approaches, in particular the fractile optimization model [77]. Using a fractile optimization model, MOP2.9$(\hat{\mathbf{p}}, \beta)$ can be transformed into the following three objective programming problem in which permissible probability levels $\hat{\mathbf{p}} = (\hat{p}_1, \hat{p}_2, \hat{p}_3)$ are set as $(\hat{p}_{1\max}, \hat{p}_{2\max}, \hat{p}_{3\max})$.

MOP 2.12 $(\hat{\mathbf{p}}_{\max}, \beta)$

$$\begin{aligned} \min \quad & f_1(\mathbf{x}, \hat{p}_{1\max}) = \Phi^{-1}(\hat{p}_{1\max}) \cdot \sqrt{\mathbf{x}^T V_1 \mathbf{x}} + \mathbf{E}[\bar{\mathbf{c}}_1]\mathbf{x} \\ \min \quad & f_2(\mathbf{x}, \hat{p}_{2\max}) = \Phi^{-1}(\hat{p}_{2\max}) \cdot \sqrt{\mathbf{x}^T V_2 \mathbf{x}} + \mathbf{E}[\bar{\mathbf{c}}_2]\mathbf{x} \\ \min \quad & f_3(\mathbf{x}, \hat{p}_{3\max}) = \Phi^{-1}(\hat{p}_{3\max}) \cdot \sqrt{\mathbf{x}^T V_3 \mathbf{x}} + \mathbf{E}[\bar{\mathbf{c}}_3]\mathbf{x} \end{aligned}$$

subject to

$$\mathbf{x} \in X(\beta) = \{\mathbf{x} \geq \mathbf{0} \mid \Pr(\omega \mid \mathbf{a}_j\mathbf{x} \leq b_j(\omega)) \geq \beta_j, j = 1,2\}$$

Table 2.3 summarizes both the optimal solution of MAXMIN2.9 and the optimal solution based on the fuzzy decision for MOP2.12$(\hat{\mathbf{p}}_{\max}, \beta)$. We observe here that a proper balance between the membership functions for permissible probability levels and the corresponding objective functions is attained at the optimal solution of our proposed method. Conversely, at the optimal solution obtained via the fractile optimization model, although permissible probability levels are fixed as $(\hat{p}_{1\max}, \hat{p}_{2\max}, \hat{p}_{3\max})$, membership functions of the corresponding objective function values become worse.

2.3 Interactive decision making for MOSPs with a special structure

In this section, we assume that a satisfactory solution based on the fuzzy decision [6, 71, 140] reflects the preference structure of the decision maker. Unfortunately,

Table 2.3: Comparing our proposed method to a fractile model in MOP2.10

	proposed method	fractile model
$f_1(\mathbf{x}, \hat{p}_{1\max})$	28.0819	33.1605 2150
$f_2(\mathbf{x}, \hat{p}_{2\max})$	–29.3565	–21.5113
$f_3(\mathbf{x}, \hat{p}_{3\max})$	–32.7991	–26.4776
p_1	0.8145	0.9
p_2	0.6577	0.95
p_3	0.6262	0.85

the decision maker does not always accept the fuzzy decision. To handle such a preference structure, in this section, MOSPs with a special structure are formulated [117]. Further, we propose an interactive linear programming algorithm to obtain a satisfactory solution in which the decision maker can update the candidate of a satisfactory solution by interacting with the computer until he or she is satisfied. In Section 2.3.1, we introduce an MP-Pareto optimal solution through a probability maximization model and formulate the minmax problem to obtain an MP-Pareto optimal solution. Similarly, in Section 2.3.2, we introduce an MF-Pareto optimal solution through a fractile optimization model and formulate the minmax problem to obtain an MF-Pareto optimal solution. In Section 2.3.3, we propose an interactive linear programming algorithm for obtaining a satisfactory solution from an MF-Pareto optimal solution set. Finally, in Section 2.3.4, we apply our interactive algorithm to a numerical example using a hypothetical decision maker.

2.3.1 A formulation using a probability model

In this subsection, we focus our efforts on MOSPs formally formulated as

MOP 2.13

$$\min_{\mathbf{x}\in X} \bar{\mathbf{z}}(\mathbf{x}) = (\bar{z}_1(\mathbf{x}), \cdots, \bar{z}_k(\mathbf{x})), \tag{2.98}$$

where $\mathbf{x} = (x_1, \cdots, x_n)^T$ is an n-dimensional decision variable column vector, X is a linear feasible set with respect to $\mathbf{x}$,

$$\bar{z}_i(\mathbf{x}) = \bar{\mathbf{c}}_i\mathbf{x} + \bar{\alpha}_i, i = 1, \cdots, k, \tag{2.99}$$

are multiobjective functions involving random variable coefficients, $\bar{\mathbf{c}}_i$ is an n-dimensional random variable row vector expressed by

$$\bar{\mathbf{c}}_i = \mathbf{c}_i^1 + \bar{t}_i\mathbf{c}_i^2, \tag{2.100}$$

where $\bar{t}_i$ is a random variable, and $\bar{\alpha}_i$ is a random variable row vector expressed by

$$\bar{\alpha}_i = \alpha_i^1 + \bar{t}_i\alpha_i^2. \tag{2.101}$$

Let us denote distribution functions of random variables $\bar{t}_i, i = 1, \cdots, k$ as $T_i(\cdot)$, which are strictly increasing and continuous.

If we adopt a probability maximization model for objective functions $\bar{z}_i(\mathbf{x}), i = 1, \cdots, k$ in MOP2.13, we can substitute the minimization of the objective functions for the maximization of the probability that each objective function $\bar{z}_i(\mathbf{x})$ is less than or equal to a certain permissible level $\hat{f}_i$. Such a probability function can be defined as

$$p_i(\mathbf{x}, \hat{f}_i) \stackrel{\text{def}}{=} \Pr(\omega \mid z_i(\mathbf{x}, \omega) \leq \hat{f}_i), i = 1, \cdots, k, \tag{2.102}$$

where

$$\hat{\mathbf{f}} = (\hat{f}_1, \cdots, \hat{f}_k), \tag{2.103}$$

is a k-dimensional vector of permissible objective levels specified subjectively by the decision maker. Then, MOP2.13 can be converted into the following form.

MOP 2.14 ($\hat{\mathbf{f}}$)

$$\max_{\mathbf{x} \in X} (p_1(\mathbf{x}, \hat{f}_1), \cdots, p_k(\mathbf{x}, \hat{f}_k)) \tag{2.104}$$

In MOP2.14($\hat{\mathbf{f}}$), assuming that $\mathbf{c}_i^2 \mathbf{x} + \alpha_i^2 > 0, i = 1, \cdots, k$ for any $\mathbf{x} \in X$, we can rewrite objective function $p_i(\mathbf{x}, \hat{f}_i)$ as follows.

$$\begin{aligned} p_i(\mathbf{x}, \hat{f}_i) &= \Pr(\omega \mid z_i(\mathbf{x}, \omega) \leq \hat{f}_i) \\ &= \Pr(\omega \mid \mathbf{c}_i(\omega)\mathbf{x} + \alpha_i(\omega) \leq \hat{f}_i) \\ &= \Pr\left(\omega \mid t_i(\omega) \leq \frac{\hat{f}_i - (\mathbf{c}_i^1 \mathbf{x} + \alpha_i^1)}{\mathbf{c}_i^2 \mathbf{x} + \alpha_i^2}\right) \\ &= T_i\left(\frac{\hat{f}_i - (\mathbf{c}_i^1 \mathbf{x} + \alpha_i^1)}{\mathbf{c}_i^2 \mathbf{x} + \alpha_i^2}\right) \end{aligned} \tag{2.105}$$

To handle MOP2.14($\hat{\mathbf{f}}$), the decision maker must specify permissible objective levels $\hat{\mathbf{f}}$ in advance; however, in general, the decision maker seems to prefer not only larger values of objective function $p_i(\mathbf{x}, \hat{f}_i)$ but also smaller values of permissible objective level $\hat{f}_i$. From this perspective, we consider the following multiobjective programming problem as a natural extension of MOP2.14($\hat{\mathbf{f}}$).

MOP 2.15

$$\max_{\mathbf{x} \in X, \hat{f}_i \in \mathrm{R}^1, i=1,\cdots,k} (p_1(\mathbf{x}, \hat{f}_1), \cdots, p_k(\mathbf{x}, \hat{f}_k), -\hat{f}_1, \cdots, -\hat{f}_k) \tag{2.106}$$

Note that in MOP2.15 that permissible objective levels $\hat{f}_i, i = 1, \cdots, k$ are not fixed values but decision variables.

Considering the imprecise nature of the decision maker's judgment, we assume that the decision maker has a fuzzy goal for each objective function in

MOP2.15. Such a fuzzy goal can be quantified via the corresponding membership function. Let us denote a membership function of objective function $p_i(\mathbf{x}, \hat{f}_i)$ as $\mu_{p_i}(p_i(\mathbf{x}, \hat{f}_i))$, and a membership function of permissible objective level $\hat{f}_i$ as $\mu_{\hat{f}_i}(\hat{f}_i)$. Then, MOP2.15 can be transformed into the following problem.

MOP 2.16

$$\max_{\mathbf{x}\in X, \hat{f}_i \in \mathrm{R}^1, i=1,\cdots,k} \left(\mu_{p_1}(p_1(\mathbf{x}, \hat{f}_1)), \cdots, \mu_{p_k}(p_k(\mathbf{x}, \hat{f}_k)), \mu_{\hat{f}_1}(\hat{f}_1), \cdots, \mu_{\hat{f}_k}(\hat{f}_k)\right) \tag{2.107}$$

Throughout this subsection, we make the following assumptions with respect to membership functions $\mu_{p_i}(p_i(\mathbf{x}, \hat{f}_i))$, $\mu_{\hat{f}_i}(\hat{f}_i), i = 1, \cdots, k$.

Assumption 2.9
$\mu_{\hat{f}_i}(\hat{f}_i), i = 1, \cdots, k$ are strictly decreasing and continuous with respect to $\hat{f}_i \in [f_{i\min}, f_{i\max}]$, where $\mu_{\hat{f}_i}(f_{i\min}) = 1, \mu_{\hat{f}_i}(f_{i\max}) = 0, i = 1, \cdots, k$.

Assumption 2.10
$\mu_{p_i}(p_i(\mathbf{x}, \hat{f}_i)), i = 1, \cdots, k$ are strictly increasing and continuous with respect to $p_i(\mathbf{x}, \hat{f}_i) \in [p_{i\min}, p_{i\max}]$, where $\mu_{p_i}(p_{i\min}) = 0, \mu_{p_i}(p_{i\max}) = 1, i = 1, \cdots, k$.

To appropriately determine these membership functions, let us assume that the decision maker sets $f_{i\min}$, $f_{i\max}$, $p_{i\min}$ and $p_{i\max}$ as follows. First, the decision maker subjectively specifies $f_{i\min}$ and $f_{i\max}$, where $f_{i\min}$ is a sufficiently satisfactory maximum value and $f_{i\max}$ is an acceptable maximum value; the decision maker then sets the corresponding membership function $\mu_{\hat{f}_i}(\hat{f}_i)$ according to Assumption 2.9. Corresponding to interval $[f_{i\min}, f_{i\max}]$, $p_{i\min}$ can be obtained by solving the following problem.

$$p_{i\max} = \max_{\mathbf{x}\in X} p_i(\mathbf{x}, \hat{f}_{i\max}), i = 1, \cdots, k \tag{2.108}$$

To obtain $p_{i\max}$, we first solve the following k linear programming problems.

$$\max_{\mathbf{x}\in X} p_i(\mathbf{x}, \hat{f}_{i\min}), i = 1, \cdots, k \tag{2.109}$$

Using optimal solutions $\mathbf{x}_i, i = 1, \cdots, k$ of (2.109), $f_{i\max}$ can be obtained as follows [139].

$$p_{i\min} = \min_{\ell=1,\cdots,k, \ell\neq i} p_i(\mathbf{x}_\ell, \hat{f}_{i\min}), i = 1, \cdots, k \tag{2.110}$$

The decision maker sets the membership function $\mu_{p_i}(p_i(\mathbf{x}, \hat{f}_i))$ defined on $[p_{i\min}, p_{i\max}]$ according to Assumption 2.10.

Note that from (2.105), $\mu_{\hat{f}_i}(\hat{f}_i)$ and $\mu_{p_i}(p_i(\mathbf{x}, \hat{f}_i))$ conflict with one another with respect to $\hat{f}_i$ for any $\mathbf{x} \in X$. Here, let us assume that the decision maker adopts the fuzzy decision [6, 71, 140] to integrate both membership functions

$\mu_{p_i}(p_i(\mathbf{x},\hat{f}_i))$ and $\mu_{\hat{f}_i}(\hat{f}_i)$. Then, integrated membership function $\mu_{D_{p_i}}(\mathbf{x},\hat{f}_i)$ can be defined as follows.

$$\mu_{D_{p_i}}(\mathbf{x},\hat{f}_i) \stackrel{\text{def}}{=} \min\{\mu_{\hat{f}_i}(\hat{f}_i), \mu_{p_i}(p_i(\mathbf{x},\hat{f}_i))\} \tag{2.111}$$

Using membership functions $\mu_{D_{p_i}}(\mathbf{x},\hat{f}_i), i = 1,\cdots,k$, MOP2.16 can be transformed into the following form.

MOP 2.17

$$\max_{\mathbf{x}\in X, \hat{f}_i = F_i, i=1,\cdots,k} \left(\mu_{D_{p_1}}(\mathbf{x},\hat{f}_1),\cdots,\mu_{D_{p_k}}(\mathbf{x},\hat{f}_k)\right) \tag{2.112}$$

where

$$F_i \stackrel{\text{def}}{=} [f_{i\min}, f_{i\max}], i = 1,\cdots,k.$$

To handle MOP2.17, we next introduce an MP-Pareto optimal solution.

Definition 2.5 $\mathbf{x}^* \in X, \hat{f}_i^* \in F_i, i = 1,\cdots,k$ is an MP-Pareto optimal solution to MOP2.17 if and only if there does not exist another $\mathbf{x} \in X, \hat{f}_i \in F_i, i = 1,\cdots,k$ such that $\mu_{D_{p_i}}(\mathbf{x},\hat{f}_i) \geq \mu_{D_{p_i}}(\mathbf{x}^*,\hat{f}_i^*)$ $i = 1,\cdots,k$, with strict inequality holding for at least one i.

To generate a candidate for a satisfactory solution that is also MP-Pareto optimal, the decision maker is asked to specify reference membership values [71]. Once reference membership values

$$\hat{\mu} = (\hat{\mu}_1,\cdots,\hat{\mu}_k) \tag{2.113}$$

are specified, the corresponding MP-Pareto optimal solution is obtained by solving the following minmax problem.

MINMAX 2.1 ($\hat{\mu}$)

$$\min_{\mathbf{x}\in X, \hat{f}_i \in F_i, i=1,\cdots,k, \lambda \in \mathrm{R}^1} \lambda \tag{2.114}$$

subject to

$$\hat{\mu}_i - \mu_{p_i}(p_i(\mathbf{x},\hat{f}_i)) \leq \lambda, i = 1,\cdots,k \tag{2.115}$$

$$\hat{\mu}_i - \mu_{\hat{f}_i}(\hat{f}_i) \leq \lambda, i = 1,\cdots,k \tag{2.116}$$

Because of Assumption 2.10 and $\mathbf{c}_i^2\mathbf{x} + \alpha_i^2 > 0$, $i = 1,\cdots,k$, constraints (2.115) can be transformed as follows.

$$\begin{aligned}
& \hat{\mu}_i - \mu_{p_i}(p_i(\mathbf{x},\hat{f}_i)) \leq \lambda \\
\Leftrightarrow\ & p_i(\mathbf{x},\hat{f}_i) \geq \mu_{p_i}^{-1}(\hat{\mu}_i - \lambda) \\
\Leftrightarrow\ & T_i\left(\frac{\hat{f}_i - (\mathbf{c}_i^1\mathbf{x} + \alpha_i^1)}{\mathbf{c}_i^2\mathbf{x} + \alpha_i^2}\right) \geq \mu_{p_i}^{-1}(\hat{\mu}_i - \lambda) \\
\Leftrightarrow\ & \hat{f}_i - (\mathbf{c}_i^1\mathbf{x} + \alpha_i^1) \\
& \geq T_i^{-1}(\mu_{p_i}^{-1}(\hat{\mu}_i - \lambda)) \cdot (\mathbf{c}_i^2\mathbf{x} + \alpha_i^2), i = 1,\cdots,k
\end{aligned} \tag{2.117}$$

where $\mu_{p_i}^{-1}(\cdot)$ is an inverse function of $\mu_{p_i}(\cdot)$. From constraints (2.116) and Assumption 2.9, it holds that $\hat{f}_i \leq \mu_{\hat{f}_i}^{-1}(\hat{\mu}_i - \lambda)$. Therefore, constraint (2.117) can be reduced to the following inequalities, where $\hat{f}_i, i = 1, \cdots, k$ have disappeared.

$$\mu_{\hat{f}_i}^{-1}(\hat{\mu}_i - \lambda) - (\mathbf{c}_i^1 \mathbf{x} + \alpha_i^1) \geq T_i^{-1}(\mu_{p_i}^{-1}(\hat{\mu}_i - \lambda)) \cdot (\mathbf{c}_i^2 \mathbf{x} + \alpha_i^2), i = 1, \cdots, k \tag{2.118}$$

Corresponding to reference membership values $\hat{\mu}_i \in [0,1], i = 1, \cdots, k$ specified by the decision maker, the feasible region of λ can be set as

$$\Lambda \stackrel{\text{def}}{=} [\max_{i=1,\cdots,k} \hat{\mu}_i - 1, \max_{i=1,\cdots,k} \hat{\mu}_i]. \tag{2.119}$$

Then, MINMAX2.1($\hat{\mu}$) can be equivalently reduced to the following problem.

MINMAX 2.2 ($\hat{\mu}$)

$$\min_{\mathbf{x} \in X, \lambda \in \Lambda} \lambda \tag{2.120}$$

subject to

$$\mu_{\hat{f}_i}^{-1}(\hat{\mu}_i - \lambda) - (\mathbf{c}_i^1 \mathbf{x} + \alpha_i^1) \geq T_i^{-1}(\mu_{p_i}^{-1}(\hat{\mu}_i - \lambda)) \cdot (\mathbf{c}_i^2 \mathbf{x} + \alpha_i^2), i = 1, \cdots, k \tag{2.121}$$

Note that constraints (2.121) can be reduced to a set of linear inequalities for some fixed value $\lambda \in \Lambda$, meaning that optimal solution $\mathbf{x}^* \in X, \lambda^* \in \Lambda$ of MINMAX2.2($\hat{\mu}$) is obtained by the combined use of the bisection method with respect to λ and the first phase of the two-phase simplex method of linear programming.

The relationship between optimal solution $\mathbf{x}^* \in X, \lambda^* \in \Lambda$ of MINMAX2.2($\hat{\mu}$) and MP-Pareto optimal solutions to MOP2.17 can be characterized by the following theorem.

Theorem 2.7

(1) If $\mathbf{x}^* \in X, \lambda^* \in \Lambda$ is a unique optimal solution of MINMAX2.2($\hat{\mu}$), then $\mathbf{x}^* \in X, \hat{f}_i^* \stackrel{\text{def}}{=} \mu_{\hat{f}_i}^{-1}(\hat{\mu}_i - \lambda^*) \in F_i, i = 1, \cdots, k$ is an MP-Pareto optimal solution to MOP2.17.

(2) If $\mathbf{x}^* \in X, \hat{f}_i^* \in F_i, i = 1, \cdots, k$ is an MP-Pareto optimal solution to MOP2.17, then $\mathbf{x}^* \in X$, $\lambda^* \stackrel{\text{def}}{=} \hat{\mu}_i - \mu_{\hat{f}_i}(\hat{f}_i^*) = \hat{\mu}_i - \mu_{p_i}(p_i(\mathbf{x}^*, \hat{f}_i^*)), i = 1, \cdots, k$ is an optimal solution of MINMAX2.2($\hat{\mu}$) for some reference membership values $\hat{\mu} = (\hat{\mu}_1, \cdots, \hat{\mu}_k)$.

(Proof)
(1) From (2.121), it holds that

$$\begin{aligned} \hat{\mu}_i - \lambda^* &\leq \mu_{p_i}(p_i(\mathbf{x}^*, \mu_{\hat{f}_i}^{-1}(\hat{\mu}_i - \lambda^*))), i = 1, \cdots, k, \\ \hat{\mu}_i - \lambda^* &= \mu_{\hat{f}_i}(\mu_{\hat{f}_i}^{-1}(\hat{\mu}_i - \lambda^*)), i = 1, \cdots, k. \end{aligned}$$

Assume that $\mathbf{x}^* \in X, \mu_{\hat{f}_i}^{-1}(\hat{\mu}_i - \lambda^*) \in F_i, i = 1, \cdots, k$ is not an MP-Pareto optimal solution to MOP2.17. Then, there exist $\mathbf{x} \in X, \hat{f}_i \in F_i, i = 1, \cdots, k$ such that

$$\begin{aligned}
\mu_{D_{p_i}}(\mathbf{x}, \hat{f}_i) &\stackrel{\text{def}}{=} \min\{\mu_{\hat{f}_i}(\hat{f}_i), \mu_{p_i}(p_i(\mathbf{x}, \hat{f}_i))\} \\
&\geq \mu_{D_{p_i}}(\mathbf{x}^*, \mu_{\hat{f}_i}^{-1}(\hat{\mu}_i - \lambda^*)) \\
&= \hat{\mu}_i - \lambda^*, i = 1, \cdots, k,
\end{aligned}$$

with strict inequality holding for at least one i. Then it holds that

$$\mu_{\hat{f}_i}(\hat{f}_i) \geq \hat{\mu}_i - \lambda^*, i = 1, \cdots, k \tag{2.122}$$

$$\mu_{p_i}(p_i(\mathbf{x}, \hat{f}_i)) \geq \hat{\mu}_i - \lambda^*, i = 1, \cdots, k. \tag{2.123}$$

From (2.105) and Assumptions 2.9 and 2.10, inequalities (2.122) and (2.123) can be transformed into the following inequalities.

$$\hat{f}_i \leq \mu_{\hat{f}_i}^{-1}(\hat{\mu}_i - \lambda^*), i = 1, \cdots, k$$

$$\hat{f}_i \geq (\mathbf{c}_i^1 \mathbf{x} + \alpha_i^1) + T_i^{-1}(\mu_{p_i}^{-1}(\hat{\mu}_i - \lambda^*)) \cdot (\mathbf{c}_i^2 \mathbf{x} + \alpha_i^2), i = 1, \cdots, k$$

This means that there exists some $\mathbf{x} \in X$ such that

$$\mu_{\hat{f}_i}^{-1}(\hat{\mu}_i - \lambda^*) - (\mathbf{c}_i^1 \mathbf{x} + \alpha_i^1) \geq T_i^{-1}(\mu_{p_i}^{-1}(\hat{\mu}_i - \lambda^*)) \cdot (\mathbf{c}_i^2 \mathbf{x} + \alpha_i^2), i = 1, \cdots, k,$$

which contradicts the fact that $\mathbf{x}^* \in X, \lambda^* \in \Lambda$ is a unique optimal solution of MINMAX2.2($\hat{\mu}$).

(2) Assume that $\mathbf{x}^* \in X, \lambda^* \in \Lambda$ is not an optimal solution of MINMAX2.2($\hat{\mu}$) for any reference membership values $\hat{\mu} = (\hat{\mu}_1, \cdots, \hat{\mu}_k)$ such that

$$\hat{\mu}_i - \lambda^* = \mu_{\hat{f}_i}(\hat{f}_i^*) = \mu_{p_i}(p_i(\mathbf{x}^*, \hat{f}_i^*)), i = 1, \cdots, k. \tag{2.124}$$

Then, there exists some $\mathbf{x} \in X, \lambda < \lambda^*$ such that

$$\begin{aligned}
& \mu_{\hat{f}_i}^{-1}(\hat{\mu}_i - \lambda) - (\mathbf{c}_i^1 \mathbf{x} + \alpha_i^1) \geq T_i^{-1}(\mu_{p_i}^{-1}(\hat{\mu}_i - \lambda)) \cdot (\mathbf{c}_i^2 \mathbf{x} + \alpha_i^2), \\
\Leftrightarrow \quad & \mu_{p_i}(p_i(\mathbf{x}, \mu_{\hat{f}_i}^{-1}(\hat{\mu}_i - \lambda)) \geq \hat{\mu}_i - \lambda, i = 1, \cdots, k.
\end{aligned} \tag{2.125}$$

Because of (2.124), (2.125) and $\hat{\mu}_i - \lambda > \hat{\mu}_i - \lambda^*, i = 1, \cdots, k$, the following inequalities hold.

$$\begin{aligned}
\mu_{\hat{f}_i}(\hat{f}_i) &> \mu_{\hat{f}_i}(\hat{f}_i^*), i = 1, \cdots, k \\
\mu_{p_i}(p_i(\mathbf{x}, \hat{f}_i)) &> \mu_{p_i}(p_i(\mathbf{x}^*, \hat{f}_i^*)), i = 1, \cdots, k
\end{aligned}$$

where $\hat{f}_i \stackrel{\text{def}}{=} \mu_{\hat{f}_i}^{-1}(\mu_i - \lambda) \in F_i$. This contradicts the fact that $\mathbf{x}^* \in X, \hat{f}_i^* \in F_i, i = 1, \cdots, k$ is an MP-Pareto optimal solution to MOP2.17. □

2.3.2 A formulation using a fractile model

In this subsection, we apply a fractile optimization model to MOP2.13. Thus, MOP2.13 can be converted into the following multiobjective programming problem, where

$$\hat{\mathbf{p}} = (\hat{p}_1, \cdots, \hat{p}_k) \tag{2.126}$$

is a k-dimensional vector of permissible probability levels specified by the decision maker and $f_i, i = 1, \cdots, k$ are target variables for the objective functions.

MOP 2.18 $(\hat{\mathbf{p}})$

$$\min_{\mathbf{x}\in X, f_i\in R^1, i=1,\cdots,k} (f_1, \cdots, f_k) \tag{2.127}$$

subject to

$$p_i(\mathbf{x}, f_i) \geq \hat{p}_i, i = 1, \cdots, k \tag{2.128}$$

In MOP2.18$(\hat{\mathbf{p}})$, since a distribution function $T_i(\cdot)$ is continuous and strictly increasing, constraints (2.128) can be transformed into the following form.

$$\begin{aligned} \hat{p}_i &\leq p_i(\mathbf{x}, f_i) = T_i\left(\frac{f_i - (\mathbf{c}_i^1\mathbf{x} + \alpha_i^1)}{\mathbf{c}_i^2\mathbf{x} + \alpha_i^2}\right) \\ \Leftrightarrow f_i &\geq T_i^{-1}(\hat{p}_i) \cdot (\mathbf{c}_i^2\mathbf{x} + \alpha_i^2) + (\mathbf{c}_i^1\mathbf{x} + \alpha_i^1), i = 1, \cdots, k \end{aligned} \tag{2.129}$$

Let us define the right-hand-side of inequality (2.129) as follows.

$$f_i(\mathbf{x}, \hat{p}_i) \stackrel{\text{def}}{=} T_i^{-1}(\hat{p}_i) \cdot (\mathbf{c}_i^2\mathbf{x} + \alpha_i^2) + (\mathbf{c}_i^1\mathbf{x} + \alpha_i^1) \tag{2.130}$$

Then, MOP2.18$(\hat{\mathbf{p}})$ can be equivalently reduced to the following simple form.

MOP 2.19 $(\hat{\mathbf{p}})$

$$\min_{\mathbf{x}\in X} (f_1(\mathbf{x}, \hat{p}_1), \cdots, f_k(\mathbf{x}, \hat{p}_k)) \tag{2.131}$$

In general, the decision maker seems to prefer not only smaller values of objective function $f_i(\mathbf{x}, \hat{p}_i)$ but also larger values of permissible probability level $\hat{p}_i$. From such a perspective, we consider the following multiobjective programming problem as a natural extension of MOP2.19$(\hat{\mathbf{p}})$.

MOP 2.20

$$\min_{\mathbf{x}\in X, \hat{p}_i\in(0,1), i=1,\cdots,k} (f_1(\mathbf{x}, \hat{p}_1), \cdots, f_k(\mathbf{x}, \hat{p}_k), -\hat{p}_1, \cdots, -\hat{p}_k) \tag{2.132}$$

Considering the imprecise nature of the decision maker's judgment, we assume that the decision maker has a fuzzy goal for each objective function in MOP2.20. Such a fuzzy goal can be quantified via the corresponding membership function. Here, let us denote a membership function of objective function

$f_i(\mathbf{x},\hat{p}_i)$ as $\mu_{f_i}(f_i(\mathbf{x},\hat{p}_i))$, and a membership function of permissible probability level $\hat{p}_i$ as $\mu_{\hat{p}_i}(\hat{p}_i)$. Then, MOP2.20 can be transformed into the following problem.

MOP 2.21

$$\max_{\mathbf{x}\in X,\hat{p}_i\in(0,1),i=1,\cdots,k} (\mu_{f_1}(f_1(\mathbf{x},\hat{p}_1)),\cdots,\mu_{f_k}(f_k(\mathbf{x},\hat{p}_k)),\mu_{\hat{p}_1}(\hat{p}_1),\cdots,\mu_{\hat{p}_k}(\hat{p}_k)) \tag{2.133}$$

Throughout this subsection, we make the following assumptions with respect to membership functions $\mu_{f_i}(f_i(\mathbf{x},\hat{p}_i))$, $\mu_{\hat{p}_i}(\hat{p}_i), i=1,\cdots,k$.

Assumption 2.11
$\mu_{\hat{p}_i}(\hat{p}_i), i=1,\cdots,k$ are strictly increasing and continuous with respect to $\hat{p}_i \in [p_{i\min}, p_{i\max}]$, where $\mu_{\hat{p}_i}(p_{i\min})=0, \mu_{\hat{p}_i}(p_{i\max})=1, i=1,\cdots,k$.

Assumption 2.12
$\mu_{f_i}(f_i(\mathbf{x},\hat{p}_i)), i=1,\cdots,k$ are strictly decreasing and continuous with respect to $f_i(\mathbf{x},\hat{p}_i)\in[f_{i\min}, f_{i\max}]$, where $\mu_{f_i}(f_{i\min})=1, \mu_{f_i}(f_{i\max})=0, i=1,\cdots,k$.

To appropriately determine these membership functions, first the decision maker subjectively specifies $p_{i\min}$ and $p_{i\max}$, where $p_{i\min}$ is an acceptable minimum value and $p_{i\max}$ is a sufficiently satisfactory minimum value, then sets membership function $\mu_{\hat{p}_i}(\hat{p}_i)$ defined on interval $[p_{i\min}, p_{i\max}]$ according to Assumption 2.11. Corresponding to interval $[p_{i\min}, p_{i\max}]$, $f_{i\min}$ can be obtained by solving the following linear programming problem.

$$f_{i\min} = \min_{\mathbf{x}\in X} f_i(\mathbf{x},\hat{p}_{i\min}), i=1,\cdots,k \tag{2.134}$$

To obtain $f_{i\max}$, we first solve the following k linear programming problems.

$$\min_{\mathbf{x}\in X} f_i(\mathbf{x},\hat{p}_{i\max}), i=1,\cdots,k \tag{2.135}$$

Let $\mathbf{x}_i, i=1,\cdots,k$ be the corresponding optimal solution. Using optimal solutions $\mathbf{x}_i, i=1,\cdots,k$, $f_{i\max}$ can be obtained as follows [139].

$$f_{i\max} = \max_{\ell=1,\cdots,k,\ell\neq i} f_i(\mathbf{x}_\ell,\hat{p}_{i\max}), i=1,\cdots,k \tag{2.136}$$

Here, the decision maker sets the membership function $\mu_{f_i}(f_i(\mathbf{x},\hat{p}_i))$ defined on interval $[f_{i\min}, f_{i\max}]$ according to Assumption 2.12.

Note that, from (2.130), $\mu_{f_i}(f_i(\mathbf{x},\hat{p}_i))$ and $\mu_{\hat{p}_i}(\hat{p}_i)$ conflict with one another with respect to $\hat{p}_i$ for any $\mathbf{x}\in X$. Here, let us assume that the decision maker adopts the fuzzy decision [6, 71, 140] to integrate both membership functions $\mu_{f_i}(f_i(\mathbf{x},\hat{p}_i))$ and $\mu_{\hat{p}_i}(\hat{p}_i)$. Then, integrated membership function $\mu_{D_{f_i}}(\mathbf{x},\hat{p}_i)$ can be defined as follows.

$$\mu_{D_{f_i}}(\mathbf{x},\hat{p}_i) \stackrel{\text{def}}{=} \min\{\mu_{\hat{p}_i}(\hat{p}_i), \mu_{f_i}(f_i(\mathbf{x},\hat{p}_i))\} \tag{2.137}$$

Using membership functions $\mu_{D_{f_i}}(\mathbf{x},\hat{p}_i), i=1,\cdots,k$, MOP2.21 can be transformed into the following form.

MOP 2.22

$$\max_{\mathbf{x}\in X,\hat{p}_i\in P_i,i=1,\cdots,k}\left(\mu_{D_{f_1}}(\mathbf{x},\hat{p}_1),\cdots,\mu_{D_{f_k}}(\mathbf{x},\hat{p}_k)\right) \tag{2.138}$$

where $P_i \stackrel{\text{def}}{=} [p_{i\min}, p_{i\max}], i = 1,\cdots,k$. In order to deal with MOP2.22, we introduce an MF-Pareto optimal solution concept.

Definition 2.6 $\mathbf{x}^* \in X, \hat{p}_i^* \in P_i, i = 1,\cdots,k$ is an MF-Pareto optimal solution to MOP2.22 if and only if there does not exist another $\mathbf{x} \in X, \hat{p}_i \in P_i, i = 1,\cdots,k$ such that $\mu_{D_{f_i}}(\mathbf{x},\hat{p}_i) \geq \mu_{D_{f_i}}(\mathbf{x}^*,\hat{p}_i^*)$ $i = 1,\cdots,k$ with strict inequality holding for at least one i.

To generate a candidate for a satisfactory solution that is also MF-Pareto optimal, the decision maker is asked to specify reference membership values [71]. Once reference membership values

$$\hat{\mu} = (\hat{\mu}_1,\cdots,\hat{\mu}_k) \tag{2.139}$$

are specified, the corresponding MF-Pareto optimal solution is obtained by solving the following minmax problem.

MINMAX 2.3 ($\hat{\mu}$)

$$\min_{\mathbf{x}\in X,\hat{p}_i\in P_i,i=1,\cdots,k,\lambda\in \mathrm{R}^1} \lambda \tag{2.140}$$

subject to

$$\hat{\mu}_i - \mu_{f_i}(f_i(\mathbf{x},\hat{p}_i)) \leq \lambda, i = 1,\cdots,k \tag{2.141}$$

$$\hat{\mu}_i - \mu_{\hat{p}_i}(\hat{p}_i) \leq \lambda, i = 1,\cdots,k \tag{2.142}$$

Because of Assumption 2.12 and $\mathbf{c}_i^2\mathbf{x} + \alpha_i^2 > 0$, $i = 1,\cdots,k$, constraints (2.141) can be transformed into

$$\begin{aligned}
& \hat{\mu}_i - \mu_{f_i}(f_i(\mathbf{x},\hat{p}_i)) \leq \lambda \\
\Leftrightarrow\ & f_i(\mathbf{x},\hat{p}_i) \leq \mu_{f_i}^{-1}(\hat{\mu}_i - \lambda) \\
\Leftrightarrow\ & T_i^{-1}(\hat{p}_i)\cdot(\mathbf{c}_i^2\mathbf{x}+\alpha_i^2) + (\mathbf{c}_i^1\mathbf{x}+\alpha_i^1) \leq \mu_{f_i}^{-1}(\hat{\mu}_i - \lambda) \\
\Leftrightarrow\ & \hat{p}_i \leq T_i\left(\frac{\mu_{f_i}^{-1}(\hat{\mu}_i-\lambda) - (\mathbf{c}_i^1\mathbf{x}+\alpha_i^1)}{\mathbf{c}_i^2\mathbf{x}+\alpha_i^2}\right), i = 1,\cdots,k.
\end{aligned} \tag{2.143}$$

From constraints (2.142) and Assumption 2.11, it holds that $\hat{p}_i \geq \mu_{\hat{p}_i}^{-1}(\hat{\mu}_i - \lambda)$. Therefore, constraint (2.143) can be reduced to the following inequality in which permissible probability levels $\hat{p}_i, i = 1,\cdots,k$ have disappeared.

$$\mu_{f_i}^{-1}(\hat{\mu}_i - \lambda) - (\mathbf{c}_i^1\mathbf{x}+\alpha_i^1) \geq T_i^{-1}(\mu_{\hat{p}_i}^{-1}(\hat{\mu}_i - \lambda))\cdot(\mathbf{c}_i^2\mathbf{x}+\alpha_i^2) \tag{2.144}$$

Corresponding to reference membership values $\hat{\mu}_i \in [0,1], i = 1, \cdots, k$ specified by the decision maker, the feasible region of λ can be set as

$$\Lambda \stackrel{\text{def}}{=} [\max_{i=1,\cdots,k} \hat{\mu}_i - 1, \max_{i=1,\cdots,k} \hat{\mu}_i] \in \mathbf{R}^1. \tag{2.145}$$

Then, MINMAX2.3($\hat{\mu}$) can be equivalently reduced to the following problem.

MINMAX 2.4 ($\hat{\mu}$)

$$\min_{\mathbf{x}\in X, \lambda\in\Lambda} \lambda \tag{2.146}$$

subject to

$$\mu_{\tilde{f}_i}^{-1}(\hat{\mu}_i - \lambda) - (\mathbf{c}_i^1\mathbf{x} + \alpha_i^1) \geq T_i^{-1}(\mu_{\tilde{p}_i}^{-1}(\hat{\mu}_i - \lambda)) \cdot (\mathbf{c}_i^2\mathbf{x} + \alpha_i^2), i = 1, \cdots, k \tag{2.147}$$

Note that constraints (2.147) can be reduced to a set of linear inequalities for some fixed value $\lambda \in \Lambda$, which means that optimal solution $\mathbf{x}^* \in X, \lambda^* \in \Lambda$ of MAXMIN2.4($\hat{\mu}$) is obtained by the combined use of the bisection method with respect to $\lambda \in \Lambda$ and the first phase of the two-phase simplex method of linear programming.

The relationship between optimal solution $\mathbf{x}^* \in X, \lambda^* \in \Lambda$ of MAXMIN2.4($\hat{\mu}$) and MF-Pareto optimal solutions to MOP2.22 can be characterized by the following theorem.

Theorem 2.8

(1) If $\mathbf{x}^* \in X, \lambda^* \in \Lambda$ is a unique optimal solution of MINMAX2.4($\hat{\mu}$), then $\mathbf{x}^* \in X, \hat{p}_i^* \stackrel{\text{def}}{=} \mu_{\tilde{p}_i}^{-1}(\hat{\mu}_i - \lambda^*) \in P_i, i = 1, \cdots, k$ is an MF-Pareto optimal solution to MOP2.22.

(2) If $\mathbf{x}^* \in X, \hat{p}_i^* \in P_i, i = 1, \cdots, k$ is an MF-Pareto optimal solution to MOP2.22, then $\mathbf{x}^* \in X$, $\lambda^* \in \Lambda$ is an optimal solution of MINMAX2.4($\hat{\mu}$) for some reference membership values $\hat{\mu} = (\hat{\mu}_1, \cdots, \hat{\mu}_k)$, where $\lambda^* \stackrel{\text{def}}{=} \hat{\mu}_i - \mu_{\tilde{p}_i}(\hat{p}_i^*) = \hat{\mu}_i - \mu_{\tilde{f}_i}(f_i(\mathbf{x}^*, \hat{p}_i^*)), i = 1, \cdots, k$.

(Proof)
(1) From (2.147), it holds that

$$\begin{aligned} \hat{\mu}_i - \lambda^* &\leq \mu_{\tilde{f}_i}(f_i(\mathbf{x}^*, \mu_{\tilde{p}_i}^{-1}(\hat{\mu}_i - \lambda^*))), i = 1, \cdots, k \\ \hat{\mu}_i - \lambda^* &= \mu_{\tilde{p}_i}(\mu_{\tilde{p}_i}^{-1}(\hat{\mu}_i - \lambda^*)), i = 1, \cdots, k. \end{aligned}$$

Assume that $\mathbf{x}^* \in X, \mu_{\tilde{p}_i}^{-1}(\hat{\mu}_i - \lambda^*) \in P_i, i = 1, \cdots, k$ is not an MF-Pareto optimal solution to MOP2.22. Then, there exist $\mathbf{x} \in X, \hat{p}_i \in P_i, i = 1, \cdots, k$ such that

$$\begin{aligned} \mu_{D_{f_i}}(\mathbf{x}, \hat{p}_i) &\stackrel{\text{def}}{=} \min\{\mu_{\hat{p}_i}(\hat{p}_i), \mu_{f_i}(f_i(\mathbf{x}, \hat{p}_i))\} \\ &\geq \mu_{D_{f_i}}(\mathbf{x}^*, \mu_{\hat{p}_i}^{-1}(\hat{\mu}_i - \lambda^*)) = \hat{\mu}_i - \lambda^*, i = 1, \cdots, k, \end{aligned}$$

with strict inequality holding for at least one i. Then it holds that

$$\mu_{\hat{p}_i}(\hat{p}_i) \geq \hat{\mu}_i - \lambda^*, i = 1, \cdots, k \tag{2.148}$$

$$\mu_{f_i}(f_i(\mathbf{x}, \hat{p}_i)) \geq \hat{\mu}_i - \lambda^*, i = 1, \cdots, k. \tag{2.149}$$

From definition (2.130) and Assumption 2.11, Assumption 2.12, inequalities (2.148) and (2.149) can be transformed into inequalities

$$\begin{aligned} \hat{p}_i &\geq \mu_{\hat{p}_i}^{-1}(\hat{\mu}_i - \lambda^*), i = 1, \cdots, k \\ \hat{p}_i &\leq T_i\left(\frac{\mu_{f_i}^{-1}(\hat{\mu}_i - \lambda^*) - (\mathbf{c}_i^1 \mathbf{x}^* + \alpha_i^1)}{\mathbf{c}_i^2 \mathbf{x}^* + \alpha_i^2}\right), i = 1, \cdots, k. \end{aligned}$$

This means that there exists some $\mathbf{x} \in X$ such that

$$\begin{aligned} & \mu_{f_i}^{-1}(\hat{\mu}_i - \lambda^*) - (\mathbf{c}_i^1 \mathbf{x} + \alpha_i^1) \\ \geq\ & T_i^{-1}(\mu_{\hat{p}_i}^{-1}(\hat{\mu}_i - \lambda^*)) \cdot (\mathbf{c}_i^2 \mathbf{x} + \alpha_i^2), i = 1, \cdots, k, \end{aligned}$$

which contradicts the fact that $\mathbf{x}^* \in X, \lambda^* \in \Lambda$ is a unique optimal solution to MINMAX2.4($\hat{\mu}$).

(2) Assume that $\mathbf{x}^* \in X, \lambda^* \in \Lambda$ is not an optimal solution to MINMAX2.4($\hat{\mu}$) for any reference membership values $\hat{\mu} = (\hat{\mu}_1, \cdots, \hat{\mu}_k)$, which satisfies inequalities

$$\hat{\mu}_i - \lambda^* = \mu_{\hat{p}_i}(\hat{p}_i^*) = \mu_{f_i}(f_i(\mathbf{x}^*, \hat{p}_i^*)), i = 1, \cdots, k. \tag{2.150}$$

Then, there exists some $\mathbf{x} \in X, \lambda < \lambda^*$ such that

$$\begin{aligned} & \mu_{f_i}^{-1}(\hat{\mu}_i - \lambda) - (\mathbf{c}_i^1 \mathbf{x} + \alpha_i^1) \geq T_i^{-1}(\mu_{\hat{p}_i}^{-1}(\hat{\mu}_i - \lambda)) \cdot (\mathbf{c}_i^2 \mathbf{x} + \alpha_i^2) \\ \Leftrightarrow\ & \mu_{f_i}(f_i(\mathbf{x}, \mu_{\hat{p}_i}^{-1}(\hat{\mu}_i - \lambda)) \geq \hat{\mu}_i - \lambda, i = 1, \cdots, k. \end{aligned} \tag{2.151}$$

Because of (2.150), (2.151), and $\hat{\mu}_i - \lambda > \hat{\mu}_i - \lambda^*, i = 1, \cdots, k$, the following inequalities hold.

$$\begin{aligned} \mu_{\hat{p}_i}(\hat{p}_i) &> \mu_{\hat{p}_i}(\hat{p}_i^*), i = 1, \cdots, k \\ \mu_{f_i}(f_i(\mathbf{x}, \hat{p}_i)) &> \mu_{f_i}(f_i(\mathbf{x}^*, \hat{p}_i^*)), i = 1, \cdots, k \end{aligned}$$

where $\hat{p}_i \stackrel{\text{def}}{=} \mu_{\hat{p}_i}^{-1}(\mu_i - \lambda)$. This contradicts the fact that $\mathbf{x}^* \in X, \hat{p}_i^* \in P_i, i = 1, \cdots, k$ is an MF-Pareto optimal solution to MOP2.22. □

2.3.3 An interactive linear programming algorithm

We begin this subsection by noting that MINMAX2.2($\hat{\mu}$) and MINMAX2.4($\hat{\mu}$) are equivalent to each another; however, to determine the membership functions, it seems to be more acceptable for the decision maker to determine membership functions $\mu_{p_i}(p_i(\mathbf{x}, \hat{f}_i)), i = 1, \cdots, k$, and corresponding domain $P_i = [p_{i\min}, p_{i\max}]$ in advance, because he or she may not be able to determine domain $F_i = [f_{i\min}, f_{i\max}]$ of the membership functions $\mu_{\hat{f}_i}(\hat{f}_i), i = 1, \cdots, k$ subjectively. From such a perspective, in this subsection, we propose an interactive algorithm based on MINMAX2.4($\hat{\mu}$) that obtains a satisfactory solution from an MF-Pareto optimal solution set.

Unfortunately, there is no guarantee that optimal solution $\mathbf{x}^* \in X, \lambda^* \in \Lambda$ of MINMAX2.4($\hat{\mu}$) is MF-Pareto optimal if $\mathbf{x}^* \in X, \lambda^* \in \Lambda$ is not unique. To guarantee MF-Pareto optimality, we first assume that k constraints (2.147) of MINMAX2.4($\hat{\mu}$) are active at optimal solution $\mathbf{x}^* \in X, \lambda^* \in \Lambda$, i.e.,

$$\mu_{\hat{f}_i}^{-1}(\hat{\mu}_i - \lambda^*) - (\mathbf{c}_i^1\mathbf{x}^* + \alpha_i^1) = T_i^{-1}(\mu_{\hat{p}_i}^{-1}(\hat{\mu}_i - \lambda^*)) \cdot (\mathbf{c}_i^2\mathbf{x}^* + \alpha_i^2), i = 1, \cdots, k. \tag{2.152}$$

If the j-th constraint of (2.147) is inactive, i.e.,

$$\begin{aligned} & \mu_{\hat{f}_j}^{-1}(\hat{\mu}_j - \lambda^*) - (\mathbf{c}_j^1\mathbf{x}^* + \alpha_j^1) > T_j^{-1}(\mu_{\hat{p}_j}^{-1}(\hat{\mu}_j - \lambda^*)) \cdot (\mathbf{c}_j^2\mathbf{x}^* + \alpha_j^2) \\ \Leftrightarrow \quad & \mu_{\hat{f}_j}^{-1}(\hat{\mu}_j - \lambda^*) > f_j(\mathbf{x}^*, \mu_{\hat{p}_j}^{-1}(\hat{\mu}_j - \lambda^*)), \end{aligned} \tag{2.153}$$

we can convert inactive constraint (2.153) into the active one by applying the bisection method for reference membership value $\hat{\mu}_j$.

For optimal solution $\mathbf{x}^* \in X, \lambda^* \in \Lambda$ of MINMAX2.4($\hat{\mu}$), where active conditions (2.152) are satisfied, we solve the MF-Pareto optimality test problem defined as follows.

Test 2.1

$$w \stackrel{\text{def}}{=} \max_{\mathbf{x} \in X, \varepsilon_i \geq 0, i=1,\cdots,k} \sum_{i=1}^{k} \varepsilon_i \tag{2.154}$$

subject to

$$\begin{aligned} & T_i^{-1}(\mu_{\hat{p}_i}^{-1}(\hat{\mu}_i - \lambda^*)) \cdot (\mathbf{c}_i^2\mathbf{x} + \alpha_i^2) + (\mathbf{c}_i^1\mathbf{x} + \alpha_i^1) + \varepsilon_i \\ = \quad & T_i^{-1}(\mu_{\hat{p}_i}^{-1}(\hat{\mu}_i - \lambda^*)) \cdot (\mathbf{c}_i^2\mathbf{x}^* + \alpha_i^2) + (\mathbf{c}_i^1\mathbf{x}^* + \alpha_i^1), \\ & i = 1, \cdots, k \end{aligned} \tag{2.155}$$

For the optimal solution of TEST2.1, the following theorem holds.

Theorem 2.9

For the optimal solution $\check{\mathbf{x}} \in X, \check{\varepsilon}_i \geq 0, i = 1, \cdots, k$ of TEST2.1, if $w = 0$ (equivalently,

$\check{\varepsilon}_i = 0, i = 1, \cdots, k$), then $\mathbf{x}^* \in X, \mu_{\hat{p}_i}^{-1}(\hat{\mu}_i - \lambda^*) \in P_i, i = 1, \cdots, k$ is an MF-Pareto optimal solution to MOP2.22.

(Proof)
From active conditions (2.152), it holds that

$$\begin{aligned} \hat{\mu}_i - \lambda^* &= \mu_{f_i}(f_i(\mathbf{x}^*, \mu_{\hat{p}_i}^{-1}(\hat{\mu}_i - \lambda^*))), i = 1, \cdots, k, \\ \hat{\mu}_i - \lambda^* &= \mu_{\hat{p}_i}(\mu_{\hat{p}_i}^{-1}(\hat{\mu}_i - \lambda^*)), i = 1, \cdots, k. \end{aligned}$$

If $\mathbf{x}^* \in X, \mu_{\hat{p}_i}^{-1}(\hat{\mu}_i - \lambda^*) \in P_i, i = 1, \cdots, k$ is not an MF-Pareto optimal solution to MOP2.22, there exists some $\mathbf{x} \in X, \hat{p}_i \in P_i, i = 1, \cdots, k$ such that

$$\begin{aligned} \mu_{D_{f_i}}(\mathbf{x}, \hat{p}_i) &= \min\{\mu_{\hat{p}_i}(\hat{p}_i), \mu_{f_i}(f_i(\mathbf{x}, \hat{p}_i))\} \\ &\geq \mu_{D_{f_i}}(\mathbf{x}^*, \mu_{\hat{p}_i}^{-1}(\hat{\mu}_i - \lambda^*)) \\ &= \hat{\mu}_i - \lambda^*, i = 1, \cdots, k, \end{aligned}$$

with strict inequality holding for at least one i. This means that the following inequalities hold.

$$\mu_{\hat{p}_i}(\hat{p}_i) \geq \hat{\mu}_i - \lambda^*, i = 1, \cdots, k \tag{2.156}$$

$$\mu_{f_i}(f_i(\mathbf{x}, \hat{p}_i)) \geq \hat{\mu}_i - \lambda^*, i = 1, \cdots, k \tag{2.157}$$

From (2.130) and Assumptions 2.11 and 2.12, (2.156) and (2.157) can be transformed into

$$\begin{aligned} \hat{p}_i &\geq \mu_{\hat{p}_i}^{-1}(\hat{\mu}_i - \lambda^*), i = 1, \cdots, k \\ \hat{p}_i &\leq T_i\left(\frac{\mu_{f_i}^{-1}(\hat{\mu}_i - \lambda^*) - (\mathbf{c}_i^1\mathbf{x} + \alpha_i^1)}{\mathbf{c}_i^2\mathbf{x} + \alpha_i^2}\right), i = 1, \cdots, k, \end{aligned}$$

indicating that there exists some $\mathbf{x} \in X$ such that

$$\begin{aligned} & \mu_{f_i}^{-1}(\hat{\mu}_i - \lambda^*) - (\mathbf{c}_i^1\mathbf{x} + \alpha_i^1) \geq T_i^{-1}(\mu_{\hat{p}_i}^{-1}(\hat{\mu}_i - \lambda^*)) \cdot (\mathbf{c}_i^2\mathbf{x} + \alpha_i^2) \\ \Leftrightarrow\ & T_i^{-1}(\mu_{\hat{p}_i}^{-1}(\hat{\mu}_i - \lambda^*)) \cdot (\mathbf{c}_i^2\mathbf{x}^* + \alpha_i^2) + (\mathbf{c}_i^1\mathbf{x}^* + \alpha_i^1) \\ & \geq T_i^{-1}(\mu_{\hat{p}_i}^{-1}(\hat{\mu}_i - \lambda^*)) \cdot (\mathbf{c}_i^2\mathbf{x} + \alpha_i^2) + (\mathbf{c}_i^1\mathbf{x} + \alpha_i^1), i = 1, \cdots, k, \end{aligned}$$

with strict inequality holding for at least one i. This contradicts the fact that $w = 0$. □

In the case of $w > 0$ at the optimal solution of TEST2.1, we can set the optimal solution of TEST2.1 as $\check{\mathbf{x}} \in X, \check{\varepsilon}_i = 0, i \in I, \check{\varepsilon}_j > 0, j \in J, I \cup J = \{1, \cdots, k\}, I \cap J = \phi$. Then, it holds that

$$\begin{aligned} & \mu_{f_j}^{-1}(\hat{\mu}_j - \lambda^*) - (\mathbf{c}_j^1\check{\mathbf{x}} + \alpha_j^1) > T_j^{-1}(\mu_{\hat{p}_j}^{-1}(\hat{\mu}_j - \lambda^*)) \cdot (\mathbf{c}_j^2\check{\mathbf{x}} + \alpha_j^2) \\ \Leftrightarrow\ & \mu_{f_j}^{-1}(\hat{\mu}_j - \lambda^*) > f_j(\check{\mathbf{x}}, \mu_{\hat{p}_j}^{-1}(\hat{\mu}_j - \lambda^*)), j \in J. \end{aligned} \tag{2.158}$$

Constraint (2.158) can be transformed into the active constraint by applying the bisection method for reference membership value $\hat{\mu}_j$.

Given the above, we next present the interactive algorithm for deriving a satisfactory solution from an MF-Pareto optimal solution set.

Algorithm 2.2

Step 1: *The decision maker subjectively specifies* $p_{i\min}(>0)$ *and* $p_{i\max}(<1), i=1,\cdots,k$. *On interval* $P_i=[p_{i\min},p_{i\max}]$, *the decision maker sets his or her membership functions* $\mu_{\hat{p}_i}(\hat{p}_i), i=1,\cdots,k$ *according to Assumption 2.11.*
Step 2: *Corresponding to interval* P_i, *compute* $f_{i\min}$ *and* $f_{i\max}$ *by solving problems (2.134) and (2.136). On interval* $F_i=[f_{i\min},f_{i\max}]$, *the decision maker sets his or her membership functions* $\mu_{f_i}(f_i(\mathbf{x},\hat{p}_i)), i=1,\cdots,k$ *according to Assumption 2.12.*
Step 3: *Set the initial reference membership values as* $\hat{\mu}_i=1, i=1,\cdots,k$.
Step 4: *Solve MINMAX2.4*($\hat{\mu}$) *by the combined use of the bisection method and the first phase of the two-phase simplex method of linear programming. For optimal solution* $(\mathbf{x}^*,\lambda^*)$, *MF-Pareto optimality test problem TEST2.1 is solved.*
Step 5: *If the decision maker is satisfied with the current values of MF-Pareto optimal solution* $\mu_{D_{f_i}}(\mathbf{x}^*,\hat{p}_i^*)$, $i=1,\cdots,k$, *where* $\hat{p}_i^* \stackrel{\text{def}}{=} \mu_{\hat{p}_i}^{-1}(\hat{\mu}_i-\lambda^*), i=1,\cdots,k$, *then stop. Otherwise, the decision maker must update his or her reference membership values* $\hat{\mu}_i, i=1,\cdots,k$, *and return to Step 4.*

2.3.4 A numerical example

In this subsection, we consider the following three objective stochastic programming problem to demonstrate the efficiency of our proposed method, formulated primarily by Sakawa et al. [73],

MOP 2.23

$$\begin{array}{ll} \min & (\mathbf{c}_1^1+\bar{t}_1\mathbf{c}_1^2)\mathbf{x}+(\alpha_1^1+\bar{t}_1\alpha_1^2) \\ \min & (\mathbf{c}_2^1+\bar{t}_2\mathbf{c}_2^2)\mathbf{x}+(\alpha_2^1+\bar{t}_2\alpha_2^2) \\ \min & (\mathbf{c}_3^1+\bar{t}_3\mathbf{c}_3^2)\mathbf{x}+(\alpha_3^1+\bar{t}_3\alpha_3^2) \end{array}$$

subject to

$$\mathbf{x}\in X \stackrel{\text{def}}{=} \{\mathbf{x}\in \mathrm{R}^{10} \mid \mathbf{a}_i\mathbf{x}\le b_i, i=1,\cdots,7, \mathbf{x}\ge \mathbf{0}\}$$

where $\mathbf{a}_i, i=1,\cdots,7, \mathbf{c}_1^1, \mathbf{c}_1^2, \mathbf{c}_2^1, \mathbf{c}_2^2, \mathbf{c}_3^1, \mathbf{c}_3^2$, are shown in Table 2.4. $\alpha_1^1=-18, \alpha_1^2=5, \alpha_2^1=-27, \alpha_2^2=6, \alpha_3^1=-10, \alpha_3^2=4$, $b_1=132.907$, $b_2=-222.897$, $b_3=-196.624$, $b_4=70.8059$, $b_5=-167.619$, $b_6=124.543$, $b_7=88.1748$, and $\bar{t}_i, i=1,2,3$ are Gaussian random variables defined as

$$\bar{t}_1\sim \mathrm{N}(4,2^2),\ \bar{t}_2\sim \mathrm{N}(3,3^2),\ \bar{t}_3\sim \mathrm{N}(3,2^2).$$

Table 2.4: Parameters of the objective functions and constraints in MOP2.23

$\mathbf{x}$	x_1	x_2	x_3	x_4	x_5	x_6	x_7	x_8	x_9	x_{10}
$\mathbf{c}_1^1$	19	48	21	10	18	35	46	11	24	33
$\mathbf{c}_1^2$	3	2	2	1	4	3	1	2	4	2
$\mathbf{c}_2^1$	12	–46	–23	–38	–33	–48	12	8	19	20
$\mathbf{c}_2^2$	1	2	4	2	2	1	2	1	2	1
$\mathbf{c}_3^1$	–18	–26	–22	–28	–15	–29	–10	–19	–17	–28
$\mathbf{c}_3^2$	2	1	3	2	1	2	3	3	2	1
$\mathbf{a}_1$	12	–2	4	–7	13	–1	–6	6	11	–8
$\mathbf{a}_2$	–2	5	3	16	6	–12	12	4	–7	–10
$\mathbf{a}_3$	3	–16	–4	–8	–8	2	–12	–12	4	–3
$\mathbf{a}_4$	–11	6	–5	9	–1	8	–4	6	–9	6
$\mathbf{a}_5$	–4	7	–6	–5	13	6	–2	–5	14	–6
$\mathbf{a}_6$	5	–3	14	–3	–9	–7	4	–4	–5	9
$\mathbf{a}_7$	–3	–4	–6	9	6	18	11	–9	–4	7

We apply Algorithm 2.2 to MOP2.23 using a hypothetical decision maker. Assume that the hypothetical decision maker sets $P_i \stackrel{\text{def}}{=} [p_{i\min}, p_{i\max}] = [0.5, 0.9]$, $i = 1,2,3$, and sets linear membership functions for the permissible probability levels (Step 1), i.e.,

$$\mu_{\hat{p}_i}(\hat{p}_i) = \frac{\hat{p}_i - p_{i\min}}{p_{i\max} - p_{i\min}}, i = 1,2,3. \tag{2.159}$$

By solving problems (2.134) and (2.136) for given values $p_{i\min}, p_{i\max}, i = 1,2,3$, $F_i \stackrel{\text{def}}{=} [f_{i\min}, f_{i\max}], i = 1,2,3$ are obtained as $[f_{1\min}, f_{1\max}] = [1855.57, 2599.30]$, $[f_{2\min}, f_{2\max}] = [340.617, 1066.26]$, $[f_{3\min}, f_{3\max}] = [-1067.25, -610.939]$. On intervals $F_i, i = 1,2,3$, the hypothetical decision maker sets linear membership functions for the objective functions as follows (Step 2).

$$\mu_{f_i}(f_i(\mathbf{x}, \hat{p}_i)) = \frac{f_{i\min} - f_i(\mathbf{x}, \hat{p}_i)}{f_{i\max} - f_{i\min}}, i = 1,2,3 \tag{2.160}$$

Next, we set initial reference membership values as $\hat{\mu}_i = 1, i = 1,2,3$ (Step 3), and solve MINMAX2.4($\hat{\mu}$) to obtain MF-Pareto optimal solution $\mathbf{x}^* \in X, \lambda^* \in \Lambda$ (Step 4). The interactive processes of the hypothetical decision maker are shown in Table 2.5. At the first iteration, optimal solution $\mu_{D_{f_i}}(\mathbf{x}^*, \hat{p}_i^*) = 0.560356, i = 1,2,3$ is obtained, where $\hat{p}_i^* \stackrel{\text{def}}{=} \mu_{\hat{p}_i}^{-1}(\hat{\mu}_i - \lambda^*), i = 1,2,3$. To improve $\mu_{D_{f_1}}(\mathbf{x}^*, \hat{p}_1^*)$ at the expense of $\mu_{D_{f_2}}(\mathbf{x}^*, \hat{p}_2^*)$, the hypothetical decision maker updates reference membership values as $(\hat{\mu}_1, \hat{\mu}_2, \hat{\mu}_3) = (0.61, 0.54, 0.56)$. In this example, at the third iteration, we obtain a satisfactory solution for the decision maker.

Below, we focus on the optimal solution at the first iteration, shown in Table 2.5. To compare our proposed method with a probability maximization model,

Table 2.5: Interactive processes in MOP2.23

	1	2	3
$\hat{\mu}_1$	1	0.61	0.585
$\hat{\mu}_2$	1	0.54	0.53
$\hat{\mu}_3$	1	0.56	0.541
$\mu_{D_{f_1}}(\mathbf{x}^*, \hat{p}_1^*)$	0.560356	0.591202	0.587493
$\mu_{D_{f_2}}(\mathbf{x}^*, \hat{p}_2^*)$	0.560356	0.521202	0.532493
$\mu_{D_{f_3}}(\mathbf{x}^*, \hat{p}_3^*)$	0.560356	0.541202	0.543493
$\hat{p}_1^*$	0.724142	0.736481	0.734997
$\hat{p}_2^*$	0.724142	0.708481	0.712997
$\hat{p}_3^*$	0.724142	0.716481	0.717397
$f_1(\mathbf{x}^*, \hat{p}_1^*)$	2182.55	2159.61	2162.37
$f_2(\mathbf{x}^*, \hat{p}_2^*)$	659.642	688.054	679.861
$f_3(\mathbf{x}^*, \hat{p}_3^*)$	–866.634	-857.894	-858.939

we first formulate the minmax problem through a probability maximization model for MOP2.23.

MINMAX 2.5 ($\hat{\mu}, \hat{\mathbf{f}}$)

$$\min_{\mathbf{x}\in X, \lambda\in\Lambda} \lambda$$

subject to

$$\hat{\mu}_i - \mu_{\hat{p}_i}(p_i(\mathbf{x}, \hat{f}_i)) \leq \lambda, i = 1,2,3$$

Table 2.6 shows optimal solutions of MINMAX2.5($\hat{\mu}, \hat{\mathbf{f}}$) in which permissible objective levels $(\hat{f}_1, \hat{f}_2, \hat{f}_3)$ are set to $(2000, 500, -900)$, (2182.55, 659.642, –866.634), and $(2300, 800, -800)$, respectively, and reference membership values $\hat{\mu}_i, i = 1,2,3$ are all set to one. In Table 2.6, optimal solution B coincides with the optimal solution at the first iteration of Table 2.5. At optimal solution A, instead of improving permissible objective levels in comparison with optimal solution B, the corresponding probability function values $p_i(\mathbf{x}^*, \hat{f}_i), i = 1,2,3$ worsen. At optimal solution C, instead of changing permissible objective levels for the worse as compared to optimal solution B, the corresponding probability function values $p_i(\mathbf{x}^*, \hat{f}_i), i = 1,2,3$ are improved. As a result, at optimal solution B, which is obtained by solving MINMAX2.4($\hat{\mu}$), a proper balance between permissible objective levels and the corresponding probability function values is attained through the fuzzy decision between membership functions (2.159) and (2.160).

Similarly, to compare our proposed method with the fractile optimization model, we formulate the minmax problem through a fractile optimization model for MOP2.23 as follows.

Table 2.6: Optimal solutions obtained via a probability model in MOP2.23

	A	B	C
$\hat{f}_1$	2000	2182.55	2300
$\hat{f}_2$	500	659.642	800
$\hat{f}_3$	–900	–866.634	–800
$\mu_{\hat{p}_1}(p_1(\mathbf{x}^*,\hat{f}_1))$	0.158237	0.560357	0.806599
$\mu_{\hat{p}_2}(p_2(\mathbf{x}^*,\hat{f}_2))$	0.158237	0.560357	0.806599
$\mu_{\hat{p}_3}(p_3(\mathbf{x}^*,\hat{f}_3))$	0.158237	0.560357	0.806599
$p_1(\mathbf{x}^*,\hat{f}_1)$	0.563295	0.724148	0.822639
$p_2(\mathbf{x}^*,\hat{f}_2)$	0.563295	0.724148	0.822639
$p_3(\mathbf{x}^*,\hat{f}_3)$	0.563295	0.724148	0.822639

MINMAX 2.6 $(\hat{\mu},\hat{\mathbf{p}})$

$$\min_{\mathbf{x}\in X,\lambda\in\Lambda} \lambda$$

subject to

$$\hat{\mu}_i - \mu_{f_i}(f_i(\mathbf{x},\hat{p}_i)) \leq \lambda, i=1,2,3$$

Table 2.7 shows optimal solutions of MINMAX2.6$(\hat{\mu},\hat{\mathbf{p}})$ in which permissible probability levels $(\hat{p}_1,\hat{p}_2,\hat{p}_3)$ are set to $(0.62,0.68,0.65)$, $(0.724142,0.724142,0.724142)$, and $(0.84,0.81,0.82)$, respectively, and reference membership values $\hat{\mu}_i, i=1,2,3$ are all set to one. In Table 2.7, optimal solution B coincides with the optimal solution at the first iteration of Table 2.5. At optimal solution A, instead of obtaining worse permissible objective levels in comparison with optimal solution B, the corresponding objective function values $f_i(\mathbf{x}^*,\hat{p}_i), i=1,2,3$ improve. At optimal solution C, instead of improving permissible probability levels in comparison with the optimal solution B, the corresponding objective function values $f_i(\mathbf{x}^*,\hat{p}_i), i=1,2,3$ worsen. As a result, at optimal solution B, which is obtained by solving MINMAX2.4$(\hat{\mu})$, a proper balance between permissible probability levels and the corresponding objective function values is attained through the fuzzy decision between membership functions (2.159) and (2.160).

2.4 Interactive decision making for MOSPs with variance covariance matrices

In this section, we formulate MOSPs with variance covariance matrices and propose an interactive convex programming algorithm to obtain a satisfactory solution [116]. In Section 2.4.1, we introduce an MP-Pareto optimal solution using a probability maximization model and formulate a minmax problem to ob-

Table 2.7: Optimal solutions obtained via a fractile model in MOP2.23

	A	B	C
$\hat{p}_1$	0.62	0.724142	0.84
$\hat{p}_2$	0.68	0.724142	0.81
$\hat{p}_3$	0.65	0.724142	0.82
$\mu_{f_1}(f_1(\mathbf{x}^*, \hat{p}_1))$	0.671099	0.560364	0.40212
$\mu_{f_2}(f_2(\mathbf{x}^*, \hat{p}_2))$	0.671099	0.560364	0.40212
$\mu_{f_3}(f_3(\mathbf{x}^*, \hat{p}_3))$	0.671099	0.560364	0.40212
$f_1(\mathbf{x}^*, \hat{p}_1)$	2100.184	2182.541	2300.232
$f_2(\mathbf{x}^*, \hat{p}_2)$	579.2821	659.6361	774.4648
$f_3(\mathbf{x}^*, \hat{p}_3)$	–917.167	–866.638	–794.43

tain an MP-Pareto optimal solution. Similarly, in Section 2.4.2, we introduce an MF-Pareto optimal solution using a fractile optimization model and formulate a minmax problem to obtain an MF-Pareto optimal solution. In Section 2.4.3, we propose an interactive algorithm for obtaining a satisfactory solution from an MF-Pareto optimal solution set. Finally, in Section 2.4.4, we apply our interactive algorithm to a numerical example using a hypothetical decision maker.

2.4.1 A formulation using a probability model

In this subsection, we focus on a multiobjective stochastic programming problem with variance covariance matrices, which we formally formulate as

MOP 2.24

$$\min_{\mathbf{x}\in X} \bar{\mathbf{C}}\mathbf{x} = (\bar{\mathbf{c}}_1\mathbf{x}, \cdots, \bar{\mathbf{c}}_k\mathbf{x}),$$

where X is a linear constraint set, $\mathbf{x} = (x_1, \cdots, x_n)^T$ is an n-dimensional decision variable column vector, $\bar{\mathbf{c}}_i \stackrel{\text{def}}{=} (\bar{c}_{i1}, \cdots, \bar{c}_{in})$ is an n-dimensional random variable row vector, $\bar{c}_{i\ell}, i = 1, \cdots, k, \ell = 1, \cdots, n$ are Gaussian random variables, i.e., $\bar{c}_{i\ell} \sim \mathrm{N}(E[\bar{c}_{i\ell}], \sigma_{i\ell\ell})$, and variance covariance matrices between Gaussian random variables $\bar{c}_{i\ell}, \ell = 1, \cdots, n$ are given as $(n \times n)$-dimensional matrices $V_i, i = 1, \cdots, k$, i.e.,

$$V_i = \begin{pmatrix} \sigma_{i11} & \sigma_{i12} & \cdots & \sigma_{i1n} \\ \sigma_{i21} & \sigma_{i22} & \cdots & \sigma_{i2n} \\ \vdots & \vdots & \ddots & \vdots \\ \sigma_{in1} & \sigma_{in2} & \cdots & \sigma_{inn} \end{pmatrix}, i = 1, \cdots, k. \tag{2.161}$$

Let us denote the vectors of the expectation for random variable row vectors $\bar{\mathbf{c}}_i, i = 1, \cdots, k$ as

$$\mathbf{E}[\bar{\mathbf{c}}_i] = (E[\bar{c}_{i1}], \cdots, E[\bar{c}_{in}]), i = 1, \cdots, k. \tag{2.162}$$

Then, using variance covariance matrices $V_i, i = 1, \cdots, k$, we can formally express vector $\bar{\mathbf{c}}_i$ as

$$\bar{\mathbf{c}}_i \sim \mathrm{N}(\mathbf{E}[\bar{\mathbf{c}}_i], V_i), i = 1, \cdots, k. \tag{2.163}$$

From the property of Gaussian random variables, objective function $\bar{\mathbf{c}}_i\mathbf{x}$ becomes Gaussian random variable.

$$\bar{\mathbf{c}}_i\mathbf{x} \sim \mathrm{N}(\mathbf{E}[\bar{\mathbf{c}}_i]\mathbf{x}, \mathbf{x}^T V_i \mathbf{x}), i = 1, \cdots, k. \tag{2.164}$$

If we adopt a probability maximization model for MOP2.24, we can substitute the minimization of objective functions $\bar{\mathbf{c}}_i\mathbf{x}, i = 1, \cdots, k$ for the maximization of the probability that each objective function $\bar{\mathbf{c}}_i\mathbf{x}$ is less than or equal to a certain permissible level $\hat{f}_i$ (called a permissible objective level). We define such a probability function as

$$p_i(\mathbf{x}, \hat{f}_i) \stackrel{\mathrm{def}}{=} \Pr(\omega \mid \mathbf{c}_i(\omega)\mathbf{x} \leq \hat{f}_i), i = 1, \cdots, k, \tag{2.165}$$

where

$$\hat{\mathbf{f}} = (\hat{f}_1, \cdots, \hat{f}_k) \tag{2.166}$$

is a k-dimensional vector of permissible objective levels specified subjectively by the decision maker. Then, MOP2.24 can be converted into the following form.

MOP 2.25 $(\hat{\mathbf{f}})$

$$\max_{\mathbf{x} \in X} (p_1(\mathbf{x}, \hat{f}_1), \cdots, p_k(\mathbf{x}, \hat{f}_k))$$

In MOP2.25$(\hat{\mathbf{f}})$, because of (2.164), we can rewrite objective function $p_i(\mathbf{x}, \hat{f}_i)$ as follows.

$$\begin{aligned} p_i(\mathbf{x}, \hat{f}_i) &= \Pr(\omega \mid \mathbf{c}_i(\omega)\mathbf{x} \leq \hat{f}_i) \\ &= \Phi\left(\frac{\hat{f}_i - \mathbf{E}[\bar{\mathbf{c}}_i]\mathbf{x}}{\sqrt{\mathbf{x}^T V_i \mathbf{x}}}\right), i = 1, \cdots, k, \end{aligned} \tag{2.167}$$

where $\Phi(\cdot)$ is a distribution function of the standard Gaussian random variable $\mathrm{N}(0,1)$.

To handle MOP2.25$(\hat{\mathbf{f}})$, the decision maker must specify his or her permissible objective levels $\hat{\mathbf{f}}$ in advance; however, in general, the decision maker seems to prefer not only larger values of probability function $p_i(\mathbf{x}, \hat{f}_i)$ but also smaller values of permissible objective level $\hat{f}_i$. From such a perspective, we consider the following multiobjective programming problem as a natural extension of MOP2.25$(\hat{\mathbf{f}})$.

MOP 2.26

$$\max_{\mathbf{x}\in X, \hat{f}_i\in R^1, i=1,\cdots,k} (p_1(\mathbf{x},\hat{f}_1),\cdots,p_k(\mathbf{x},\hat{f}_k),-\hat{f}_1,\cdots,-\hat{f}_k)$$

where permissible objective levels $\hat{f}_i, i=1,\cdots,k$ are not fixed values, but rather decision variables.

Considering the imprecise nature of the decision maker's judgment, it is natural to assume that the decision maker has a fuzzy goal for each objective function $p_i(\mathbf{x},\hat{f}_i)$ in MOP2.26. In this subsection, we assume that such fuzzy goals can be quantified via the corresponding membership functions. Let us denote a membership function of a probability function $p_i(\mathbf{x},\hat{f}_i)$ as $\mu_{p_i}(p_i(\mathbf{x},\hat{f}_i))$ and a membership function of permissible objective level $\hat{f}_i$ as $\mu_{\hat{f}_i}(\hat{f}_i)$. Then, MOP2.26 can be transformed into the following multiobjective programming problem.

MOP 2.27

$$\max_{\mathbf{x}\in X, \hat{f}_i, i=1,\cdots,k} \left(\mu_{p_1}(p_1(\mathbf{x},\hat{f}_1)),\cdots,\mu_{p_k}(p_k(\mathbf{x},\hat{f}_k)),\mu_{\hat{f}_1}(\hat{f}_1),\cdots,\mu_{\hat{f}_k}(\hat{f}_k)\right)$$

Throughout this subsection, we make the following assumptions.

Assumption 2.13
$\mu_{\hat{f}_i}(\hat{f}_i), i=1,\cdots,k$ are strictly decreasing and continuous with respect to

$$\hat{f}_i \in F_i \stackrel{\text{def}}{=} [\hat{f}_{i\min},\hat{f}_{i\max}], i=1,\cdots,k, \tag{2.168}$$

where $\hat{f}_{i\min}$ is a sufficiently satisfactory maximum value and $\hat{f}_{i\max}$ is an acceptable maximum value, that is, $\mu_{\hat{f}_i}(\hat{f}_{i\min})=1, \mu_{\hat{f}_i}(\hat{f}_{i\max})=0, i=1,\cdots,k$.

Assumption 2.14
$\mu_{p_i}(p_i(\mathbf{x},\hat{f}_i)), i=1,\cdots,k$ are strictly increasing and continuous with respect to

$$p_i(\mathbf{x},\hat{f}_i) \in [p_{i\min},p_{i\max}], i=1,\cdots,k, \tag{2.169}$$

where $p_{i\min}$ is an acceptable minimum value and $p_{i\max}$ is a sufficiently satisfactory minimum value, that is, $\mu_{p_i}(p_{i\min})=0, \mu_{p_i}(p_{i\max})=1, i=1,\cdots,k$.

To appropriately determine these membership functions, let us assume that the decision maker sets $\hat{f}_{i\min}$, $\hat{f}_{i\max}$, $p_{i\min}$, $p_{i\max}$ as follows.

Considering the individual minimum and maximum of $\mathbf{E}[\bar{\mathbf{c}}_i]\mathbf{x}$, i.e.,

$$\min_{\mathbf{x}\in X} \mathbf{E}[\bar{\mathbf{c}}_i]\mathbf{x}, i=1,\cdots,k, \tag{2.170}$$

$$\max_{\mathbf{x}\in X} \mathbf{E}[\bar{\mathbf{c}}_i]\mathbf{x}, i=1,\cdots,k, \tag{2.171}$$

the decision maker subjectively specifies a sufficiently satisfactory maximum value $\hat{f}_{i\min}$ and an acceptable maximum value $\hat{f}_{i\max}$. Corresponding to $\hat{f}_{i\min}$ and $\hat{f}_{i\max}$, $p_{i\max}$ can be obtained by solving the following problem.

$$p_{i\max} \stackrel{\text{def}}{=} \max_{\mathbf{x}\in X} p_i(\mathbf{x}, \hat{f}_{i\max}) = \Phi\left(\frac{\hat{f}_{i\max} - \mathbf{E}[\bar{\mathbf{c}}_i]\mathbf{x}}{\sqrt{\mathbf{x}^T V_i \mathbf{x}}}\right), i = 1, \cdots, k \tag{2.172}$$

We denote the optimal solutions to

$$\max_{\mathbf{x}\in X} p_i(\mathbf{x}, \hat{f}_{i\min}), i = 1, \cdots, k, \tag{2.173}$$

as $\mathbf{x}_i \in X, i = 1, \cdots, k$. Using the optimal solutions $\mathbf{x}_i, i = 1, \cdots, k$, $p_{i\min}$ can be obtained as follows.

$$p_{i\min} \stackrel{\text{def}}{=} \min_{\ell=1,\cdots,k,\ell\neq i} p_i(\mathbf{x}_\ell, \hat{f}_{i\min}), i = 1, \cdots, k \tag{2.174}$$

Note that $\mu_{\hat{f}_i}(\hat{f}_i)$ and $\mu_{p_i}(p_i(\mathbf{x}, \hat{f}_i))$, $i = 1, \cdots, k$ perfectly conflict with respect to $\hat{f}_i$. Here, we assume that the decision maker adopts the fuzzy decision [6, 71, 140] to integrate both membership functions $\mu_{\hat{f}_i}(\hat{f}_i)$ and $\mu_{p_i}(p_i(\mathbf{x}, \hat{f}_i))$. Then, the integrated membership function can be defined as

$$\mu_{D_{p_i}}(\mathbf{x}, \hat{f}_i) \stackrel{\text{def}}{=} \min\{\mu_{\hat{f}_i}(\hat{f}_i), \mu_{p_i}(p_i(\mathbf{x}, \hat{f}_i))\}. \tag{2.175}$$

Using membership functions $\mu_{D_{p_i}}(\mathbf{x}, \hat{f}_i), i = 1, \cdots, k$, MOP2.27 can be transformed into the following form.

MOP 2.28

$$\max_{\mathbf{x}\in X, \hat{f}_i \in \mathrm{R}^1, i=1,\cdots,k} \left(\mu_{D_{p_1}}(\mathbf{x}, \hat{f}_1), \cdots, \mu_{D_{p_k}}(\mathbf{x}, \hat{f}_k)\right)$$

To handle MOP2.28, we next introduce an MP-Pareto optimal solution.

Definition 2.7 $\mathbf{x}^* \in X, \hat{f}_i^* \in \mathrm{R}^1, i = 1, \cdots, k$ is an MP-Pareto optimal solution to MOP2.28 if and only if there does not exist another $\mathbf{x} \in X, \hat{f}_i \in \mathrm{R}^1, i = 1, \cdots, k$ such that $\mu_{D_{p_i}}(\mathbf{x}, \hat{f}_i) \geq \mu_{D_{p_i}}(\mathbf{x}^*, \hat{f}_i^*)$ $i = 1, \cdots, k$, with strict inequality holding for at least one i.

To generate a candidate of a satisfactory solution that is also MP-Pareto optimal, the decision maker is asked to specify reference membership values [71]. Once reference membership values

$$\hat{\mu} = (\hat{\mu}_1, \cdots, \hat{\mu}_k) \tag{2.176}$$

are specified, the corresponding MP-Pareto optimal solution to MOP2.28 is obtained by solving the following minmax problem.

MINMAX 2.7 ($\hat{\mu}$)

$$\min_{\mathbf{x}\in X, \hat{f}_i\in F_i, i=1,\cdots,k, \lambda\in\Lambda} \lambda \tag{2.177}$$

subject to

$$\hat{\mu}_i - \mu_{p_i}(p_i(\mathbf{x}, \hat{f}_i)) \leq \lambda, i=1,\cdots,k \tag{2.178}$$

$$\hat{\mu}_i - \mu_{\hat{f}_i}(\hat{f}_i) \leq \lambda, i=1,\cdots,k \tag{2.179}$$

where

$$\Lambda \stackrel{\text{def}}{=} [\max_{i=1,\cdots,k} \hat{\mu}_i - 1, \max_{i=1,\cdots,k} \hat{\mu}_i] = [\lambda_{\min}, \lambda_{\max}].$$

Because of Assumption 2.14, constraints (2.178) can be transformed as

$$\begin{aligned} & \hat{\mu}_i - \mu_{p_i}(p_i(\mathbf{x}, \hat{f}_i)) \leq \lambda \\ \Leftrightarrow \quad & \hat{f}_i - \mathbf{E}[\bar{\mathbf{c}}_i]\mathbf{x} \geq \Phi^{-1}(\mu_{p_i}^{-1}(\hat{\mu}_i - \lambda)) \cdot \sqrt{\mathbf{x}^T V_i \mathbf{x}}. \end{aligned} \tag{2.180}$$

From constraints (2.179) and Assumption 2.13, it holds that

$$\hat{f}_i \leq \mu_{\hat{f}_i}^{-1}(\hat{\mu}_i - \lambda).$$

Therefore, two constraints (2.178) and (2.179) can be reduced to inequalities

$$\mu_{\hat{f}_i}^{-1}(\hat{\mu}_i - \lambda) \geq \mathbf{E}[\bar{\mathbf{c}}_i]\mathbf{x} + \Phi^{-1}(\mu_{p_i}^{-1}(\hat{\mu}_i - \lambda)) \cdot \sqrt{\mathbf{x}^T V_i \mathbf{x}}, i=1,\cdots,k. \tag{2.181}$$

Then, MINMAX2.7($\hat{\mu}$) is equivalently transformed into the following min-max problem in which permissible objective levels $\hat{f}_i, i=1,\cdots,k$ have disappeared.

MINMAX 2.8 ($\hat{\mu}$)

$$\min_{\mathbf{x}\in X, \lambda\in\Lambda} \lambda \tag{2.182}$$

subject to

$$\mu_{\hat{f}_i}^{-1}(\hat{\mu}_i - \lambda) - \mathbf{E}[\bar{\mathbf{c}}_i]\mathbf{x} \geq \Phi^{-1}(\mu_{p_i}^{-1}(\hat{\mu}_i - \lambda)) \cdot \sqrt{\mathbf{x}^T V_i \mathbf{x}}, \quad i=1,\cdots,k \tag{2.183}$$

To solve MINMAX2.8($\hat{\mu}$), which is a nonlinear programming problem, we first define function $g_i(\mathbf{x}, \lambda)$ for constraints (2.183) as

$$\begin{aligned} g_i(\mathbf{x}, \lambda) \stackrel{\text{def}}{=} & \mu_{\hat{f}_i}^{-1}(\hat{\mu}_i - \lambda) - \mathbf{E}[\bar{\mathbf{c}}_i]\mathbf{x} \\ & -\Phi^{-1}(\mu_{p_i}^{-1}(\hat{\mu}_i - \lambda)) \cdot \sqrt{\mathbf{x}^T V_i \mathbf{x}}, i=1,\cdots,k. \end{aligned} \tag{2.184}$$

Note that $g_i(\mathbf{x}, \lambda), i=1,\cdots,k$ are concave functions for any fixed $\lambda \in \Lambda$, if $\Phi^{-1}(\mu_{p_i}^{-1}(\hat{\mu}_i - \lambda)) > 0, i=1,\cdots,k$ for any $\lambda \in \Lambda$. Therefore, we make the following assumption.

Assumption 2.15

$$0.5 < p_{i\min} < p_{i\max} < 1, i = 1, \cdots, k.$$

It seems to be acceptable for the decision maker that $0.5 < p_{i\min}$. Under Assumption 2.15, it is clear that constraint set $G(\lambda)$ is a convex set for any $\lambda \in \Lambda$, which is defined as

$$G(\lambda) \stackrel{\text{def}}{=} \{\mathbf{x} \in X \mid g_i(\mathbf{x}, \lambda) \geq 0, i = 1, \cdots, k\}. \tag{2.185}$$

Constraint set $G(\lambda)$ satisfies the following property for $\lambda \in \Lambda$.

Property 2.2
If $\lambda_1, \lambda_2 \in \Lambda$ and $\lambda_1 < \lambda_2$, then it holds that $G(\lambda_1) \subset G(\lambda_2)$.
(Proof)
From Assumptions 2.13 and 2.14, it holds that

$$\mu_{\hat{f}_i}^{-1}(\hat{\mu}_i - \lambda_1) < \mu_{\hat{f}_i}^{-1}(\hat{\mu}_i - \lambda_2)$$
$$\Phi^{-1}(\mu_{p_i}^{-1}(\hat{\mu}_i - \lambda_1)) > \Phi^{-1}(\mu_{p_i}^{-1}(\hat{\mu}_i - \lambda_2)).$$

This means that $g_i(\mathbf{x}, \lambda_1) < g_i(\mathbf{x}, \lambda_2)$ for any $\mathbf{x} \in X$. Therefore, $G(\lambda_1) \subset G(\lambda_2)$ for any $\lambda_1 < \lambda_2$. □

From Property 2.2, we easily obtain optimal solution $\mathbf{x}^* \in X, \lambda^* \in \Lambda$ of MINMAX2.8($\hat{\mu}$) by applying Algorithm 2.3 based on the bisection method for $\lambda \in \Lambda$ in which we assume that $G(\lambda_{\max}) \neq \phi, G(\lambda_{\min}) = \phi$.

Algorithm 2.3

Step 1: *Set* $\lambda_0 = \lambda_{\max}, \lambda_1 = \lambda_{\min}$, $\lambda \leftarrow (\lambda_0 + \lambda_1)/2$.
Step 2: *Solve the following convex programming problem for* $\lambda \in \Lambda$, *and let us denote the optimal solution as* $\mathbf{x}^* \in X$.

$$\max_{\mathbf{x} \in X} g_j(\mathbf{x}, \lambda)$$

subject to

$$g_i(\mathbf{x}, \lambda) \geq 0, i = 1, \cdots, k, i \neq j$$

Step 3: *If* $|\lambda_1 - \lambda_0| < \varepsilon$, *then go to Step 4, where* ε *is a sufficiently small positive constant. If* $g_j(\mathbf{x}^*, \lambda) < 0$ *or there exists index* $i \neq j$ *such that* $g_i(\mathbf{x}^*, \lambda) < 0$, *then set* $\lambda_1 \leftarrow \lambda, \lambda \leftarrow (\lambda_0 + \lambda_1)/2$, *and go to Step 2. Otherwise, if* $g_j(\mathbf{x}^*, \lambda) \geq 0$ *and* $g_i(\mathbf{x}^*, \lambda) \geq 0$ *for any* $i = 1, \cdots, k, i \neq j$, *then set* $\lambda_0 \leftarrow \lambda, \lambda \leftarrow (\lambda_0 + \lambda_1)/2$, *and go to Step 2.*
Step 4: *Set* $\lambda^* \leftarrow \lambda$ *and obtain optimal solution* $\mathbf{x}^* \in X, \lambda^* \in \Lambda$ *of MINMAX2.8(*$\hat{\mu}$*).*

The relationships between optimal solution $\mathbf{x}^* \in X, \lambda^* \in \Lambda$ of MINMAX2.8 ($\hat{\mu}$) and MP-Pareto optimal solutions to MOP2.28 can be characterized by the following theorem.

Theorem 2.10

(1) If $\mathbf{x}^* \in X, \lambda^* \in \Lambda$ is a unique optimal solution of MINMAX2.8($\hat{\mu}$), then $\mathbf{x}^* \in X, \hat{f}_i^* \stackrel{\text{def}}{=} \mu_{\hat{f}_i}^{-1}(\hat{\mu}_i - \lambda^*) \in F_i, i = 1, \cdots, k$ is an MP-Pareto optimal solution to MOP2.28.

(2) If $\mathbf{x}^* \in X, \hat{f}_i^* \in F_i, i = 1, \cdots, k$ is an MP-Pareto optimal solution to MOP2.28, then $\mathbf{x}^* \in X$, $\lambda^* \in \Lambda$ is an optimal solution of MINMAX2.8($\hat{\mu}$) for some reference membership values $\hat{\mu} = (\hat{\mu}_1, \cdots, \hat{\mu}_k)$, where $\lambda^* \stackrel{\text{def}}{=} \hat{\mu}_i - \mu_{\hat{f}_i}(\hat{f}_i^*) = \hat{\mu}_i - \mu_{p_i}(p_i(\mathbf{x}^*, \hat{f}_i^*)), i = 1, \cdots, k$.

(Proof)

(1) Assume that $\mathbf{x}^* \in X, \mu_{\hat{f}_i}^{-1}(\hat{\mu}_i - \lambda^*) \in F_i, i = 1, \cdots, k$ is not an MP-Pareto optimal solution to MOP2.28. Since it holds that $\hat{\mu}_i - \lambda^* \leq \mu_{p_i}(p_i(\mathbf{x}^*, \mu_{\hat{f}_i}^{-1}(\hat{\mu}_i - \lambda^*))), i = 1, \cdots, k$ at the optimal solution $\mathbf{x}^* \in X, \lambda^* \in \Lambda$, there exist $\mathbf{x} \in X, \hat{f}_i \in F_i, i = 1, \cdots, k$ such that

$$\begin{aligned} \mu_{D_{p_i}}(\mathbf{x}, \hat{f}_i) &\stackrel{\text{def}}{=} \min\{\mu_{\hat{f}_i}(\hat{f}_i), \mu_{p_i}(p_i(\mathbf{x}, \hat{f}_i))\} \\ &\geq \mu_{D_{p_i}}(\mathbf{x}^*, \mu_{\hat{f}_i}^{-1}(\hat{\mu}_i - \lambda^*)) = \hat{\mu}_i - \lambda^*, i = 1, \cdots, k, \end{aligned}$$

with strict inequality holding for at least one i. Then it holds that

$$\begin{aligned} \mu_{\hat{f}_i}(\hat{f}_i) &\geq \hat{\mu}_i - \lambda^*, i = 1, \cdots, k \\ \mu_{p_i}(p_i(\mathbf{x}, \hat{f}_i)) &\geq \hat{\mu}_i - \lambda^*, i = 1, \cdots, k. \end{aligned}$$

From (2.167) and Assumptions 2.13 and 2.14, these inequalities can be transformed into inequalities

$$\begin{aligned} \hat{f}_i &\leq \mu_{\hat{f}_i}^{-1}(\hat{\mu}_i - \lambda^*) \\ \hat{f}_i &\geq \mathbf{E}[\bar{\mathbf{c}}_i]\mathbf{x} + \Phi^{-1}(\mu_{p_i}^{-1}(\hat{\mu}_i - \lambda^*)) \cdot \sqrt{\mathbf{x}^T V_i \mathbf{x}}, \end{aligned}$$

respectively. This means that there exists some $\mathbf{x} \in X$ such that

$$\mu_{\hat{f}_i}^{-1}(\hat{\mu}_i - \lambda^*) - \mathbf{E}[\bar{\mathbf{c}}_i]\mathbf{x} \geq \Phi^{-1}(\mu_{p_i}^{-1}(\hat{\mu}_i - \lambda^*)) \cdot \sqrt{\mathbf{x}^T V_i \mathbf{x}}, i = 1, \cdots, k,$$

with strict inequality holding for at least one i, which contradicts the fact that $\mathbf{x}^* \in X, \lambda^* \in \Lambda$ is a unique optimal solution of MINMAX2.8($\hat{\mu}$).

(2) Assume that $\mathbf{x}^* \in X, \lambda^* \in \Lambda$ is not an optimal solution of MINMAX2.8($\hat{\mu}$) for any reference membership values $\hat{\mu} = (\hat{\mu}_1, \cdots, \hat{\mu}_k)$ that satisfy inequalities

$$\hat{\mu}_i - \lambda^* = \mu_{\hat{f}_i}(\hat{f}_i^*) = \mu_{p_i}(p_i(\mathbf{x}^*, \hat{f}_i^*)), i = 1, \cdots, k.$$

Then, there exists some $\mathbf{x} \in X, \lambda < \lambda^*$ such that

$$\begin{aligned} & \mu_{\hat{f}_i}^{-1}(\hat{\mu}_i - \lambda) - \mathbf{E}[\bar{\mathbf{c}}_i]\mathbf{x} \geq \Phi^{-1}(\mu_{\hat{p}_i}^{-1}(\hat{\mu}_i - \lambda)) \cdot \sqrt{\mathbf{x}^T V_i \mathbf{x}} \\ \Leftrightarrow \quad & \mu_{p_i}(p_i(\mathbf{x}, \mu_{\hat{f}_i}^{-1}(\hat{\mu}_i - \lambda)) \geq \hat{\mu}_i - \lambda, i = 1, \cdots, k. \end{aligned}$$

Because of $\hat{\mu}_i - \lambda > \hat{\mu}_i - \lambda^*, i = 1, \cdots, k$, the following inequalities hold.

$$\mu_{\hat{f}_i}(\mu_{\hat{f}_i}^{-1}(\mu_i - \lambda)) > \mu_{\hat{f}_i}(\hat{f}_i^*), i = 1, \cdots, k$$

$$\mu_{p_i}(p_i(\mathbf{x}, \mu_{\hat{p}_i}^{-1}(\mu_i - \lambda))) > \mu_{p_i}(p_i(\mathbf{x}^*, \hat{f}_i^*)), i = 1, \cdots, k$$

This contradicts the fact that $\mathbf{x}^* \in X, \hat{f}_i^* \in F_i, i = 1, \cdots, k$ is an MP-Pareto optimal solution to MOP2.28. □

2.4.2 A formulation using a fractile model

Similar to the previous subsection, in this subsection, we apply a fractile optimization model [77] to MOP2.24. If the decision maker specifies permissible probability levels

$$\hat{\mathbf{p}} = (\hat{p}_1, \cdots, \hat{p}_k), \tag{2.186}$$

then, MOP2.24 can be transformed into the following multiobjective programming problem, where $f_i, i = 1, \cdots, k$ are target variables for the objective functions.

MOP 2.29 ($\hat{\mathbf{p}}$)

$$\min_{\mathbf{x} \in X, f_i \in R^1, i=1,\cdots,k} (f_1, \cdots, f_k) \tag{2.187}$$

subject to

$$p_i(\mathbf{x}, f_i) \geq \hat{p}_i, i = 1, \cdots, k \tag{2.188}$$

In MOP2.29($\hat{\mathbf{p}}$), since distribution function $\Phi(\cdot)$ is continuous and strictly increasing, constraints (2.188) can be transformed into the following form.

$$\begin{aligned} \hat{p}_i \quad & \leq \quad p_i(\mathbf{x}, f_i) \stackrel{\text{def}}{=} \Phi\left(\frac{f_i - \mathbf{E}[\bar{\mathbf{c}}_i]\mathbf{x}}{\sqrt{\mathbf{x}^T V_i \mathbf{x}}} \right) \\ \Leftrightarrow f_i \quad & \geq \quad \mathbf{E}[\bar{\mathbf{c}}_i]\mathbf{x} + \Phi^{-1}(\hat{p}_i) \cdot \sqrt{\mathbf{x}^T V_i \mathbf{x}} \end{aligned} \tag{2.189}$$

Let us define the right-hand-side of inequality (2.189) as follows.

$$f_i(\mathbf{x}, \hat{p}_i) \stackrel{\text{def}}{=} \mathbf{E}[\bar{\mathbf{c}}_i]\mathbf{x} + \Phi^{-1}(\hat{p}_i) \cdot \sqrt{\mathbf{x}^T V_i \mathbf{x}} \tag{2.190}$$

Then, MOP2.29($\hat{\mathbf{p}}$) can be equivalently reduced to the following simple form.

MOP 2.30 ($\hat{\mathbf{p}}$)

$$\min_{\mathbf{x}\in X}(f_1(\mathbf{x},\hat{p}_1),\cdots,f_k(\mathbf{x},\hat{p}_k)) \tag{2.191}$$

To handle MOP2.30($\hat{\mathbf{p}}$), the decision maker must specify permissible probability levels $\hat{\mathbf{p}}$ in advance; however, in general, the decision maker seems to prefer not only smaller values of objective function $f_i(\mathbf{x},\hat{p}_i)$ but also larger value of the permissible probability level $\hat{p}_i$. From such a perspective, we consider the following multiobjective programming problem as a natural extension of MOP2.30($\hat{\mathbf{p}}$).

MOP 2.31

$$\min_{\mathbf{x}\in X,\hat{p}_i\in(0,1),i=1,\cdots,k}(f_1(\mathbf{x},\hat{p}_1),\cdots,f_k(\mathbf{x},\hat{p}_k),-\hat{p}_1,\cdots,-\hat{p}_k)$$

Here, permissible probability levels $\hat{p}_i, i=1,\cdots,k$ are not fixed values, but rather decision variables.

Considering the imprecise nature of the decision maker's judgment, we assume that the decision maker has a fuzzy goal for each objective function in MOP2.31. Such a fuzzy goal can be quantified via the corresponding membership function. Let us denote a membership function of objective function $f_i(\mathbf{x},\hat{p}_i)$ as $\mu_{f_i}(f_i(\mathbf{x},\hat{p}_i))$, and a membership function of permissible probability level $\hat{p}_i$ as $\mu_{\hat{p}_i}(\hat{p}_i)$ respectively. Then, MOP2.31 can be transformed into the following problem.

MOP 2.32

$$\max_{\mathbf{x}\in X,\hat{p}_i\in(0,1),i=1,\cdots,k}(\mu_{f_1}(f_1(\mathbf{x},\hat{p}_1)),\cdots,\mu_{f_k}(f_k(\mathbf{x},\hat{p}_k)),\mu_{\hat{p}_1}(\hat{p}_1),\cdots,\mu_{\hat{p}_k}(\hat{p}_k)) \tag{2.192}$$

Throughout this subsection, we make the following assumptions with respect to membership functions $\mu_{f_i}(f_i(\mathbf{x},\hat{p}_i))$, $\mu_{\hat{p}_i}(\hat{p}_i), i=1,\cdots,k$.

Assumption 2.16
$\mu_{\hat{p}_i}(\hat{p}_i), i=1,\cdots,k$ are strictly increasing and continuous with respect to $\hat{p}_i \in P_i \stackrel{\text{def}}{=} [\hat{p}_{i\min},\hat{p}_{i\max}] \subset (0,1)$, where $\mu_{\hat{p}_i}(\hat{p}_{i\min})=0$, $\mu_{\hat{p}_i}(\hat{p}_{i\max})=1$.

Assumption 2.17
$\mu_{f_i}(f_i(\mathbf{x},\hat{p}_i)), i=1,\cdots,k$ are strictly decreasing and continuous with respect to $f_i(\mathbf{x},\hat{p}_i) \in [f_{i\min},f_{i\max}]$, where $\mu_{f_i}(f_{i\min})=1, \mu_{f_i}(f_{i\max})=0, i=1,\cdots,k$.

To appropriately determine these membership functions, let us assume that the decision maker sets $[\hat{p}_{i\min},\hat{p}_{i\max}]$, $[f_{i\min},f_{i\max}]$ as follows.

The decision maker subjectively specifies $\hat{p}_{i\min}$ and $\hat{p}_{i\max}$, where $\hat{p}_{i\min}$ is an acceptable minimum value and $\hat{p}_{i\max}$ is a sufficiently satisfactory minimum value, and sets intervals $P_i=[\hat{p}_{i\min},\hat{p}_{i\max}], i=1,\cdots,k$. Corresponding to interval

P_i, we compute interval $[f_{i\min}, f_{i\max}]$ as follows. First, $f_{i\min}, i = 1, \cdots, k$ can be obtained by solving the following problems.

$$f_{i\min} \stackrel{\text{def}}{=} \min_{\mathbf{x} \in X} f_i(\mathbf{x}, \hat{p}_{i\min}), i = 1, \cdots, k \tag{2.193}$$

To obtain $f_{i\max}, i = 1, \cdots, k$, we solve the following k programming problems.

$$\min_{\mathbf{x} \in X} f_i(\mathbf{x}, \hat{p}_{i\max}), i = 1, \cdots, k \tag{2.194}$$

Let $\mathbf{x}_i, i = 1, \cdots, k$ be the corresponding optimal solution. Using optimal solutions $\mathbf{x}_i, i = 1, \cdots, k$, $f_{i\max}$ can be obtained as

$$f_{i\max} \stackrel{\text{def}}{=} \max_{\ell = 1, \cdots, k, \ell \neq i} f_i(\mathbf{x}_\ell, \hat{p}_{i\max}), i = 1, \cdots, k. \tag{2.195}$$

Note that from (2.190), $\mu_{f_i}(f_i(\mathbf{x}, \hat{p}_i))$ and $\mu_{\hat{p}_i}(\hat{p}_i)$ perfectly conflict with one another with respect to $\hat{p}_i$. Here, let us assume that the decision maker adopts the fuzzy decision [6, 71, 140] to integrate both membership functions $\mu_{f_i}(f_i(\mathbf{x}, \hat{p}_i))$ and $\mu_{\hat{p}_i}(\hat{p}_i)$. Then, the integrated membership function $\mu_{D_{f_i}}(\mathbf{x}, \hat{p}_i)$ can be defined as

$$\mu_{D_{f_i}}(\mathbf{x}, \hat{p}_i) \stackrel{\text{def}}{=} \min\{\mu_{\hat{p}_i}(\hat{p}_i), \mu_{f_i}(f_i(\mathbf{x}, \hat{p}_i))\}. \tag{2.196}$$

Using integrated membership functions $\mu_{D_{f_i}}(\mathbf{x}, \hat{p}_i), i = 1, \cdots, k$, MOP2.32 can be transformed into the following form.

MOP 2.33

$$\max_{\mathbf{x} \in X, \hat{p}_i \in P_i, i = 1, \cdots, k} \left(\mu_{D_{f_1}}(\mathbf{x}, \hat{p}_1), \cdots, \mu_{D_{f_k}}(\mathbf{x}, \hat{p}_k)\right)$$

To handle MOP2.33, we next introduce an MF-Pareto optimal solution.

Definition 2.8 $\mathbf{x}^* \in X, \hat{p}_i^* \in P_i, i = 1, \cdots, k$ is an MF-Pareto optimal solution to MOP2.33 if and only if there does not exist another $\mathbf{x} \in X, \hat{p}_i \in P_i, i = 1, \cdots, k$ such that $\mu_{D_{f_i}}(\mathbf{x}, \hat{p}_i) \geq \mu_{D_{f_i}}(\mathbf{x}^*, \hat{p}_i^*)$ $i = 1, \cdots, k$, with strict inequality holding for at least one i.

To generate a candidate for a satisfactory solution that is also MF-Pareto optimal, the decision maker is asked to specify reference membership values [71]. Once reference membership values $\hat{\mu} = (\hat{\mu}_1, \cdots, \hat{\mu}_k)$ are specified, the corresponding MF-Pareto optimal solution is obtained by solving the following minmax problem.

MINMAX 2.9 $(\hat{\mu})$

$$\min_{\mathbf{x}\in X, \hat{p}_i \in P_i, i=1,\cdots,k, \lambda\in\Lambda} \lambda \tag{2.197}$$

subject to

$$\hat{\mu}_i - \mu_{f_i}(f_i(\mathbf{x},\hat{p}_i)) \leq \lambda, i=1,\cdots,k \tag{2.198}$$

$$\hat{\mu}_i - \mu_{\hat{p}_i}(\hat{p}_i) \leq \lambda, i=1,\cdots,k \tag{2.199}$$

Because of Assumption 2.17, constraints (2.198) can be transformed as follows.

$$\hat{p}_i \leq \Phi\left(\frac{\mu_{f_i}^{-1}(\hat{\mu}_i - \lambda) - \mathbf{E}[\bar{\mathbf{c}}_i]\mathbf{x}}{\sqrt{\mathbf{x}^T V_i \mathbf{x}}}\right), i=1,\cdots,k \tag{2.200}$$

From constraints (2.199) and Assumption 2.16, it holds that $\hat{p}_i \geq \mu_{\hat{p}_i}^{-1}(\hat{\mu}_i - \lambda)$. Therefore, constraints (2.198) and (2.199) can be reduced to the following inequality in which permissible probability levels $\hat{p}_i, i=1,\cdots,k$ have disappeared.

$$\mu_{f_i}^{-1}(\hat{\mu}_i - \lambda) - \mathbf{E}[\bar{\mathbf{c}}_i]\mathbf{x} \geq \Phi^{-1}(\mu_{\hat{p}_i}^{-1}(\hat{\mu}_i - \lambda)) \cdot \sqrt{\mathbf{x}^T V_i \mathbf{x}} \tag{2.201}$$

Then, MINMAX2.9$(\hat{\mu})$ can be equivalently reduced to the following problem.

MINMAX 2.10 $(\hat{\mu})$

$$\min_{\mathbf{x}\in X, \lambda\in\Lambda} \lambda \tag{2.202}$$

subject to

$$\mu_{f_i}^{-1}(\hat{\mu}_i - \lambda) - \mathbf{E}[\bar{\mathbf{c}}_i]\mathbf{x} \geq \Phi^{-1}(\mu_{\hat{p}_i}^{-1}(\hat{\mu}_i - \lambda)) \cdot \sqrt{\mathbf{x}^T V_i \mathbf{x}}, \quad i=1,\cdots,k \tag{2.203}$$

Similar to the previous subsection, we make the following assumption for interval $P_i = [\hat{p}_{i\min}, \hat{p}_{i\max}]$.

Assumption 2.18

$$0.5 < \hat{p}_{i\min} < \hat{p}_{i\max} < 1, i=1,\cdots,k.$$

Under Assumption 2.18, we can easily solve MINMAX2.10$(\hat{\mu})$ by applying Algorithm 2.3 from the previous subsection.

Finally, the relationships between optimal solution $\mathbf{x}^* \in X, \lambda^* \in \Lambda$ of MINMAX2.10$(\hat{\mu})$ and MF-Pareto optimal solutions to MOP2.33 can be characterized by the following theorem.

Theorem 2.11

(1) If $\mathbf{x}^* \in X, \lambda^* \in \Lambda$ is a unique optimal solution of MINMAX2.10($\hat{\mu}$), then $\mathbf{x}^* \in X, \hat{p}_i^* \stackrel{\text{def}}{=} \mu_{\tilde{p}_i}^{-1}(\hat{\mu}_i - \lambda^*) \in P_i, i = 1, \cdots, k$ is an MF-Pareto optimal solution to MOP2.33.

(2) If $\mathbf{x}^* \in X, \hat{p}_i^* \in P_i, i = 1, \cdots, k$ is an MF-Pareto optimal solution to MOP2.33, then $\mathbf{x}^* \in X$, $\lambda^* \in \Lambda$ is an optimal solution of MINMAX2.10($\hat{\mu}$) for some reference membership values $\hat{\mu} = (\hat{\mu}_1, \cdots, \hat{\mu}_k)$, where $\lambda^* \stackrel{\text{def}}{=} \hat{\mu}_i - \mu_{\tilde{p}_i}(\hat{p}_i^*) = \hat{\mu}_i - \mu_{\tilde{f}_i}(f_i(\mathbf{x}^*, \hat{p}_i^*)), i = 1, \cdots, k$.

2.4.3 An interactive convex programming algorithm

In this subsection, we propose an interactive convex programming algorithm to obtain a satisfactory solution from an MF-Pareto optimal solution set. Unfortunately, there is no guarantee that optimal solution $\mathbf{x}^* \in X$, $\lambda^* \in \Lambda$ of MINMAX2.10($\hat{\mu}$) is MF-Pareto optimal, if $\mathbf{x}^* \in X$, $\lambda^* \in \Lambda$ is not unique. To guarantee MF-Pareto optimality, we first assume that k constraints (2.203) of MINMAX2.10($\hat{\mu}$) are active at optimal solution $\mathbf{x}^* \in X$, $\lambda^* \in \Lambda$, i.e.,

$$\mu_{\tilde{f}_i}^{-1}(\hat{\mu}_i - \lambda^*) = f_i(\mathbf{x}^*, \mu_{\tilde{p}_i}^{-1}(\hat{\mu}_i - \lambda^*)), i = 1, \cdots, k.$$

If the j-th constraint of (2.203) is inactive, i.e.,

$$\mu_{\tilde{f}_j}^{-1}(\hat{\mu}_j - \lambda^*) > f_j(\mathbf{x}^*, \mu_{\tilde{p}_j}^{-1}(\hat{\mu}_j - \lambda^*)),$$

we can convert the inactive constraint into the active one by applying Algorithm 2.4, shown below, which is based on the bisection method with respect to reference membership value $\hat{\mu}_j$.

Algorithm 2.4

Step 1: *Set $\mu_j^L \leftarrow \hat{\mu}_j, \mu_j^R \leftarrow \lambda^* + 1$.*
Step 2: *Set $\mu_j \leftarrow (\mu_j^L + \mu_j^R)/2$.*
Step 3: *If $h_j(\mu_j) > 0$, then $\mu_j^L \leftarrow \mu_j$ and go to Step 2, else if $h_j(\mu_j) < 0$, then $\mu_j^R \leftarrow \mu_j$ and go to Step 2, else update the reference membership value as $\hat{\mu}_j \leftarrow \mu_j$ and stop, where*

$$h_j(\hat{\mu}_j) \stackrel{\text{def}}{=} \mu_{\tilde{f}_j}^{-1}(\hat{\mu}_j - \lambda^*) - f_j(\mathbf{x}^*, \mu_{\tilde{p}_j}^{-1}(\hat{\mu}_j - \lambda^*)).$$

For optimal solution $\mathbf{x}^* \in X$, $\lambda^* \in \Lambda$ of MINMAX2.10($\hat{\mu}$), where the active conditions for (2.203) are satisfied, we solve MF-Pareto optimality test problems for each $j = 1, \cdots, k$.

Test 2.2

$$w_j \stackrel{\text{def}}{=} \min_{\mathbf{x}\in X} f_j(\mathbf{x}, \mu_{\hat{p}_j}^{-1}(\hat{\mu}_j - \lambda^*))$$

subject to

$$f_i(\mathbf{x}, \mu_{\hat{p}_i}^{-1}(\hat{\mu}_i - \lambda^*)) \leq f_i(\mathbf{x}^*, \mu_{\hat{p}_i}^{-1}(\hat{\mu}_i - \lambda^*)), i = 1, \cdots, k$$

Note that the test problems shown above are convex programming problems. For these test problems, the theorem below holds.

Theorem 2.12
If $w_j = f_j(\mathbf{x}^*, \mu_{\hat{p}_j}^{-1}(\hat{\mu}_j - \lambda^*)), j = 1, \cdots, k$ in Test2.2, then $\mathbf{x}^* \in X, \mu_{\hat{p}_i}^{-1}(\hat{\mu}_i - \lambda^*) \in P_i, i = 1, \cdots, k$ is an MF-Pareto optimal solution to MOP2.33.

Following the above discussion, we present the following interactive algorithm to derive a satisfactory solution from an MF-Pareto optimal solution set.

Algorithm 2.5

Step 1: *The decision maker subjectively specifies $\hat{p}_{i\min}(> 0.5)$, $\hat{p}_{i\max}(< 1)$, $i = 1, \cdots, k$. On interval $P_i = [\hat{p}_{i\min}, \hat{p}_{i\max}]$, the decision maker sets his or her membership functions $\mu_{\hat{p}_i}(\hat{p}_i), i = 1, \cdots, k$ according to Assumption 2.16.*
Step 2: *Corresponding to interval P_i, compute $f_{i\min}$ and $f_{i\max}$ by solving problems (2.193) and (2.195). On interval $[f_{i\min}, f_{i\max}]$, the decision maker sets his or her membership functions $\mu_{f_i}(f_i(\mathbf{x}, \hat{p}_i)), i = 1, \cdots, k$ according to Assumption 2.17.*
Step 3: *Set the initial reference membership values as $\hat{\mu}_i = 1, i = 1, \cdots, k$.*
Step 4: *Solve MINMAX2.10($\hat{\mu}$) by applying Algorithm 2.3. For optimal solution $\mathbf{x}^* \in X$, $\lambda^* \in \Lambda$ of MINMAX2.10($\hat{\mu}$), MF-Pareto optimality test problems TEST2.2 are solved.*
Step 5: *If the decision maker is satisfied with the current values of MF-Pareto optimal solution $\mu_{D_{f_i}}(\mathbf{x}^*, \mu_{\hat{p}_i}^{-1}(\hat{\mu}_i - \lambda^*)), i = 1, \cdots, k$, then stop. Otherwise, the decision maker must update his or her reference membership values $\hat{\mu}_i, i = 1, \cdots, k$, and return to Step 4.*

2.4.4 A numerical example

Next, we consider the following three objective stochastic programming problem to demonstrate the feasibility of our proposed method with a hypothetical decision maker formulated by Sakawa et al. [77].

MOP 2.34

$$\min \ (\bar{\mathbf{c}}_1\mathbf{x}, \bar{\mathbf{c}}_2\mathbf{x}, \bar{\mathbf{c}}_3\mathbf{x})$$

subject to

$$\mathbf{x} \in X \stackrel{\text{def}}{=} \{(x_1,x_2,x_3,x_4)^T \geq \mathbf{0} \mid 7x_1+3x_2+4x_3+6x_4 \leq 30.146,\ -5x_1-6x_2-7x_3-9x_4 \leq -11.329\}$$

Here, each of the elements of $\bar{\mathbf{c}}_i, i=1,2,3$ is a Gaussian random variable, where the corresponding mean vectors $\mathbf{E}[\bar{\mathbf{c}}_i], i=1,2,3$ and the variance covariance matrices $V_i, i=1,2,3$ are defined as

$$\begin{aligned}
\mathbf{E}[\bar{\mathbf{c}}_1] &= (E[\bar{c}_{11}],E[\bar{c}_{12}],E[\bar{c}_{13}],E[\bar{c}_{14}]) = (2,3,2,4)\\
\mathbf{E}[\bar{\mathbf{c}}_2] &= (E[\bar{c}_{21}],E[\bar{c}_{22}],E[\bar{c}_{23}],E[\bar{c}_{24}]) = (10,-7,1,-2)\\
\mathbf{E}[\bar{\mathbf{c}}_3] &= (E[\bar{c}_{31}],E[\bar{c}_{32}],E[\bar{c}_{33}],E[\bar{c}_{34}]) = (-8,-5,-7,-14)
\end{aligned}$$

$$\begin{aligned}
V_1 &= \begin{pmatrix} 25 & -1 & 0.8 & -2 \\ -1 & 4 & -1.2 & 1.2 \\ 0.8 & -1.2 & 4 & 2 \\ -2 & 1.2 & 2 & 9 \end{pmatrix}\\
V_2 &= \begin{pmatrix} 16 & 1.4 & -1.2 & 1.4 \\ 1.4 & 1 & 1.5 & -0.8 \\ -1.2 & 1.5 & 25 & -0.6 \\ 1.4 & -0.8 & -0.6 & 4 \end{pmatrix}\\
V_3 &= \begin{pmatrix} 4 & -1.9 & 1.5 & 1.8 \\ -1.9 & 25 & 0.8 & -0.4 \\ 1.5 & 0.8 & 9 & 2.5 \\ 1.8 & -0.4 & 2.5 & 36 \end{pmatrix}.
\end{aligned}$$

We assume that the hypothetical decision maker sets the interval of permissible probability levels as $[\hat{p}_{1\min}, \hat{p}_{1\max}] = [0.6, 0.9]$, $[\hat{p}_{2\min}, \hat{p}_{2\max}] = [0.65, 0.95]$, and $[\hat{p}_{3\min}, \hat{p}_{3\max}] = [0.6, 0.85]$, and adopts linear membership functions for permissible probability levels (i.e., Step 1) as

$$\mu_{\hat{p}_i}(\hat{p}_i) = \frac{\hat{p}_{i\min} - \hat{p}_i}{\hat{p}_{i\min} - \hat{p}_{i\max}}, i = 1,2,3.$$

By solving problems (2.193) and (2.195), we obtain $[f_{1\min}, f_{1\max}] = [4.05698, 55.9025]$, $[f_{2\min}, f_{2\max}] = [-66.4696, 13.6200]$, and $[f_{3\min}, f_{3\max}] = [-62.7041, 1.83056]$. On these intervals, we assume that the hypothetical decision maker establishes linear membership functions (i.e., Step 2) as

$$\mu_{f_i}(f_i(\mathbf{x}, \hat{p}_i)) = \frac{f_i(\mathbf{x}, \hat{p}_i) - f_{i\max}}{f_{i\min} - f_{i\max}}, i = 1,2,3.$$

Table 2.8: Interactive processes in MOP2.34

	1	2	3
$\hat{\mu}_1$	1	0.58	0.566
$\hat{\mu}_2$	1	0.537	0.523
$\hat{\mu}_3$	1	0.51	0.51
$\mu_{D_{f_1}}(\mathbf{x}^*, \hat{p}_1^*)$	0.53661	0.56645	0.56327
$\mu_{D_{f_2}}(\mathbf{x}^*, \hat{p}_2^*)$	0.53661	0.52345	0.52027
$\mu_{D_{f_3}}(\mathbf{x}^*, \hat{p}_3^*)$	0.53661	0.49645	0.50727
$\hat{p}_1^*$	0.76098	0.76993	0.76898
$\hat{p}_2^*$	0.81098	0.80703	0.80608
$\hat{p}_3^*$	0.73415	0.72411	0.726818
$f_1(\mathbf{x}^*, \hat{p}_1^*)$	28.082	26.535	26.699
$f_2(\mathbf{x}^*, \hat{p}_2^*)$	–29.356	–28.303	–28.048
$f_3(\mathbf{x}^*, \hat{p}_3^*)$	–32.799	–30.207	–30.906

Next, we set the initial reference membership values as $\hat{\mu}_i = 1, i = 1, \cdots, k$ (i.e., Step 3), and solve MINMAX2.10($\hat{\mu}$) to obtain optimal solution $\mathbf{x}^* \in X, \lambda^* \in \Lambda$ (i.e., Step 4). The interactive processes with the hypothetical decision maker are shown in Table 2.8.

Chapter 3

Multiobjective Fuzzy Random Programming Problems (MOFRPs)

CONTENTS

The concept of fuzzy random variable was first introduced by Kwakernaak [46], with its definition in an n-dimensional Euclidean space given by Puri and Ralescu [65]. From a practical point of view, Wang and Zhang [102] also defined a fuzzy random variable in a one-dimensional Euclidean space. Roughly speaking, fuzzy random variables defined by Wang and Zhang [102] can be interpreted as random variables whose realized values are not real values, but rather are fuzzy sets [77].

From the perspective that both randomness and fuzziness are often involved simultaneously in real-world decision making problems, fuzzy random linear programming problems were formulated and corresponding solutions have been proposed in numerous studies [35, 53, 103]. As a natural extension, Katagiri et al. [38, 39] formulated multiobjective fuzzy random programming problems (MOFRPs) and proposed interactive decision making methods to obtain satisfactory solutions. In their methods, to handle the randomness of fuzzy random variable coefficients in objective functions, either a probability maximization model or a fractile optimization model is adopted. In such models, the decision maker is requested to specify permissible levels in advance. Unfortunately, since permissible levels have a strong influence on the corresponding solution, the decision maker must carefully select appropriate values.

In practice, it seems to be difficult for decision makers to identify appropriate values of permissible levels. To circumvent such difficulties, in Sections 3.1 and 3.2, we propose interactive decision making methods for MOFRPs. More specifically, in Section 3.1, we formulate MOFRPs with a special structure and propose an interactive linear programming algorithm [131, 132, 133, 134] to obtain a satisfactory solution in which the decision maker sets not permissible levels, but rather their membership functions. Similarly, in Section 3.2, we formulate MOFRPs with variance covariance matrices and propose an interactive convex programming algorithm [114, 119, 120] to obtain a satisfactory solution. In Section 3.3, we formulate MOFRPs using both expectation and variance models and propose an interactive convex programming algorithm [127, 128] to obtain a satisfactory solution. Finally, in Section 3.4, we focus on MOFRPs with simple recourse in which we formulate equality constraints involving fuzzy random variable coefficients by applying a simple recourse programming technique [7, 33]. We also propose an interactive algorithm [135] to obtain a satisfactory solution in which a decision maker iteratively updates not only the reference membership values but also a permissible possibility level for fuzzy random variable coefficients.

3.1 Interactive decision making for MOFRPs with a special structure

In Section 3.1, we formulate MOFRPs with a special structure and propose an interactive linear programming algorithm [131, 132, 133, 134] to obtain a satisfactory solution. In Section 3.1.1, by introducing a possibility measure [18], MOFRPs with a special structure are transformed into multiobjective stochastic programming problems. Next, in Section 3.1.2, transformed multiobjective stochastic programming problems are formulated using a probability maximization model and the fuzzy decision [6, 71, 140]; using this approach, the decision maker specifies membership functions rather than permissible objective levels. Further, we define an MP-Pareto optimal solution to MOFRPs with a special structure and investigate relationships between the minmax problem and an MP-Pareto optimal solution set. Similarly, in Section 3.1.3, transformed multiobjective stochastic programming problems are formulated using a fractile optimization model and the fuzzy decision; using this approach, the decision maker again specifies membership functions rather than permissible probability levels. Here, we define an MF-Pareto optimal solution to MOFRPs with a special structure and investigate relationships between the minmax problem and an MF-Pareto optimal solution set. Next, in Section 3.1.4, we propose an interactive linear programming algorithm to obtain a satisfactory solution from an MF-Pareto optimal solution set. Finally, in Section 3.1.5, the interactive algorithm is applied to a numerical example with a hypothetical decision maker.

3.1.1 MOFRPs and a possibility measure

In this subsection, we focus on multiobjective programming problems involving fuzzy random variable coefficients [46, 65, 77] in objective functions; as noted above, such problems are called multiobjective fuzzy random programming problems. We propose an interactive linear programming algorithm to obtain a satisfactory solution for the decision maker. To begin, a multiobjective fuzzy random programming problem is formally formulated as

MOP 3.1

$$\min \tilde{\bar{\mathbf{C}}}\mathbf{x} = (\tilde{\bar{\mathbf{c}}}_1\mathbf{x}, \cdots, \tilde{\bar{\mathbf{c}}}_k\mathbf{x}) \tag{3.1}$$

subject to

$$\mathbf{x} \in X \stackrel{\text{def}}{=} \{\mathbf{x} \in \mathrm{R}^n \mid A\mathbf{x} \leq \mathbf{b}, \mathbf{x} \geq \mathbf{0}\}, \tag{3.2}$$

where $\mathbf{x} = (x_1, \cdots, x_n)^T$ is an n-dimensional decision variable column vector, A is an $(m \times n)$-dimensional coefficient matrix, $\mathbf{b} = (b_1, \cdots, b_m)^T$ is an m-dimensional column vector. $\tilde{\bar{\mathbf{c}}}_i = (\tilde{\bar{c}}_{i1}, \cdots, \tilde{\bar{c}}_{in}), i = 1, \cdots, k$, are n-dimensional coefficient vectors of objective function $\tilde{\bar{\mathbf{c}}}_i\mathbf{x}$, whose elements $\tilde{\bar{c}}_{ij}, i = 1, \cdots, k, j = 1, \cdots, n$ are fuzzy random variables [46, 65, 77], and symbols "-" and "~" indicate randomness and fuzziness, respectively.

To handle objective functions $\tilde{\bar{\mathbf{c}}}_i\mathbf{x}, i = 1, \cdots, k$, Katagiri et al. [38, 39] proposed an LR-type fuzzy random variable, which can be viewed as a special version of a fuzzy random variable. Given the occurrence of each elementary event ω, $\tilde{\bar{c}}_{ij}(\omega)$ is a realization of LR-type fuzzy random variable $\tilde{\bar{c}}_{ij}$, which is an LR fuzzy number [18] with membership function defined as

$$\mu_{\tilde{\bar{c}}_{ij}(\omega)}(s) = \begin{cases} L\left(\frac{d_{ij}(\omega)-s}{\alpha_{ij}(\omega)}\right), & s \leq d_{ij}(\omega) \\ R\left(\frac{s-d_{ij}(\omega)}{\beta_{ij}(\omega)}\right), & s > d_{ij}(\omega). \end{cases}$$

Here, function $L(t) \stackrel{\text{def}}{=} \max\{0, l(t)\}$ is a real-valued continuous function from $[0, \infty)$ to $[0, 1]$, $l(t)$ is a strictly decreasing continuous function satisfying $l(0) = 1$. Also, $R(t) \stackrel{\text{def}}{=} \max\{0, r(t)\}$ satisfies the same conditions, and $\bar{d}_{ij}, \bar{\alpha}_{ij}, \bar{\beta}_{ij}$ are random variables expressed by

$$\begin{aligned} \bar{d}_{ij} &= d_{ij}^1 + \bar{t}_i d_{ij}^2 \\ \bar{\alpha}_{ij} &= \alpha_{ij}^1 + \bar{t}_i \alpha_{ij}^2 \\ \bar{\beta}_{ij} &= \beta_{ij}^1 + \bar{t}_i \beta_{ij}^2. \end{aligned}$$

Here, $\bar{t}_i$ is a random variable whose distribution function is denoted by $T_i(\cdot)$, a strictly increasing and continuous function, and d_{ij}^1, d_{ij}^2, $\alpha_{ij}^1, \alpha_{ij}^2, \beta_{ij}^1, \beta_{ij}^2$ are constants. Given the occurrence of each elementary event ω, we assume that $\alpha_{ij}^1 + t_i(\omega)\alpha_{ij}^2 > 0$, $\beta_{ij}^1 + t_i(\omega)\beta_{ij}^2 > 0$.

Katagiri et al. [38, 39] transformed MOP3.1 into a multiobjective stochastic programming problem via the concept of a possibility measure [18]. As shown by Katagiri et al. [38, 39], realization $\tilde{\bar{\mathbf{c}}}_i(\omega)\mathbf{x}$ becomes an LR fuzzy number characterized by the following membership functions on the basis of the extension principle [18], i.e.,

$$\mu_{\tilde{\bar{\mathbf{c}}}_i(\omega)\mathbf{x}}(y) = \begin{cases} L\left(\frac{\mathbf{d}_i(\omega)\mathbf{x}-y}{\alpha_i(\omega)\mathbf{x}}\right), & y \leq \mathbf{d}_i(\omega)\mathbf{x} \\ R\left(\frac{y-\mathbf{d}_i(\omega)\mathbf{x}}{\beta_i(\omega)\mathbf{x}}\right), & y > \mathbf{d}_i(\omega)\mathbf{x}, \end{cases}$$

where

$$\begin{aligned} \mathbf{d}_i(\omega) &= (d_{i1}(\omega), \cdots, d_{in}(\omega)) \\ \alpha_i(\omega) &= (\alpha_{i1}(\omega), \cdots, \alpha_{in}(\omega)) \\ \beta_i(\omega) &= (\beta_{i1}(\omega), \cdots, \beta_{in}(\omega)). \end{aligned}$$

For realizations $\tilde{\bar{\mathbf{c}}}_i(\omega)\mathbf{x}, i = 1, \cdots, k$ of the objective functions to MOP3.1, we assume that the decision maker has fuzzy goals $\tilde{G}_i, i = 1, \cdots, k$ [71, 140], whose membership functions $\mu_{\widetilde{G}_i}(y), i = 1, \cdots, k$ are continuous and strictly decreasing for minimization problems. Using the concept of a possibility measure [18], the degree of possibility that objective function value $\widetilde{\overline{\mathbf{c}}}_i\mathbf{x}$ satisfies fuzzy goal $\widetilde{G}_i$ is expressed as [35]

$$\Pi_{\widetilde{\overline{\mathbf{c}}}_i\mathbf{x}}(\widetilde{G}_i) \stackrel{\text{def}}{=} \sup_y \min\{\mu_{\widetilde{\overline{\mathbf{c}}}_i\mathbf{x}}(y), \mu_{\widetilde{G}_i}(y)\}.$$

Using a possibility measure, MOP3.1 can be transformed into the following multiobjective stochastic programming problem [38, 39].

MOP 3.2

$$\max_{\mathbf{x}\in X}(\Pi_{\widetilde{\overline{\mathbf{c}}}_1\mathbf{x}}(\widetilde{G}_1), \cdots, \Pi_{\widetilde{\overline{\mathbf{c}}}_k\mathbf{x}}(\widetilde{G}_k)) \tag{3.3}$$

Katagiri et al. [38, 39] transformed MOP3.2 into the usual multiobjective programming problem using a probability maximization model and a fractile maximization model, proposing interactive algorithms to obtain a satisfactory solution. In their methods, the decision maker specifies permissible probability levels or permissible possibility levels for objective functions in advance; however, in practice, it is difficult to specify appropriate permissible levels, because they have a strong influence on objective function or distribution function values. In the following subsections, by assuming that the decision maker has fuzzy goals for permissible probability and permissible possibility levels, and specifies the corresponding membership functions, we propose an interactive fuzzy decision making method for MOP3.2 to obtain a satisfactory solution.

3.1.2 A formulation using a probability model

For the objective function of MOP3.2, if the decision maker specifies permissible possibility levels $h_i \in [0,1], i = 1, \cdots, k$, then MOP3.2 can be formulated as the following multiobjective programming problem using a probability maximization model.

MOP 3.3 (h)

$$\max_{\mathbf{x}\in X}(\Pr(\omega \mid \Pi_{\widetilde{\overline{\mathbf{c}}}_1(\omega)\mathbf{x}}(\widetilde{G}_1) \geq h_1), \cdots, \Pr(\omega \mid \Pi_{\widetilde{\overline{\mathbf{c}}}_k(\omega)\mathbf{x}}(\widetilde{G}_k) \geq h_k)) \tag{3.4}$$

Here, $\Pr(\cdot)$ is a probability measure, $\mathbf{h} = (h_1, \cdots, h_k)$ is a k-dimensional vector of permissible possibility levels. In MOP3.3($\mathbf{h}$), inequality $\Pi_{\widetilde{\overline{\mathbf{c}}}_i(\omega)\mathbf{x}}(\widetilde{G}_i) \geq h_i$ can be equivalently transformed into

$$\begin{aligned} & \sup_y \min\{\mu_{\widetilde{\overline{\mathbf{c}}}_i\mathbf{x}}(y), \mu_{\widetilde{G}_i}(y)\} \geq h_i \\ \Leftrightarrow \quad & (\mathbf{d}_i(\omega) - L^{-1}(h_i)\alpha_i(\omega))\mathbf{x} \leq \mu_{\widetilde{G}_i}^{-1}(h_i), \end{aligned}$$

where $L^{-1}(\cdot)$, $R^{-1}(\cdot)$, and $\mu_{\widetilde{G}_i}^{-1}(\cdot)$ are inverse functions of $L(\cdot)$, $R(\cdot)$, and $\mu_{\widetilde{G}_i}(\cdot)$, respectively. Therefore, using distribution function $T_i(\cdot)$ of random variable $\bar{t}_i$, the objective functions in MOP3.3($\mathbf{h}$) can be expressed as

$$\begin{aligned} & \Pr(\omega \mid \Pi_{\widetilde{\bar{\mathbf{c}}}_i(\omega)\mathbf{x}}(\widetilde{G}_i) \geq h_i) \\ = \; & \Pr\left(\omega \mid (\mathbf{d}_i(\omega) - L^{-1}(h_i)\alpha_i(\omega))\mathbf{x} \leq \mu_{\widetilde{G}_i}^{-1}(h_i)\right) \\ = \; & T_i\left(\frac{\mu_{\widetilde{G}_i}^{-1}(h_i) - (\mathbf{d}_i^1\mathbf{x} - L^{-1}(h_i)\alpha_i^1\mathbf{x})}{\mathbf{d}_i^2\mathbf{x} - L^{-1}(h_i)\alpha_i^2\mathbf{x}}\right) \\ \stackrel{\text{def}}{=} \; & p_i(\mathbf{x}, h_i), \end{aligned} \tag{3.5}$$

where we assume that $(\mathbf{d}_i^2 - L^{-1}(h_i)\alpha_i^2)\mathbf{x} > 0$, $i = 1, \cdots, k$, for any $\mathbf{x} \in X$ and $h_i \in [0, 1]$. As a result, using $p_i(\mathbf{x}, h_i), i = 1, \cdots, k$, MOP3.3($\mathbf{h}$) can be transformed into the following simple form [38].

MOP 3.4 ($\mathbf{h}$)

$$\max_{\mathbf{x} \in X} (p_1(\mathbf{x}, h_1), \cdots, p_k(\mathbf{x}, h_k)) \tag{3.6}$$

In MOP3.4($\mathbf{h}$), the decision maker prefers not only larger values of permissible possibility level h_i but also larger values of corresponding distribution function $p_i(\mathbf{x}, h_i)$. Since these values conflict with one another, larger values of permissible possibility level h_i result in smaller values of corresponding distribution function $p_i(\mathbf{x}, h_i)$. From such a perspective, we consider the following multiobjective programming problem, which can be viewed as a natural extension of MOP3.4($\mathbf{h}$).

MOP 3.5

$$\max_{\mathbf{x} \in X, h_i \in [0,1], i=1,\cdots,k} (p_1(\mathbf{x}, h_1), \cdots, p_k(\mathbf{x}, h_k), h_1, \cdots, h_k) \tag{3.7}$$

Note that in MOP3.5, permissible possibility levels $h_i, i = 1, \cdots, k$ are not fixed values but rather decision variables. Considering the imprecise nature of the decision maker's judgment, it is natural to assume that the decision maker has fuzzy goals for $p_i(\mathbf{x}, h_i), i = 1, \cdots, k$. In this subsection, we assume that such fuzzy goals can be quantified by eliciting the corresponding membership functions. Let us denote a membership function of a distribution function $p_i(\mathbf{x}, h_i)$ as $\mu_{p_i}(p_i(\mathbf{x}, h_i))$. Then, MOP3.5 can be transformed into the following multiobjective programming problem.

MOP 3.6

$$\max_{\mathbf{x} \in X, h_i \in [0,1], i=1,\cdots,k} (\mu_{p_1}(p_1(\mathbf{x}, h_1)), \cdots, \mu_{p_k}(p_k(\mathbf{x}, h_k)), h_1, \cdots, h_k) \tag{3.8}$$

To appropriately elicit membership functions $\mu_{p_i}(p_i(\mathbf{x},h_i)), i = 1,\cdots,k$, we suggest the following procedure. First, the decision maker sets intervals

$$H_i \stackrel{\text{def}}{=} [h_{i\min}, h_{i\max}], i = 1,\cdots,k, \tag{3.9}$$

for permissible possibility levels, where $h_{i\min}$ is a maximum value of unacceptable levels and $h_{i\max}$ is a minimum value of sufficiently satisfactory levels. For interval H_i, the corresponding domain of $p_i(\mathbf{x},h_i)$ can then be defined as

$$P_i(H_i) \stackrel{\text{def}}{=} [p_{i\min}, p_{i\max}] = \{p_i(\mathbf{x},h_i) \mid \mathbf{x} \in X, h_i \in H_i\}.$$

$p_{i\max}, i = 1,\cdots,k$ can be obtained by solving the following optimization problems.

$$p_{i\max} \stackrel{\text{def}}{=} \max_{\mathbf{x}\in X} p_i(\mathbf{x},h_{i\min}), i = 1,\cdots,k \tag{3.10}$$

To obtain $p_{i\min}, i = 1,\cdots,k$, we first solve optimization problems

$$\max_{\mathbf{x}\in X} p_i(\mathbf{x},h_{i\max}), i = 1,\cdots,k, \tag{3.11}$$

and denote the corresponding optimal solutions as $\mathbf{x}_i, i = 1,\cdots,k$. Using optimal solution $\mathbf{x}_i, i = 1,\cdots,k$, $p_{i\min}, i = 1,\cdots,k$ can be obtained as the following minimum values.

$$p_{i\min} \stackrel{\text{def}}{=} \min_{\ell=1,\cdots,k,\ell\neq i} p_i(\mathbf{x}_\ell,h_{i\max}), i = 1,\cdots,k \tag{3.12}$$

For membership functions $\mu_{p_i}(p_i(\mathbf{x},h_i)), i = 1,\cdots,k$ defined on $P_i(H_i)$, we make the following assumption.

Assumption 3.1
$\mu_{p_i}(p_i(\mathbf{x},h_i)), i = 1,\cdots,k$ are strictly increasing and continuous with respect to $p_i(\mathbf{x},h_i) \in P_i(H_i)$, and $\mu_{p_i}(p_{i\min}) = 0, \mu_{p_i}(p_{i\max}) = 1, i = 1,\cdots,k$.

From definition (3.5), we note that $\mu_{p_i}(p_i(\mathbf{x},h_i))$ is strictly decreasing with respect to $h_i \in H_i$. If the decision maker adopts the fuzzy decision [6, 71, 140] to integrate $\mu_{p_i}(p_i(\mathbf{x},h_i))$ and h_i, the integrated membership function can be defined as

$$\mu_{D_{p_i}}(\mathbf{x},h_i) \stackrel{\text{def}}{=} \min\{h_i, \mu_{p_i}(p_i(\mathbf{x},h_i))\}. \tag{3.13}$$

Then, MOP3.6 can be transformed into the following form.

MOP 3.7

$$\max_{\mathbf{x}\in X, h_i = H_i, i=1,\cdots,k} \left(\mu_{D_{p_1}}(\mathbf{x},h_1),\cdots,\mu_{D_{p_k}}(\mathbf{x},h_k)\right) \tag{3.14}$$

In MOP3.7, permissible possibility levels $h_i, i = 1,\cdots,k$ are automatically adjusted via the fuzzy decision. The decision maker does not need to specify such parameters in advance. Further, to handle MOP3.7, we introduce an MP-Pareto optimal solution in the following definition.

Definition 3.1 $\mathbf{x}^* \in X, h_i^* \in H_i, i = 1, \cdots, k$ is an MP-Pareto optimal solution to MOP3.7 if and only if there does not exist another $\mathbf{x} \in X, h_i \in H_i, i = 1, \cdots, k$ such that $\mu_{D_{p_i}}(\mathbf{x}, h_i) \geq \mu_{D_{p_i}}(\mathbf{x}^*, h_i^*)$ $i = 1, \cdots, k$ with strict inequality holding for at least one i.

To generate a candidate for a satisfactory solution that is also MP-Pareto optimal, the decision maker is asked to specify reference membership values [71]. Once reference membership values

$$\hat{\mu} = (\hat{\mu}_1, \cdots, \hat{\mu}_k) \tag{3.15}$$

are specified, the corresponding MP-Pareto optimal solution is obtained by solving the following minmax problem.

MINMAX 3.1 ($\hat{\mu}$)

$$\min_{\mathbf{x}\in X, h_i \in H_i, i=1,\cdots,k, \lambda \in \Lambda} \lambda \tag{3.16}$$

subject to

$$\begin{aligned} \hat{\mu}_i - \mu_{p_i}(p_i(\mathbf{x}, h_i)) &\leq \lambda, i = 1, \cdots, k &(3.17)\\ \hat{\mu}_i - h_i &\leq \lambda, i = 1, \cdots, k &(3.18) \end{aligned}$$

where

$$\Lambda \stackrel{\text{def}}{=} [\max_{i=1,\cdots,k} \hat{\mu}_i - 1, \max_{i=1,\cdots,k} \hat{\mu}_i]. \tag{3.19}$$

From definition (3.5), constraints (3.18) can be expressed as follows by using $p_i(\mathbf{x}, h_i)$.

$$\begin{aligned} h_i &= \mu_{\tilde{G}_i}\left((\mathbf{d}_i^1\mathbf{x} - L^{-1}(h_i)\alpha_i^1\mathbf{x}) + T_i^{-1}(p_i(\mathbf{x}, h_i)) \cdot (\mathbf{d}_i^2\mathbf{x} - L^{-1}(h_i)\alpha_i^2\mathbf{x})\right) \\ &\geq \hat{\mu}_i - \lambda \end{aligned} \tag{3.20}$$

From Assumption 3.1, constraints (3.17) can be transformed into

$$\begin{aligned} & \hat{\mu}_i - \mu_{p_i}(p_i(\mathbf{x}, h_i)) \leq \lambda \\ \Leftrightarrow\ & p_i(\mathbf{x}, h_i) \geq \mu_{p_i}^{-1}(\hat{\mu}_i - \lambda) \\ \Leftrightarrow\ & \mu_{\tilde{G}_i}^{-1}(h_i) \geq (\mathbf{d}_i^1\mathbf{x} - L^{-1}(h_i)\alpha_i^1\mathbf{x}) + T_i^{-1}(\mu_{p_i}^{-1}(\hat{\mu}_i - \lambda)) \cdot (\mathbf{d}_i^2\mathbf{x} - L^{-1}(h_i)\alpha_i^2\mathbf{x}) \\ \Leftrightarrow\ & \mu_{\tilde{G}_i}^{-1}(h_i) \geq (\mathbf{d}_i^1\mathbf{x} + T_i^{-1}(\mu_{p_i}^{-1}(\hat{\mu}_i - \lambda))\mathbf{d}_i^2\mathbf{x}) \\ & \quad -L^{-1}(h_i)(\alpha_i^1\mathbf{x} + T_i^{-1}(\mu_{p_i}^{-1}(\hat{\mu}_i - \lambda))\alpha_i^2\mathbf{x}). \end{aligned} \tag{3.21}$$

Because of $\hat{\mu}_i - \lambda \leq h_i$, it holds that $\mu_{\tilde{G}_i}^{-1}(h_i) \leq \mu_{\tilde{G}_i}^{-1}(\hat{\mu}_i - \lambda)$ and $L^{-1}(h_i) \leq L^{-1}(\hat{\mu}_i - \lambda)$. Since it is guaranteed that $(\alpha_i^1\mathbf{x} + T_i^{-1}(\mu_{p_i}^{-1}(\hat{\mu}_i - \lambda))\ \alpha_i^2\mathbf{x}) > 0$, the following inequalities can be derived.

$$\begin{aligned}
& (\mathbf{d}_i^1\mathbf{x} + T_i^{-1}(\mu_{p_i}^{-1}(\hat{\mu}_i - \lambda))\mathbf{d}_i^2\mathbf{x}) \\
& \quad -L^{-1}(h_i)(\alpha_i^1\mathbf{x} + T_i^{-1}(\mu_{p_i}^{-1}(\hat{\mu}_i - \lambda))\alpha_i^2\mathbf{x}) \\
\geq \; & (\mathbf{d}_i^1\mathbf{x} + T_i^{-1}(\mu_{p_i}^{-1}(\hat{\mu}_i - \lambda))\mathbf{d}_i^2\mathbf{x}) \\
& \quad -L^{-1}(\hat{\mu}_i - \lambda)(\alpha_i^1\mathbf{x} + T_i^{-1}(\mu_{p_i}^{-1}(\hat{\mu}_i - \lambda))\alpha_i^2\mathbf{x}) \\
= \; & (\mathbf{d}_i^1\mathbf{x} - L^{-1}(\hat{\mu}_i - \lambda)\alpha_i^1\mathbf{x}) \\
& \quad +T_i^{-1}(\mu_{p_i}^{-1}(\hat{\mu}_i - \lambda)) \cdot (\mathbf{d}_i^2\mathbf{x} - L^{-1}(\hat{\mu}_i - \lambda)\alpha_i^2\mathbf{x})
\end{aligned} \tag{3.22}$$

From (3.21) and (3.22), it holds that

$$\begin{aligned}
& \mu_{\tilde{G}_i}^{-1}(\hat{\mu}_i - \lambda) \geq \mu_{\tilde{G}_i}^{-1}(h_i) \\
\geq \; & (\mathbf{d}_i^1\mathbf{x} - L^{-1}(\hat{\mu}_i - \lambda)\alpha_i^1\mathbf{x}) + T_i^{-1}(\mu_{p_i}^{-1}(\hat{\mu}_i - \lambda)) \cdot (\mathbf{d}_i^2\mathbf{x} - L^{-1}(\hat{\mu}_i - \lambda)\alpha_i^2\mathbf{x}).
\end{aligned}$$

Therefore, MINMAX3.1($\hat{\mu}$) can be reduced to the following minmax problem, where permissible possibility levels $h_i, i = 1, \cdots, k$ have disappeared.

MINMAX 3.2 ($\hat{\mu}$)

$$\min_{\mathbf{x} \in X, \lambda \in \Lambda} \lambda \tag{3.23}$$

subject to

$$\begin{aligned}
\mu_{\tilde{G}_i}^{-1}(\hat{\mu}_i - \lambda) \; \geq \; & (\mathbf{d}_i^1\mathbf{x} - L^{-1}(\hat{\mu}_i - \lambda)\alpha_i^1\mathbf{x}) \\
& +T_i^{-1}(\mu_{p_i}^{-1}(\hat{\mu}_i - \lambda)) \cdot (\mathbf{d}_i^2\mathbf{x} - L^{-1}(\hat{\mu}_i - \lambda)\alpha_i^2\mathbf{x}), \\
& i = 1, \cdots, k
\end{aligned} \tag{3.24}$$

Note that constraints (3.24) can be reduced to a set of linear inequalities for some fixed value $\lambda \in \Lambda$, meaning that optimal solution $\mathbf{x}^* \in X, \lambda^* \in \Lambda$ of MINMAX3.2($\hat{\mu}$) is obtained via the combined use of the bisection method with respect to $\lambda \in \Lambda$ and the first phase of the two-phase simplex method of linear programming. Relationships between optimal solution $\mathbf{x}^* \in X, \lambda^* \in \Lambda$ of MINMAX3.2($\hat{\mu}$) and MP-Pareto optimal solutions are characterized by the following theorem.

Theorem 3.1

(1) If $\mathbf{x}^* \in X, \lambda^* \in \Lambda$ is a unique optimal solution of MINMAX3.2($\hat{\mu}$), then $\mathbf{x}^* \in X, h_i^* \stackrel{\text{def}}{=} \hat{\mu}_i - \lambda^* \in H_i, i = 1, \cdots, k$ is an MP-Pareto optimal solution to MOP3.7.

(2) If $\mathbf{x}^* \in X, h_i^* \in H_i, i = 1,\cdots,k$ is an MP-Pareto optimal solution to MOP3.7, then $\mathbf{x}^* \in X$, $\lambda^* \stackrel{\text{def}}{=} \hat{\mu}_i - h_i^* = \hat{\mu}_i - \mu_{p_i}(p_i(\mathbf{x}^*, h_i^*)), i = 1,\cdots,k$ is an optimal solution of MINMAX3.2$(\hat{\mu})$ for some reference membership values $\hat{\mu} = (\hat{\mu}_1,\cdots,\hat{\mu}_k)$.

(Proof)
(1) Inequalities of (3.24) can be equivalently expressed as follows at $\mathbf{x}^* \in X, \lambda^* \in \Lambda$.

$$\hat{\mu}_i - \lambda^* \leq \mu_{p_i}(p_i(\mathbf{x}^*, \hat{\mu}_i - \lambda^*)), i = 1,\cdots,k$$

Assume that $\mathbf{x}^* \in X, \hat{\mu}_i - \lambda^* \in H_i, i = 1,\cdots,k$ is not an MP-Pareto optimal solution to MOP3.7. Then, there exists $\mathbf{x} \in X, h_i \in H_i$ such that

$$\begin{aligned} \mu_{D_{p_i}}(\mathbf{x}, h_i) &= \min\{h_i, \mu_{p_i}(p_i(\mathbf{x}, h_i))\} \\ &\geq \mu_{D_{p_i}}(\mathbf{x}^*, \hat{\mu}_i - \lambda^*) \\ &= \hat{\mu}_i - \lambda^*, i = 1,\cdots,k, \end{aligned}$$

with strict inequality holding for at least one i. Then it holds that

$$h_i \geq \hat{\mu}_i - \lambda^*, i = 1,\cdots,k, \tag{3.25}$$

$$\mu_{p_i}(p_i(\mathbf{x}, h_i)) \geq \hat{\mu}_i - \lambda^*, i = 1,\cdots,k. \tag{3.26}$$

From Assumption 3.1, (3.5) and $L^{-1}(h_i) \leq L^{-1}(\hat{\mu}_i - \lambda^*)$, (3.25) and (3.26) can be transformed as follows.

$$\begin{aligned} \mu_{\tilde{G}_i}^{-1}(h_i) &\leq \mu_{\tilde{G}_i}^{-1}(\hat{\mu}_i - \lambda^*), i = 1,\cdots,k \\ \mu_{\tilde{G}_i}^{-1}(h_i) &\geq (\mathbf{d}_i^1\mathbf{x} - L^{-1}(\hat{\mu}_i - \lambda^*)\alpha_i^1\mathbf{x}) \\ &\quad + T_i^{-1}(\mu_{p_i}^{-1}(\hat{\mu}_i - \lambda^*)) \cdot (\mathbf{d}_i^2\mathbf{x} - L^{-1}(\hat{\mu}_i - \lambda^*)\alpha_i^2\mathbf{x}), i = 1,\cdots,k \end{aligned}$$

As a result, there exists $\mathbf{x} \in X$ such that

$$\begin{aligned} & \mu_{\tilde{G}_i}^{-1}(\hat{\mu}_i - \lambda^*) - (\mathbf{d}_i^1\mathbf{x} - L^{-1}(\hat{\mu}_i - \lambda^*)\alpha_i^1\mathbf{x}), \\ \geq\ & T_i^{-1}(\mu_{p_i}^{-1}(\hat{\mu}_i - \lambda^*)) \cdot (\mathbf{d}_i^2\mathbf{x} - L^{-1}(\hat{\mu}_i - \lambda^*)\alpha_i^2\mathbf{x}), i = 1,\cdots,k, \end{aligned}$$

which contradicts the fact that $\mathbf{x}^* \in X, \lambda^* \in \Lambda$ is a unique optimal solution of MINMAX3.2$(\hat{\mu})$.

(2) From Assumption 3.1, it holds that $h_i^* = \mu_{p_i}(p_i(\mathbf{x}^*, h_i^*)), i = 1,\cdots,k$. Assume that $\mathbf{x}^* \in X, \lambda^* \in \Lambda$ is not an optimal solution of MINMAX3.2$(\hat{\mu})$ for any reference membership values $\hat{\mu} = (\hat{\mu}_1,\cdots,\hat{\mu}_k)$, which satisfy equalities

$$\hat{\mu}_i - \lambda^* = h_i^* = \mu_{p_i}(p_i(\mathbf{x}^*, h_i^*)), i = 1,\cdots,k. \tag{3.27}$$

Then, there exists some $\mathbf{x} \in X, \lambda < \lambda^*$ such that

$$\begin{aligned} & \mu_{\tilde{G}_i}^{-1}(\hat{\mu}_i - \lambda) - (\mathbf{d}_i^1\mathbf{x} - L^{-1}(\hat{\mu}_i - \lambda)\alpha_i^1\mathbf{x}), \\ & \quad \geq T_i^{-1}(\mu_{p_i}^{-1}(\hat{\mu}_i - \lambda)) \cdot (\mathbf{d}_i^2\mathbf{x} - L^{-1}(\hat{\mu}_i - \lambda)\alpha_i^2\mathbf{x}), \\ \Leftrightarrow\ & \mu_{p_i}(p_i(\mathbf{x}, \hat{\mu}_i - \lambda)) \geq \hat{\mu}_i - \lambda, i = 1,\cdots,k. \end{aligned} \tag{3.28}$$

Because of (3.27), (3.28) and $\hat{\mu}_i - \lambda > \hat{\mu}_i - \lambda^*, i = 1, \cdots, k$, the following inequalities hold.

$$\mu_{p_i}(p_i(\mathbf{x}, h_i)) \quad > \quad \mu_{p_i}(p_i(\mathbf{x}^*, h_i^*)), i = 1, \cdots, k$$

where $h_i \stackrel{\text{def}}{=} \hat{\mu}_i - \lambda \in H_i$. Then, because of $h_i > h_i^*$, there exists $\mathbf{x} \in X, h_i \in H_i, i = 1, \cdots, k$ such that

$$\mu_{D_{p_i}}(\mathbf{x}, h_i) > \mu_{D_{p_i}}(\mathbf{x}^*, h_i^*), i = 1, \cdots, k.$$

This contradicts the fact that $\mathbf{x}^* \in X, h_i^* \in H_i, i = 1, \cdots, k$ is an MP-Pareto optimal solution to MOP3.7. □

3.1.3 A formulation using a fractile model

If we adopt a fractile optimization model for objective functions $\Pi_{\tilde{\bar{\mathbf{c}}}_i\mathbf{x}}(\widetilde{G}_i)$, $i = 1, \cdots, k$ to MOP3.2, we can convert MOP3.2 into the following multiobjective programming problem, where the decision maker subjectively specifies permissible probability levels $\hat{\mathbf{p}} = (\hat{p}_1, \cdots, \hat{p}_k)$.

MOP 3.8 $(\hat{\mathbf{p}})$

$$\max_{\mathbf{x} \in X, h_i \in [0,1], i=1,\cdots,k} (h_1, \cdots, h_k) \tag{3.29}$$

subject to

$$p_i(\mathbf{x}, h_i) \geq \hat{p}_i, i = 1, \cdots, k \tag{3.30}$$

Since distribution functions $T_i(\cdot), i = 1, \cdots, k$ are continuous and strictly increasing, constraints (3.30) can be transformed into

$$\begin{aligned} & \hat{p}_i \leq p_i(\mathbf{x}, h_i) \\ & \quad = T_i\left(\frac{\mu_{\widetilde{G}_i}^{-1}(h_i) - (\mathbf{d}_i^1\mathbf{x} - L^{-1}(h_i)\alpha_i^1\mathbf{x})}{\mathbf{d}_i^2\mathbf{x} - L^{-1}(h_i)\alpha_i^2\mathbf{x}}\right) \\ \Leftrightarrow \quad & \mu_{\widetilde{G}_i}^{-1}(h_i) \geq (\mathbf{d}_i^1\mathbf{x} - L^{-1}(h_i)\alpha_i^1\mathbf{x}) + T_i^{-1}(\hat{p}_i) \cdot (\mathbf{d}_i^2\mathbf{x} - L^{-1}(h_i)\alpha_i^2\mathbf{x}), \\ & i = 1, \cdots, k. \end{aligned} \tag{3.31}$$

Let us define the right-hand-side of inequality (3.31) as follows.

$$f_i(\mathbf{x}, h_i, \hat{p}_i) \stackrel{\text{def}}{=} (\mathbf{d}_i^1\mathbf{x} - L^{-1}(h_i)\alpha_i^1\mathbf{x}) + T_i^{-1}(\hat{p}_i) \cdot (\mathbf{d}_i^2\mathbf{x} - L^{-1}(h_i)\alpha_i^2\mathbf{x}) \tag{3.32}$$

Then, MOP3.8$(\hat{\mathbf{p}})$ can be equivalently transformed into the following form, because $\mu_{\widetilde{G}_i}(\cdot), i = 1, \cdots, k$ are continuous and strictly decreasing.

MOP 3.9 ($\hat{\mathbf{p}}$)

$$\max_{\mathbf{x}\in X, h_i\in[0,1], i=1,\cdots,k} (h_1,\cdots,h_k) \tag{3.33}$$

subject to

$$\mu_{\tilde{G}_i}(f_i(\mathbf{x},h_i,\hat{p}_i)) \geq h_i, i=1,\cdots,k \tag{3.34}$$

In MOP3.9($\hat{\mathbf{p}}$), let us pay attention to inequalities (3.34). Because of definition (3.32), $f_i(\mathbf{x},h_i,\hat{p}_i)$ can be transformed into

$$\begin{aligned} f_i(\mathbf{x},h_i,\hat{p}_i) &= (\mathbf{d}_i^1\mathbf{x} - L^{-1}(h_i)\alpha_i^1\mathbf{x}) + T_i^{-1}(\hat{p}_i)\cdot(\mathbf{d}_i^2\mathbf{x} - L^{-1}(h_i)\alpha_i^2\mathbf{x}) \\ &= (\mathbf{d}_i^1\mathbf{x} + T_i^{-1}(\hat{p}_i)\mathbf{d}_i^2\mathbf{x}) - L^{-1}(h_i)(\alpha_i^1\mathbf{x} + T_i^{-1}(\hat{p}_i)\alpha_i^2\mathbf{x}). \end{aligned}$$

Here, $L^{-1}(h_i)$ is continuous and strictly decreasing with respect to h_i, and we assume that $(\alpha_i^1\mathbf{x} + T_i^{-1}(\hat{p}_i)\alpha_i^2\mathbf{x}) > 0$ from the property of spread parameters [18]. As a result, $f_i(\mathbf{x},h_i,\hat{p}_i)$ is continuous and strictly increasing with respect to h_i for any $\mathbf{x}\in X$, meaning that the left-hand-side $\mu_{\tilde{G}_i}(f_i(\mathbf{x},h_i,\hat{p}_i))$ of (3.34) is continuous and strictly decreasing with respect to h_i for any $\mathbf{x}\in X$. Since the right-hand-side of (3.34) is continuous and strictly increasing with respect to h_i, inequalities (3.34) must always satisfy the active condition, that is, $\mu_{\tilde{G}_i}(f_i(\mathbf{x},h_i,\hat{p}_i)) = h_i$, $i=1,\cdots,k$ at the optimal solution. From such a perspective, MOP3.9($\hat{\mathbf{p}}$) is equivalently expressed as follows.

MOP 3.10 ($\hat{\mathbf{p}}$)

$$\max_{\mathbf{x}\in X, h_i\in[0,1], i=1,\cdots,k} (\mu_{\tilde{G}_1}(f_1(\mathbf{x},h_1,\hat{p}_1)),\cdots,\mu_{\tilde{G}_k}(f_k(\mathbf{x},h_k,\hat{p}_k))) \tag{3.35}$$

subject to

$$\mu_{\tilde{G}_i}(f_i(\mathbf{x},h_i,\hat{p}_i)) = h_i, i=1,\cdots,k \tag{3.36}$$

To handle MOP3.10($\hat{\mathbf{p}}$), the decision maker must specify permissible probability levels $\hat{\mathbf{p}}$ in advance; however, in general, the decision maker prefers not only larger values of a permissible probability level but also larger values of corresponding membership function $\mu_{\tilde{G}_i}(\cdot)$. From such a perspective, we consider the following multiobjective programming problem, which can be viewed as a natural extension of MOP3.10($\hat{\mathbf{p}}$).

MOP 3.11

$$\begin{aligned} \max_{\mathbf{x}\in X, h_i\in[0,1], \hat{p}_i\in(0,1), i=1,\cdots,k} \quad & (\mu_{\tilde{G}_1}(f_1(\mathbf{x},h_1,\hat{p}_1)), \\ & \cdots,\mu_{\tilde{G}_k}(f_k(\mathbf{x},h_k,\hat{p}_k)),\hat{p}_1,\cdots,\hat{p}_k) \end{aligned} \tag{3.37}$$

subject to

$$\mu_{\tilde{G}_i}(f_i(\mathbf{x},h_i,\hat{p}_i)) = h_i, i=1,\cdots,k \tag{3.38}$$

It should be noted in MOP3.11 that permissible probability levels $(\hat{p}_1, \cdots, \hat{p}_k)$ are not the fixed values but the decision variables.

Considering the imprecise nature of the decision maker's judgment, we assume that the decision maker has a fuzzy goal for each permissible probability level. Such a fuzzy goal can be quantified by eliciting the corresponding membership function. Let us denote a membership function of a permissible probability level $\hat{p}_i$ as $\mu_{\hat{p}_i}(\hat{p}_i)$. Then, MOP3.11 can be transformed into the following multiobjective programming problem.

MOP 3.12

$$\max_{\mathbf{x}\in X, h_i\in[0,1], \hat{p}_i\in(0,1), i=1,\cdots,k} \quad (\mu_{\tilde{G}_1}(f_1(\mathbf{x}, h_1, \hat{p}_1)), \cdots, \mu_{\tilde{G}_k}(f_k(\mathbf{x}, h_k, \hat{p}_k)),$$

$$\mu_{\hat{p}_1}(\hat{p}_1), \cdots, \mu_{\hat{p}_k}(\hat{p}_k)) \tag{3.39}$$

subject to

$$\mu_{\tilde{G}_i}(f_i(\mathbf{x}, h_i, \hat{p}_i)) = h_i, i = 1, \cdots, k \tag{3.40}$$

To appropriately elicit the membership functions, we suggest the following procedure. First, the decision maker sets intervals

$$P_i \stackrel{\text{def}}{=} [p_{i\min}, p_{i\max}], i = 1, \cdots, k,$$

where $p_{i\min}$ is an unacceptable maximum value of $\hat{p}_i$ and $p_{i\max}$ is a sufficiently satisfactory minimum value of $\hat{p}_i$. Throughout this subsection, we make the following assumption.

Assumption 3.2
$\mu_{\hat{p}_i}(\hat{p}_i), i = 1, \cdots, k$ are strictly increasing and continuous with respect to $\hat{p}_i \in P_i$, and $\mu_{\hat{p}_i}(p_{i\min}) = 0$, $\mu_{\hat{p}_i}(p_{i\max}) = 1$.

Corresponding to interval P_i, interval

$$F_i(P_i) \stackrel{\text{def}}{=} [f_{i\min}, f_{i\max}], i = 1, \cdots, k,$$

can be computed as follows. Minimum values $f_{i\min}, i = 1, \cdots, k$ can be obtained by solving the following problem.

$$f_{i\min} \stackrel{\text{def}}{=} \min_{\mathbf{x}\in X, h_i\in[0,1]} f_i(\mathbf{x}, h_i, p_{i\min}), i = 1, \cdots, k \tag{3.41}$$

$$\text{subject to } h_i = \mu_{\tilde{G}_i}(f_i(\mathbf{x}, h_i, p_{i\min})) \tag{3.42}$$

This is equivalent to the following problem.

$$f_{i\min} \stackrel{\text{def}}{=} \min_{\mathbf{x}\in X, h_i\in[0,1]} \mu_{\tilde{G}_i}^{-1}(h_i), i = 1, \cdots, k \tag{3.43}$$

subject to

$$\mu_{\tilde{G}_i}^{-1}(h_i) = (\mathbf{d}_i^1\mathbf{x} - L^{-1}(h_i)\alpha_i^1\mathbf{x}) + T_i^{-1}(p_{i\min}) \cdot (\mathbf{d}_i^2\mathbf{x} - L^{-1}(h_i)\alpha_i^2\mathbf{x}) \tag{3.44}$$

Optimal solution $\mathbf{x}^* \in X, h_i^* \in [0,1], i = 1,\cdots,k$ of the above problem can be obtained via the combined use of the bisection method with respect to $h_i \in [0,1]$ and the first phase of the two-phase simplex method of linear programming. To obtain $f_{i\max}$, we first solve the following k linear programming problems.

$$\min_{\mathbf{x}\in X, h_i\in[0,1]} f_i(\mathbf{x}, h_i, p_{i\max}), i = 1,\cdots,k \tag{3.45}$$

$$\text{subject to } h_i = \mu_{\tilde{G}_i}(f_i(\mathbf{x}, h_i, p_{i\max})) \tag{3.46}$$

Let $\mathbf{x}^* \in X, h_i^* \in [0,1], i = 1,\cdots,k$ be the above optimal solution. Using optimal solutions $\mathbf{x}^* \in X, h_i^* \in [0,1], i = 1,\cdots,k$, $f_{i\max}, i = 1,\cdots,k$ can be obtained as follows.

$$f_{i\max} \stackrel{\text{def}}{=} \max_{\ell=1,\cdots,k,\ell\neq i} f_i(\mathbf{x}_\ell^*, h_\ell^*, p_{i\max}), i = 1,\cdots,k \tag{3.47}$$

Corresponding to $F_i(P_i)$, the interval of $h_i, i = 1,\cdots,k$ can be obtained as follows.

$$H_i(P_i) \stackrel{\text{def}}{=} [h_{i\min}, h_{i\max}] = [\mu_{\tilde{G}_i}(f_{i\max}), \mu_{\tilde{G}_i}(f_{i\min})], i = 1,\cdots,k \tag{3.48}$$

Note that $\mu_{\tilde{G}_i}(f_i(\mathbf{x}, h_i, \hat{p}_i))$ is strictly decreasing with respect to $\hat{p}_i$. If the decision maker adopts the fuzzy decision [6, 71, 140] to integrate $\mu_{\tilde{G}_i}(f_i(\mathbf{x}, h_i, \hat{p}_i))$ and $\mu_{\hat{p}_i}(\hat{p}_i)$, integrated membership function $\mu_{D_{G_i}}(\mathbf{x}, h_i, \hat{p}_i)$ can be defined as

$$\mu_{D_{G_i}}(\mathbf{x}, h_i, \hat{p}_i) \stackrel{\text{def}}{=} \min\{\mu_{\hat{p}_i}(\hat{p}_i), \mu_{\tilde{G}_i}(f_i(\mathbf{x}, h_i, \hat{p}_i))\}. \tag{3.49}$$

Using integrated membership function $\mu_{D_{G_i}}(\mathbf{x}, h_i, \hat{p}_i)$, MOP3.12 is then transformed into the following form.

MOP 3.13

$$\max_{\mathbf{x}\in X, \hat{p}_i\in P_i, h_i\in H_i(P_i), i=1,\cdots,k} \left(\mu_{D_{G_1}}(\mathbf{x}, h_1, \hat{p}_1), \cdots, \mu_{D_{G_k}}(\mathbf{x}, h_k, \hat{p}_k)\right) \tag{3.50}$$

subject to

$$\mu_{\tilde{G}_i}(f_i(\mathbf{x}, h_i, \hat{p}_i)) = h_i, i = 1,\cdots,k \tag{3.51}$$

In order to deal with MOP3.13, we introduce an MF-Pareto optimal solution concept.

Definition 3.2 $\mathbf{x}^* \in X, \hat{p}_i^* \in P_i, h_i^* \in H_i(P_i), i = 1,\cdots,k$ is an MF-Pareto optimal solution to MOP3.13 if and only if there does not exist another $\mathbf{x} \in X, \hat{p}_i \in P_i, h_i \in H_i(P_i), i = 1,\cdots,k$ such that $\mu_{D_{G_i}}(\mathbf{x}, h_i, \hat{p}_i) \geq \mu_{D_{G_i}}(\mathbf{x}^*, h_i^*, \hat{p}_i^*), i = 1,\cdots,k$

with strict inequality holding for at least one i, where $\mu_{\tilde{G}_i}(f_i(\mathbf{x}^*, h_i^*, \hat{p}_i^*)) = h_i^*$, $\mu_{\tilde{G}_i}(f_i(\mathbf{x}, h_i, \hat{p}_i)) = h_i, i = 1, \cdots, k$.

To generate a candidate for a satisfactory solution that is also MF-Pareto optimal, the decision maker is asked to specify reference membership values [71].

$$\hat{\mu} = (\hat{\mu}_1, \cdots, \hat{\mu}_k) \tag{3.52}$$

Once these reference membership values are specified, the corresponding MF-Pareto optimal solution is obtained by solving the following minmax problem.

MINMAX 3.3 ($\hat{\mu}$)

$$\min_{\mathbf{x} \in X, \hat{p}_i \in P_i, h_i \in H_i(P_i), i=1,\cdots,k, \lambda \in \Lambda} \lambda \tag{3.53}$$

subject to

$$\hat{\mu}_i - \mu_{\hat{p}_i}(\hat{p}_i) \leq \lambda, i = 1, \cdots, k \tag{3.54}$$
$$\hat{\mu}_i - h_i \leq \lambda, i = 1, \cdots, k \tag{3.55}$$
$$\mu_{\tilde{G}_i}(f_i(\mathbf{x}, h_i, \hat{p}_i)) = h_i, i = 1, \cdots, k \tag{3.56}$$

where

$$\Lambda \stackrel{\text{def}}{=} [\max_{i=1,\cdots,k} \hat{\mu}_i - 1, \max_{i=1,\cdots,k} \hat{\mu}_i]. \tag{3.57}$$

In constraints (3.55) and (3.56), it holds that

$$\begin{aligned} & h_i = \mu_{\tilde{G}_i}(f_i(\mathbf{x}, h_i, \hat{p}_i)) \geq \hat{\mu}_i - \lambda \\ \Leftrightarrow \quad & \mu_{\tilde{G}_i}^{-1}(h_i) = f_i(\mathbf{x}, h_i, \hat{p}_i) \leq \mu_{\tilde{G}_i}^{-1}(\hat{\mu}_i - \lambda) \\ \Leftrightarrow \quad & \mu_{\tilde{G}_i}^{-1}(h_i) = (\mathbf{d}_i^1\mathbf{x} - L^{-1}(h_i)\alpha_i^1\mathbf{x}) \\ & + T_i^{-1}(\hat{p}_i) \cdot (\mathbf{d}_i^2\mathbf{x} - L^{-1}(h_i)\alpha_i^2\mathbf{x}) \leq \mu_{\tilde{G}_i}^{-1}(\hat{\mu}_i - \lambda). \end{aligned} \tag{3.58}$$

In the right-hand-side of (3.58), because of $L^{-1}(h_i) \leq L^{-1}(\hat{\mu}_i - \lambda)$ and $\alpha_i^1\mathbf{x} + T_i^{-1}(\hat{p}_i)\alpha_i^2\mathbf{x} > 0$, it holds that

$$\begin{aligned} & (\mathbf{d}_i^1\mathbf{x} - L^{-1}(h_i)\alpha_i^1\mathbf{x}) + T_i^{-1}(\hat{p}_i) \cdot (\mathbf{d}_i^2\mathbf{x} - L^{-1}(h_i)\alpha_i^2\mathbf{x}), \\ = \quad & (\mathbf{d}_i^1\mathbf{x} + T_i^{-1}(\hat{p}_i)\mathbf{d}_i^2\mathbf{x}) - L^{-1}(h_i)\left(\alpha_i^1\mathbf{x} + T_i^{-1}(\hat{p}_i)\alpha_i^2\mathbf{x}\right), \\ \geq \quad & (\mathbf{d}_i^1\mathbf{x} + T_i^{-1}(\hat{p}_i)\mathbf{d}_i^2\mathbf{x}) - L^{-1}(\hat{\mu}_i - \lambda)\left(\alpha_i^1\mathbf{x} + T_i^{-1}(\hat{p}_i)\alpha_i^2\mathbf{x}\right). \end{aligned} \tag{3.59}$$

Using (3.58) and (3.59), it holds that

$$\begin{aligned} & \mu_{\tilde{G}_i}^{-1}(\hat{\mu}_i - \lambda) \\ \geq \quad & (\mathbf{d}_i^1\mathbf{x} + T_i^{-1}(\hat{p}_i)\mathbf{d}_i^2\mathbf{x}) - L^{-1}(\hat{\mu}_i - \lambda)\left(\alpha_i^1\mathbf{x} + T_i^{-1}(\hat{p}_i)\alpha_i^2\mathbf{x}\right), \\ = \quad & (\mathbf{d}_i^1\mathbf{x} - L^{-1}(\hat{\mu}_i - \lambda)\alpha_i^1\mathbf{x}) + T_i^{-1}(\hat{p}_i) \cdot (\mathbf{d}_i^2\mathbf{x} - L^{-1}(\hat{\mu}_i - \lambda)\alpha_i^2\mathbf{x}). \end{aligned} \tag{3.60}$$

Moreover, from (3.54) and Assumption 3.2, it holds that $\hat{p}_i \geq \mu_{\hat{p}_i}^{-1}(\hat{\mu}_i - \lambda)$. Therefore, inequality (3.60) can be transformed into the following form.

$$T_i\left(\frac{\mu_{\tilde{G}_i}^{-1}(\hat{\mu}_i-\lambda)-(\mathbf{d}_i^1\mathbf{x}-L^{-1}(\hat{\mu}_i-\lambda)\alpha_i^1\mathbf{x})}{\mathbf{d}_i^2\mathbf{x}-L^{-1}(\hat{\mu}_i-\lambda)\alpha_i^2\mathbf{x}}\right)\geq \hat{p}_i\geq \mu_{\hat{p}_i}^{-1}(\hat{\mu}_i-\lambda)$$

$$\begin{aligned}\Leftrightarrow\quad & \mu_{\tilde{G}_i}^{-1}(\hat{\mu}_i-\lambda)\\ & \geq(\mathbf{d}_i^1\mathbf{x}-L^{-1}(\hat{\mu}_i-\lambda)\alpha_i^1\mathbf{x})\\ & +T_i^{-1}(\mu_{\hat{p}_i}^{-1}(\hat{\mu}_i-\lambda))\cdot(\mathbf{d}_i^2\mathbf{x}-L^{-1}(\hat{\mu}_i-\lambda)\alpha_i^2\mathbf{x})\end{aligned}\tag{3.61}$$

Then, MINMAX3.3($\hat{\mu}$) can be reduced to the following minmax problem, in which both permissible probability levels $\hat{p}_i$, $i = 1,\cdots,k$ and permissible possibility levels $h_i, i = 1,\cdots,k$ have disappeared.

MINMAX 3.4 ($\hat{\mu}$)

$$\min_{\mathbf{x}\in X,\lambda\in\Lambda}\ \lambda \tag{3.62}$$

subject to

$$\begin{aligned}&\mu_{\tilde{G}_i}^{-1}(\hat{\mu}_i-\lambda)\geq(\mathbf{d}_i^1\mathbf{x}-L^{-1}(\hat{\mu}_i-\lambda)\alpha_i^1\mathbf{x})\\ &+T_i^{-1}(\mu_{\hat{p}_i}^{-1}(\hat{\mu}_i-\lambda))\cdot(\mathbf{d}_i^2\mathbf{x}-L^{-1}(\hat{\mu}_i-\lambda)\alpha_i^2\mathbf{x}), i=1,\cdots,k\end{aligned}\tag{3.63}$$

Note that MINMAX3.4($\hat{\mu}$) is equivalent to MINMAX3.2($\hat{\mu}$).

Similar to Theorem 3.1, the relationships between optimal solution $(\mathbf{x}^*, \lambda^*)$ of MINMAX3.4($\hat{\mu}$) and MF-Pareto optimal solutions can be characterized by the following theorem.

Theorem 3.2

(1) If $\mathbf{x}^* \in X, \lambda^* \in \Lambda$ is a unique optimal solution of MINMAX3.4($\hat{\mu}$), then $\mathbf{x}^* \in X, \hat{p}_i^* \stackrel{\text{def}}{=} \mu_{\hat{p}_i}^{-1}(\hat{\mu}_i - \lambda^*) \in P_i, h_i^* \stackrel{\text{def}}{=} \hat{\mu}_i - \lambda^* \in H_i(P_i), i = 1,\cdots,k$ is an MF-Pareto optimal solution to MOP3.13.

(2) If $\mathbf{x}^* \in X, \hat{p}_i^* \in P_i, h_i^* \in H_i(P_i), i = 1,\cdots,k$ is an MF-Pareto optimal solution to MOP3.13, then $\mathbf{x}^* \in X$, $\lambda^* \stackrel{\text{def}}{=} \hat{\mu}_i - \mu_{\hat{p}_i}(\hat{p}_i^*) = \hat{\mu}_i - \mu_{\tilde{G}_i}(f_i(\mathbf{x}^*, h_i^*, \hat{p}_i^*)), i = 1,\cdots,k$ is an optimal solution of MINMAX3.4($\hat{\mu}$) for some reference membership values $\hat{\mu} = (\hat{\mu}_1,\cdots,\hat{\mu}_k)$.

(Proof)
(1) From inequality constraints (3.63), it holds that

$$\hat{\mu}_i-\lambda^*\leq\mu_{\tilde{G}_i}(f_i(\mathbf{x}^*,\hat{\mu}_i-\lambda^*,\mu_{\hat{p}_i}^{-1}(\hat{\mu}_i-\lambda^*))), i=1,\cdots,k,$$

and it is obvious that $\hat{\mu}_i - \lambda^* = \mu_{\hat{p}_i}(\mu_{\hat{p}_i}^{-1}(\hat{\mu}_i - \lambda^*))$. Assume that $\mathbf{x}^* \in X, \hat{\mu}_i - \lambda^* \in H_i(P_i), \mu_{\hat{p}_i}^{-1}(\hat{\mu}_i - \lambda^*) \in P_i, i = 1, \cdots, k$ is not an MF-Pareto optimal solution to MOP3.13. Then, there exist $\mathbf{x} \in X, \hat{p}_i \in P_i, h_i \in H_i(P_i), i = 1, \ldots, k$ such that

$$\begin{aligned} \mu_{D_{G_i}}(\mathbf{x}, h_i, \hat{p}_i) &= \min\{\mu_{\hat{p}_i}(\hat{p}_i), \mu_{\tilde{G}_i}(f_i(\mathbf{x}, h_i, \hat{p}_i))\} \\ &\geq \mu_{D_{G_i}}(\mathbf{x}^*, \hat{\mu}_i - \lambda^*, \mu_{\hat{p}_i}^{-1}(\hat{\mu}_i - \lambda^*)) \\ &= \hat{\mu}_i - \lambda^*, i = 1, \cdots, k, \end{aligned}$$

with strict inequality holding for at least one i, and $\mu_{\tilde{G}_i}(f_i(\mathbf{x}, h_i, \hat{p}_i)) = h_i, i = 1, \ldots, k$. Then it holds that

$$\mu_{\hat{p}_i}(\hat{p}_i) \geq \hat{\mu}_i - \lambda^*, i = 1, \cdots, k, \tag{3.64}$$

$$\mu_{\tilde{G}_i}(f_i(\mathbf{x}, h_i, \hat{p}_i)) \geq \hat{\mu}_i - \lambda^*, i = 1, \cdots, k. \tag{3.65}$$

From (3.32) and Assumption 3.2, (3.64) and (3.65) can be transformed as follows.

$$\begin{aligned} \hat{p}_i &\geq \mu_{\hat{p}_i}^{-1}(\hat{\mu}_i - \lambda^*), i = 1, \cdots, k \\ \hat{p}_i &\leq T_i\left(\frac{\mu_{\tilde{G}_i}^{-1}(\hat{\mu}_i - \lambda^*) - (\mathbf{d}_i^1\mathbf{x} - L^{-1}(h_i)\alpha_i^1\mathbf{x})}{\mathbf{d}_i^2\mathbf{x} - L^{-1}(h_i)\alpha_i^2\mathbf{x}}\right), i = 1, \cdots, k \end{aligned}$$

Because of $L^{-1}(h_i) \leq L^{-1}(\hat{\mu}_i - \lambda^*), i = 1, \cdots, k$, there exists $\mathbf{x} \in X$ such that

$$\begin{aligned} & \mu_{\tilde{G}_i}^{-1}(\hat{\mu}_i - \lambda^*) - (\mathbf{d}_i^1\mathbf{x} - L^{-1}(h_i)\alpha_i^1\mathbf{x}) \\ & \geq T_i^{-1}(\mu_{\hat{p}_i}^{-1}(\hat{\mu}_i - \lambda^*)) \cdot (\mathbf{d}_i^2\mathbf{x} - L^{-1}(h_i)\alpha_i^2\mathbf{x}), \\ \Leftrightarrow\ & \mu_{\tilde{G}_i}^{-1}(\hat{\mu}_i - \lambda^*) \geq (\mathbf{d}_i^1\mathbf{x} + T_i^{-1}(\mu_{\hat{p}_i}^{-1}(\hat{\mu}_i - \lambda^*)) \cdot \mathbf{d}_i^2\mathbf{x}) \\ & -L^{-1}(h_i)(\alpha_i^1\mathbf{x} + T_i^{-1}(\mu_{\hat{p}_i}^{-1}(\hat{\mu}_i - \lambda^*)) \cdot \alpha_i^2\mathbf{x}), \\ \Leftrightarrow\ & \mu_{\tilde{G}_i}^{-1}(\hat{\mu}_i - \lambda^*) \geq (\mathbf{d}_i^1\mathbf{x} + T_i^{-1}(\mu_{\hat{p}_i}^{-1}(\hat{\mu}_i - \lambda^*)) \cdot \mathbf{d}_i^2\mathbf{x}) \\ & -L^{-1}(\hat{\mu}_i - \lambda^*)(\alpha_i^1\mathbf{x} + T_i^{-1}(\mu_{\hat{p}_i}^{-1}(\hat{\mu}_i - \lambda^*)) \cdot \alpha_i^2\mathbf{x}), i = 1, \cdots, k. \end{aligned}$$

This contradicts the fact that $\mathbf{x}^* \in X, \lambda^* \in \Lambda$ is a unique optimal solution of MINMAX3.4($\hat{\mu}$).

(2) At $\mathbf{x}^* \in X, \hat{p}_i^* \in P_i, h_i^* \in H_i(P_i), i = 1, \cdots, k$, it holds that $\mu_{\hat{p}_i}(\hat{p}_i^*) = \mu_{\tilde{G}_i}(f_i(\mathbf{x}^*, h_i^*, \hat{p}_i^*)), i = 1, \cdots, k$, since $\mu_{\tilde{G}_i}(f_i(\mathbf{x}, h_i, \hat{p}_i))$ is continuous and strictly monotone decreasing with respect to $\hat{p}_i$ for any $\mathbf{x} \in X, h_i \in H_i(P_i)$. Assume that $\mathbf{x}^* \in X, \lambda^* \in \Lambda$ is not an optimal solution of MINMAX3.4($\hat{\mu}$) for any reference membership values $\hat{\mu} = (\hat{\mu}_1, \ldots, \hat{\mu}_k)$ that satisfy equalities

$$\hat{\mu}_i - \lambda^* = \mu_{\hat{p}_i}(\hat{p}_i^*) = \mu_{\tilde{G}_i}(f_i(\mathbf{x}^*, h_i^*, \hat{p}_i^*)), i = 1, \cdots, k. \tag{3.66}$$

Then, there exists some $\mathbf{x} \in X, \lambda < \lambda^*$ such that

$$\begin{aligned} & \mu_{\tilde{G}_i}^{-1}(\hat{\mu}_i - \lambda) - (\mathbf{d}_i^1\mathbf{x} - L^{-1}(\hat{\mu}_i - \lambda)\alpha_i^1\mathbf{x}) \\ & \geq T_i^{-1}(\mu_{\hat{p}_i}^{-1}(\hat{\mu}_i - \lambda)) \cdot (\mathbf{d}_i^2\mathbf{x} - L^{-1}(\hat{\mu}_i - \lambda)\alpha_i^2\mathbf{x}), \\ \Leftrightarrow \quad & \mu_{\tilde{G}_i}(f_i(\mathbf{x}, \hat{\mu}_i - \lambda, \mu_{\hat{p}_i}^{-1}(\hat{\mu}_i - \lambda)) \geq \hat{\mu}_i - \lambda, i = 1, \cdots, k. \end{aligned} \tag{3.67}$$

Because of (3.66), (3.67) and $\hat{\mu}_i - \lambda > \hat{\mu}_i - \lambda^*, i = 1, \cdots, k$, the following inequalities hold.

$$\begin{aligned} \mu_{\hat{p}_i}(\hat{p}_i) &> \mu_{\hat{p}_i}(\hat{p}_i^*), i = 1, \cdots, k \\ \mu_{\tilde{G}_i}(f_i(\mathbf{x}, h_i, \hat{p}_i)) &> \mu_{\tilde{G}_i}(f_i(\mathbf{x}^*, h_i^*, \hat{p}_i^*)), i = 1, \cdots, k \end{aligned}$$

where $\hat{p}_i \stackrel{\text{def}}{=} \mu_{\hat{p}_i}^{-1}(\mu_i - \lambda) \in P_i$, $h_i \stackrel{\text{def}}{=} \hat{\mu}_i - \lambda \in H_i(P_i), i = 1, \cdots, k$. This means that there exists some $\mathbf{x} \in X, \hat{p}_i \in P_i, h_i \in H_i(P_i), i = 1, \cdots, k$ such that

$$\mu_{D_{f_i}}(\mathbf{x}, h_i, \hat{p}_i) > \mu_{D_{f_i}}(\mathbf{x}^*, h_i^*, \hat{p}_i^*), i = 1, \cdots, k.$$

This contradicts the fact that $\mathbf{x}^* \in X, \hat{p}_i^* \in P_i, h_i^* \in H_i(P_i), i = 1, \ldots, k$ is an MF-Pareto optimal solution to MOP3.13. □

3.1.4 An interactive linear programming algorithm

In this subsection, we propose an interactive algorithm for obtaining a satisfactory solution from an MF-Pareto optimal solution set to MOP3.13. From Theorem 3.2, it is not guaranteed that optimal solution $\mathbf{x}^* \in X, \lambda^* \in \Lambda$ of MINMAX3.4($\hat{\mu}$) is MF-Pareto optimal if it is not unique. To guarantee MF-Pareto optimality, we first assume that k constraints (3.63) of MINMAX3.4($\hat{\mu}$) are active at optimal solution $(\mathbf{x}^*, \lambda^*)$, i.e.,

$$\begin{aligned} & \mu_{\tilde{G}_i}^{-1}(\hat{\mu}_i - \lambda^*) - (\mathbf{d}_i^1\mathbf{x}^* - L^{-1}(\hat{\mu}_i - \lambda^*)\alpha_i^1\mathbf{x}^*) \\ = \quad & T_i^{-1}(\mu_{\hat{p}_i}^{-1}(\hat{\mu}_i - \lambda^*)) \cdot (\mathbf{d}_i^2\mathbf{x}^* - L^{-1}(\hat{\mu}_i - \lambda^*)\alpha_i^2\mathbf{x}^*), \\ & i = 1, \cdots, k. \end{aligned} \tag{3.68}$$

If the j-th constraint of (3.63) is inactive, i.e.,

$$\begin{aligned} & \mu_{\tilde{G}_j}^{-1}(\hat{\mu}_j - \lambda^*) - (\mathbf{d}_i^1\mathbf{x}^* - L^{-1}(\hat{\mu}_i - \lambda^*)\alpha_i^1\mathbf{x}^*) \\ & > T_j^{-1}(\mu_{\hat{p}_j}^{-1}(\hat{\mu}_j - \lambda^*)) \cdot (\mathbf{d}_i^2\mathbf{x}^* - L^{-1}(\hat{\mu}_i - \lambda^*)\alpha_i^2\mathbf{x}^*), \\ \Leftrightarrow \quad & \mu_{\tilde{G}_j}^{-1}(\hat{\mu}_j - \lambda^*) > f_j(\mathbf{x}^*, \hat{\mu}_j - \lambda^*, \mu_{\hat{p}_j}^{-1}(\hat{\mu}_j - \lambda^*)), \end{aligned} \tag{3.69}$$

we can convert inactive constraint (3.69) into the active one by applying the bisection method for reference membership value $\hat{\mu}_j \in [\lambda^*, \lambda^* + 1]$.

For optimal solution $\mathbf{x}^* \in X, \lambda^* \in \Lambda$ of MINMAX3.4($\hat{\mu}$), where active conditions (3.68) are satisfied, we solve the MF-Pareto optimality test problem defined as

Test 3.1

$$w \stackrel{\text{def}}{=} \max_{\mathbf{x}\in X, \varepsilon_i \geq 0, i=1,\cdots,k} \sum_{i=1}^{k} \varepsilon_i \tag{3.70}$$

subject to

$$\begin{aligned} & T_i^{-1}(\mu_{\tilde{p}_i}^{-1}(\hat{\mu}_i - \lambda^*)) \cdot (\mathbf{d}_i^2\mathbf{x} - L^{-1}(\hat{\mu}_i - \lambda^*)\alpha_i^2\mathbf{x}) \\ & +(\mathbf{d}_i^1\mathbf{x} - L^{-1}(\hat{\mu}_i - \lambda^*)\alpha_i^1\mathbf{x}) + \varepsilon_i \\ = \; & T_i^{-1}(\mu_{\tilde{p}_i}^{-1}(\hat{\mu}_i - \lambda^*)) \cdot (\mathbf{d}_i^2\mathbf{x}^* - L^{-1}(\hat{\mu}_i - \lambda^*)\alpha_i^2\mathbf{x}^*) \\ & +(\mathbf{d}_i^1\mathbf{x}^* - L^{-1}(\hat{\mu}_i - \lambda^*)\alpha_i^1\mathbf{x}^*), i = 1, \cdots, k. \end{aligned} \tag{3.71}$$

For the optimal solution of the above test problem, the following theorem holds.

Theorem 3.3

For optimal solution $\check{\mathbf{x}} \in X, \check{\varepsilon}_i \geq 0, i = 1, \cdots, k$ of TEST3.1, if $w = 0$ (equivalently, $\check{\varepsilon}_i = 0, i = 1, \cdots, k$), $\mathbf{x}^* \in X, \mu_{\tilde{p}_i}^{-1}(\hat{\mu}_i - \lambda^*) \in P_i, \hat{\mu}_i - \lambda^* \in H_i(P_i), i = 1, \cdots, k$ is an MF-Pareto optimal solution to MOP3.13.

(Proof)
From active conditions (3.68), it holds that

$$\hat{\mu}_i - \lambda^* = \mu_{\tilde{G}_i}(f_i(\mathbf{x}^*, \hat{\mu}_i - \lambda^*, \mu_{\tilde{p}_i}^{-1}(\hat{\mu}_i - \lambda^*))), i = 1, \cdots, k,$$

at optimal solution $\mathbf{x}^* \in X, \lambda^* \in \Lambda$ of MINMAX3.4($\hat{\mu}$). It is obvious that $\hat{\mu}_i - \lambda^* = \mu_{\tilde{p}_i}(\mu_{\tilde{p}_i}^{-1}(\hat{\mu}_i - \lambda^*)), i = 1, \ldots, k$. Assume that $\mathbf{x}^* \in X, \mu_{\tilde{p}_i}^{-1}(\hat{\mu}_i - \lambda^*) \in P_i, \hat{\mu}_i - \lambda^* \in H_i(P_i), i = 1, \cdots, k$ is not an MF-Pareto optimal solution to MOP3.13. Then, there exists some $\mathbf{x} \in X, \hat{p}_i \in P_i, h_i \in H_i(P_i), i = 1, \cdots, k$ such that

$$\begin{aligned} \mu_{D_{G_i}}(\mathbf{x}, h_i, \hat{p}_i) &= \min\{\mu_{\tilde{p}_i}(\hat{p}_i), \mu_{\tilde{G}_i}(f_i(\mathbf{x}, h_i, \hat{p}_i))\}, \\ &\geq \mu_{D_{f_i}}(\mathbf{x}^*, \hat{\mu}_i - \lambda^*, \mu_{\tilde{p}_i}^{-1}(\hat{\mu}_i - \lambda^*)), \\ &= \hat{\mu}_i - \lambda^*, i = 1, \cdots, k, \end{aligned}$$

with strict inequality holding for at least one i, and

$$\mu_{\tilde{G}_i}(f_i(\mathbf{x}, h_i, \hat{p}_i)) = h_i, i = 1, \cdots, k.$$

This means that the following inequalities hold.

$$\mu_{\hat{p}_i}(\hat{p}_i) \geq \hat{\mu}_i - \lambda^*, i = 1, \cdots, k \tag{3.72}$$

$$\mu_{\tilde{G}_i}(f_i(\mathbf{x}, h_i, \hat{p}_i)) \geq \hat{\mu}_i - \lambda^*, i = 1, \cdots, k \tag{3.73}$$

Because of Assumption 3.2, (3.32), and $L^{-1}(h_i) \leq L^{-1}(\hat{\mu}_i - \lambda^*)$, (3.72) and (3.73) can be transformed into the following forms.

$$\hat{p}_i \geq \mu_{\hat{p}_i}^{-1}(\hat{\mu}_i - \lambda^*), i = 1, \cdots, k$$

$$\hat{p}_i \leq T_i\left(\frac{\mu_{\tilde{G}_i}^{-1}(\hat{\mu}_i - \lambda^*) - (\mathbf{d}_i^1\mathbf{x} - L^{-1}(\hat{\mu}_i - \lambda^*)\alpha_i^1\mathbf{x})}{\mathbf{d}_i^2\mathbf{x} - L^{-1}(\hat{\mu}_i - \lambda^*)\alpha_i^2\mathbf{x}}\right), i = 1, \cdots, k$$

Therefore, there exists some $\mathbf{x} \in X$ such that

$$\begin{aligned}
& \mu_{\tilde{G}_i}^{-1}(\hat{\mu}_i - \lambda^*) - (\mathbf{d}_i^1\mathbf{x} - L^{-1}(\hat{\mu}_i - \lambda^*)\alpha_i^1\mathbf{x}) \\
& \geq T_i^{-1}(\mu_{\hat{p}_i}^{-1}(\hat{\mu}_i - \lambda^*)) \cdot (\mathbf{d}_i^2\mathbf{x} - L^{-1}(\hat{\mu}_i - \lambda^*)\alpha_i^2\mathbf{x}) \\
\Leftrightarrow\ & T_i^{-1}(\mu_{\hat{p}_i}^{-1}(\hat{\mu}_i - \lambda^*)) \cdot (\mathbf{d}_i^2\mathbf{x}^* - L^{-1}(\hat{\mu}_i - \lambda^*)\alpha_i^2\mathbf{x}^*) \\
& +(\mathbf{d}_i^1\mathbf{x}^* - L^{-1}(\hat{\mu}_i - \lambda^*)\alpha_i^1\mathbf{x}^*) \\
& \geq T_i^{-1}(\mu_{\hat{p}_i}^{-1}(\hat{\mu}_i - \lambda^*)) \cdot (\mathbf{d}_i^2\mathbf{x} - L^{-1}(\hat{\mu}_i - \lambda)\alpha_i^2\mathbf{x}) \\
& +(\mathbf{d}_i^1\mathbf{x} - L^{-1}(\hat{\mu}_i - \lambda)\alpha_i^1\mathbf{x}), i = 1, \cdots, k.
\end{aligned} \tag{3.74}$$

with strict inequality holding for at least one i. This contradicts the fact that $w = 0$. □

Given the above discussion, we present an interactive algorithm to derive a satisfactory solution from an MF-Pareto optimal solution set to MOP3.13.

Algorithm 3.1

Step 1: *The decision maker specifies intervals* $P_i \stackrel{\text{def}}{=} [p_{i\min}, p_{i\max}], i = 1, \ldots, k$, *of permissible probability levels* $\hat{\mathbf{p}}$ *of MOP3.11, where* $p_{i\min}(> 0)$ *is an unacceptable maximum value of* $\hat{p}_i$ *and* $p_{i\max}(< 1)$ *is a sufficiently satisfactory minimum value of* $\hat{p}_i$*. According to Assumption 3.2, the decision maker sets membership functions* $\mu_{\hat{p}_i}(\hat{p}_i), i = 1, \cdots, k$ *on interval* $P_i, i = 1, \cdots, k$.
Step 2: *Corresponding to intervals* $P_i, i = 1, \ldots, k$, *compute* $f_{i\min}$ *and* $f_{i\max}$ *by solving optimization problems (3.41) and (3.47), respectively. The decision maker sets membership functions* $\mu_{\tilde{G}_i}(y), i = 1, \ldots, k$ *on interval* $F_i(P_i) \stackrel{\text{def}}{=} [f_{i\min}, f_{i\max}]$, *in which* $\mu_{\tilde{G}_i}(y), i = 1, \ldots, k$ *are strictly monotone decreasing and continuous.*
Step 3: *Set the initial reference membership values as* $\hat{\mu}_i = 1, i = 1, \ldots, k$.
Step 4: *Solve MINMAX3.4(*$\hat{\mu}$*) via the combined use of the bisection method with respect to* $\lambda \in \Lambda$ *and the first phase of the two-phase simplex method of linear programming, thus obtaining optimal solution* $\mathbf{x}^* \in X, \lambda^* \in \Lambda$. *For optimal solution*

$\mathbf{x}^* \in X, \lambda^* \in \Lambda$, the corresponding MF-Pareto optimality test problem TEST3.1 is formulated and solved.

Step 5: *If the decision maker is satisfied with the current values of MF-Pareto optimal solution $\mu_{D_{G_i}}(\mathbf{x}^*, h_i^*, \hat{p}_i^*), i = 1, \cdots, k$, where $\hat{p}_i^* \stackrel{\text{def}}{=} \mu_{\tilde{p}_i}^{-1}(\hat{\mu}_i - \lambda^*), h_i^* \stackrel{\text{def}}{=} \hat{\mu}_i - \lambda^*, i = 1, \cdots, k$, then stop. Otherwise, the decision maker updates reference membership values $\hat{\mu}_i, i = 1, \ldots, k$ and returns to Step 4.*

3.1.5 A numerical example

To demonstrate the feasibility of Algorithm 3.1, consider the following two objective fuzzy random linear programming problem formulated by Katagiri et al. [38].

MOP 3.14

$$\begin{array}{ll} \min & \tilde{\bar{\mathbf{c}}}_1\mathbf{x} = \tilde{\bar{c}}_{11}x_1 + \tilde{\bar{c}}_{12}x_2 + \tilde{\bar{c}}_{13}x_3 \\ \min & \tilde{\bar{\mathbf{c}}}_2\mathbf{x} = \tilde{\bar{c}}_{21}x_1 + \tilde{\bar{c}}_{22}x_2 + \tilde{\bar{c}}_{23}x_3 \end{array}$$

subject to

$$\begin{aligned} \mathbf{x} \in X \;=\; & \{\mathbf{x} = (x_1, x_2, x_3) \geq \mathbf{0} \mid 2x_1 + 6x_2 + 3x_3 \leq 150, 6x_1 + 3x_2 + 5x_3 \leq 175, \\ & 5x_1 + 4x_2 + 2x_3 \leq 160, 2x_1 + 2x_2 + 3x_3 \geq 90\} \end{aligned}$$

We assume here that realization $\tilde{\bar{c}}_{ij}(\omega)$ of fuzzy random variable $\tilde{\bar{c}}_{ij}$ is an LR fuzzy number whose membership function is defined as

$$\mu_{\tilde{\bar{c}}_{ij}(\omega)}(s) = \begin{cases} L\left(\dfrac{d_{ij}^1 + t_i(\omega)d_{ij}^2 - s}{\alpha_{ij}^1 + t_i(\omega)\alpha_{ij}^2}\right), & s \leq d_{ij}(\omega) \\ R\left(\dfrac{s - d_{ij}^1 + t_i(\omega)d_{ij}^2}{\beta_{ij}^1 + t_i(\omega)\beta_{ij}^2}\right), & s > d_{ij}(\omega), \end{cases}$$

where $L(t) = R(t) = \max\{0, 1 - t\}$, and parameters $d_{ij}^1, d_{ij}^2, \alpha_{ij}^1, \alpha_{ij}^2, \beta_{ij}^1$, and β_{ij}^2 are given in Table 3.1 [38]. Moreover, $\bar{t}_i, i = 1, 2$ are Gaussian random variables defined as $\bar{t}_i \sim \mathrm{N}(0, 1)$.

We assume that the decision maker adopts linear membership functions $\mu_{\tilde{G}_i}(f_i(\mathbf{x}, h_i, \hat{p}_i)), i = 1, 2$ and $\mu_{\hat{p}_i}(\hat{p}_i), i = 1, 2$ defined as follows (i.e., Step 1 and Step 2).

Table 3.1: Parameters of LR-type fuzzy random variables in MOP3.14

j	1	2	3	j	1	2	3
d_{1j}^1	2	1	3	d_{1j}^2	1.3	1.1	1.2
d_{2j}^1	−7	−7	−9	d_{2j}^2	1.1	1.2	1.1
α_{1j}^1	0.5	0.4	0.5	α_{1j}^2	0.05	0.04	0.05
α_{2j}^1	0.3	0.5	0.4	α_{2j}^2	0.05	0.04	0.05
β_{1j}^1	0.6	0.5	0.6	β_{1j}^2	0.06	0.05	0.06
β_{2j}^1	0.4	0.5	0.5	β_{2j}^2	0.06	0.06	0.05

$$\begin{aligned}
\mu_{\tilde{G}_1}(f_1(\mathbf{x},h_1,\hat{p}_1)) &= \frac{96.42857 - f_1(\mathbf{x},h_1,\hat{p}_1)}{96.42857 - 75} \\
\mu_{\tilde{G}_2}(f_2(\mathbf{x},h_2,\hat{p}_2)) &= \frac{(-285) - f_2(\mathbf{x},h_2,\hat{p}_2)}{(-285) - (-332.143)} \\
\mu_{\hat{p}_1}(\hat{p}_1) &= \frac{\hat{p}_1 - 0.401066}{(0.714968 - 0.401066)} \\
\mu_{\hat{p}_2}(\hat{p}_2) &= \frac{\hat{p}_2 - 0.213304}{(0.812859 - 0.213304)}
\end{aligned}$$

At Step 3, set the initial reference membership values as $(\hat{\mu}_1, \hat{\mu}_2) = (1,1)$. Next, we solve MINMAX3.4($\hat{\mu}$) via the combined use of the bisection method with respect to λ and the first phase of the two-phase simplex method of linear programming to obtain corresponding MF-Pareto optimal solution $\mathbf{x}^* \in X, \lambda^* \in \Lambda$ (i.e., Step 4).

$$\begin{aligned}
\mu_{\tilde{G}_1}(f_1(\mathbf{x}^*,h_1^*,\hat{p}_1^*)) &= \mu_{\hat{p}_1}(\hat{p}_1^*) = 0.564271 \\
\mu_{\tilde{G}_2}(f_2(\mathbf{x}^*,h_2^*,\hat{p}_2^*)) &= \mu_{\hat{p}_2}(\hat{p}_2^*) = 0.564271
\end{aligned}$$

The hypothetical decision maker is not satisfied with the current value of the MF-Pareto optimal solution $\mathbf{x}^* \in X, \lambda^* \in \Lambda$; thus, to improve $\mu_{D_{G_2}}(\cdot)$ at the expense of $\mu_{D_{G_1}}(\cdot)$, he or she updates his or her reference membership values as $(\hat{\mu}_1, \hat{\mu}_2) = (0.5, 0.6)$ (i.e., Step 5). Next, the corresponding MF-Pareto optimal solution is again obtained by solving MINMAX3.4($\hat{\mu}$) (i.e., Step 4).

$$\begin{aligned}
\mu_{\tilde{G}_1}(f_1(\mathbf{x}^*,h_1^*,\hat{p}_1^*)) &= \mu_{\hat{p}_1}(\hat{p}_1^*) = 0.514421 \\
\mu_{\tilde{G}_2}(f_2(\mathbf{x}^*,h_2^*,\hat{p}_2^*)) &= \mu_{\hat{p}_2}(\hat{p}_2^*) = 0.614421
\end{aligned}$$

For the current value of the MF-Pareto optimal solution, the hypothetical decision maker updates his or her reference membership values $(\hat{\mu}_1, \hat{\mu}_2) = (0.52, 0.59)$ to improve $\mu_{D_{G_1}}(\cdot)$ at the expense of $\mu_{D_{G_2}}(\cdot)$ (i.e., Step 5). The corresponding MF-Pareto optimal solution is obtained by solving MINMAX3.4($\hat{\mu}$) (i.e., Step 4).

Table 3.2: Interactive processes in MOP3.14

	1	2	3
$\hat{\mu}_1$	1	0.5	0.52
$\hat{\mu}_2$	1	0.6	0.59
$\mu_{D_{G_1}}(\mathbf{x}^*, h_1^*, \hat{p}_1^*)$	0.564271	0.514421	0.529412
$\mu_{D_{G_2}}(\mathbf{x}^*, h_2^*, \hat{p}_2^*)$	0.564271	0.614421	0.599412
$\hat{p}_1^*$	0.578193	0.562545	0.567250
$\hat{p}_2^*$	0.551616	0.581684	0.572685
$f_1(\mathbf{x}^*, h_1^*, \hat{p}_1^*)$	84.3370	85.4053	85.0840
$f_2(\mathbf{x}^*, h_2^*, \hat{p}_2^*)$	–311.601	–313.966	–313.258

$$\begin{aligned} \mu_{\tilde{G}_1}(f_1(\mathbf{x}^*, h_1^*, \hat{p}_1^*)) &= \mu_{\hat{p}_1}(\hat{p}_1^*) = 0.529412 \\ \mu_{\tilde{G}_2}(f_2(\mathbf{x}^*, h_2^*, \hat{p}_2^*)) &= \mu_{\hat{p}_2}(\hat{p}_2^*) = 0.599412 \end{aligned}$$

Finally, since the hypothetical decision maker is satisfied with the current value of the MF-Pareto optimal solution, we stop (i.e., Step 5). These interactive processes under the hypothetical decision maker are summarized in Table 3.2.

3.2 Interactive decision making for MOFRPs with variance covariance matrices

In this section, we formulate MOFRPs with variance covariance matrices and propose an interactive convex programming algorithm [114, 119, 120] to obtain satisfactory solutions to such problems. First, in Section 3.2.1, by introducing a possibility measure [18], we transform MOFRPs with variance covariance matrices into multiobjective stochastic programming problems. In Section 3.2.2, we then formulate transformed multiobjective stochastic programming problems through a probability maximization model and the fuzzy decision [6, 71, 140]; in our approach, the decision maker specifies membership functions rather than permissible objective levels. We also define an MP-Pareto optimal solution for MOFRPs with variance covariance matrices and investigate the relationships between the minmax problem and an MP-Pareto optimal solution set. Similarly, in Section 3.2.3, we formulate transformed multiobjective stochastic programming problems through a fractile optimization model and the fuzzy decision; in this approach, the decision maker specifies membership functions rather than permissible probability levels. We define an MF-Pareto optimal solution for MOFRPs with variance covariance matrices and investigate the relationships between the minmax problem and an MF-Pareto optimal solution set. Next, in Section 3.2.4, we propose an interactive convex programming algorithm to obtain satisfactory solutions from an MF-Pareto optimal solution set. Finally, in Section 3.2.5, we apply our interactive algorithm to a numerical example with a hypothetical decision maker.

3.2.1 MOFRPs and a possibility measure

In this subsection, we focus on MOFRPs with variance covariance matrices defined as

MOP 3.15

$$\min_{\mathbf{x}\in X} \tilde{\bar{\mathbf{C}}}\mathbf{x} = (\tilde{\bar{\mathbf{c}}}_1\mathbf{x}, \cdots, \tilde{\bar{\mathbf{c}}}_k\mathbf{x}),$$

where $\mathbf{x} = (x_1, \cdots, x_n)^T$ is an n-dimensional decision variable column vector and X is a feasible set of linear constraints. Further, $\tilde{\bar{\mathbf{c}}}_i = (\tilde{\bar{c}}_{i1}, \cdots, \tilde{\bar{c}}_{in}), i = 1, \cdots, k$ are coefficient vectors of objective function $\tilde{\bar{\mathbf{c}}}_i\mathbf{x}$, whose elements are fuzzy random variables. The concept of fuzzy random variable $\tilde{\bar{c}}_{ij}$ is defined precisely in [46, 65, 77], and symbols "-" and "~" indicate randomness and fuzziness respectively.

To handle objective functions $\tilde{\bar{\mathbf{c}}}_i\mathbf{x}, i = 1, \cdots, k$, Katagiri et al. [38] proposed an LR-type fuzzy random variable, which can be viewed as a special version of a fuzzy random variable. Given the occurrence of each elementary event ω, $\tilde{\bar{c}}_{ij}(\omega)$ is a realization of LR-type fuzzy random variable $\tilde{\bar{c}}_{ij}$, which is an LR fuzzy number [18] whose membership function is defined as

$$\mu_{\tilde{\bar{c}}_{ij}(\omega)}(s) = \begin{cases} L\left(\frac{d_{ij}(\omega)-s}{\alpha_{ij}}\right), & s \le d_{ij}(\omega) \\ R\left(\frac{s-d_{ij}(\omega)}{\beta_{ij}}\right), & s > d_{ij}(\omega), \end{cases}$$

where function $L(t) \overset{\text{def}}{=} \max\{0, l(t)\}$ is a real-valued continuous function from $[0, \infty)$ to $[0, 1]$ and $l(t)$ is a strictly decreasing continuous function satisfying $l(0) = 1$. Further, $R(t) \overset{\text{def}}{=} \max\{0, r(t)\}$ satisfies the same conditions. Let us assume $\alpha_{ij} > 0$ and $\beta_{ij} > 0$ are spread parameters [18], mean value [18] $\bar{d}_{ij}$ is a Gaussian random variable, i.e.,

$$\bar{d}_{ij} \sim \mathrm{N}(E[\bar{d}_{ij}], \sigma_{ijj}),$$

and positive-definite variance covariance matrices $V_i, i = 1, \cdots, k$ between Gaussian random variables $\bar{d}_{ij}, j = 1, \cdots, n$ are given as

$$V_i = \begin{pmatrix} \sigma_{i11} & \sigma_{i12} & \cdots & \sigma_{i1n} \\ \sigma_{i21} & \sigma_{i22} & \cdots & \sigma_{i2n} \\ \vdots & \vdots & \ddots & \vdots \\ \sigma_{in1} & \sigma_{in2} & \cdots & \sigma_{inn} \end{pmatrix}, \ i = 1, \cdots, k. \tag{3.75}$$

Let us denote vectors of the expectation for the random variable row vectors $\bar{\mathbf{d}}_i \stackrel{\text{def}}{=} (\bar{d}_{i1}, \cdots, \bar{d}_{in}), i = 1, \cdots, k$ as

$$\mathbf{E}[\bar{\mathbf{d}}_i] = (E[\bar{d}_{i1}], \cdots, E[\bar{d}_{in}]), i = 1, \cdots, k.$$

Then, from the property of Gaussian random variables, $\bar{\mathbf{d}}_i\mathbf{x}, i = 1, \cdots, k$ also become Gaussian random variables

$$\bar{\mathbf{d}}_i\mathbf{x} \sim \mathrm{N}(\mathbf{E}[\bar{\mathbf{d}}_i]\mathbf{x}, \mathbf{x}^T V_i \mathbf{x}), i = 1, \cdots, k. \tag{3.76}$$

Katagiri et al. [38] transformed MOP3.15 into a multiobjective stochastic programming problem by using the concept of a possibility measure [18]. As shown in [38], realizations $\tilde{\bar{\mathbf{c}}}_i(\omega)\mathbf{x}$ become LR fuzzy numbers characterized by the following membership functions based on the extension principle [18].

$$\mu_{\tilde{\bar{\mathbf{c}}}_i(\omega)\mathbf{x}}(y) = \begin{cases} L\left(\frac{\mathbf{d}_i(\omega)\mathbf{x} - y}{\alpha_i \mathbf{x}}\right), & y \leq \mathbf{d}_i(\omega)\mathbf{x} \\ R\left(\frac{y - \mathbf{d}_i(\omega)\mathbf{x}}{\beta_i \mathbf{x}}\right), & y > \mathbf{d}_i(\omega)\mathbf{x} \end{cases}$$

For realizations $\tilde{\bar{\mathbf{c}}}_i(\omega)\mathbf{x}$, we assume the decision maker has fuzzy goals $\widetilde{G}_i, i = 1, \cdots, k$ [71, 140], whose membership functions $\mu_{\widetilde{G}_i}(y)$, $i = 1, \cdots, k$ are continuous and strictly decreasing for minimization problems. By using the concept of a possibility measure [18], the degree of possibility that objective function value $\tilde{\bar{\mathbf{c}}}_i\mathbf{x}$ satisfies fuzzy goal $\widetilde{G}_i$ is expressed as [35, 38]

$$\Pi_{\tilde{\bar{\mathbf{c}}}_i\mathbf{x}}(\widetilde{G}_i) \stackrel{\text{def}}{=} \sup_y \min\{\mu_{\tilde{\bar{\mathbf{c}}}_i\mathbf{x}}(y), \mu_{\widetilde{G}_i}(y)\}, i = 1, \cdots, k. \tag{3.77}$$

Using a possibility measure, MOP3.15 can be transformed into the following multiobjective stochastic programming problem.

MOP 3.16

$$\max_{\mathbf{x} \in X} (\Pi_{\tilde{\bar{\mathbf{c}}}_1\mathbf{x}}(\widetilde{G}_1), \cdots, \Pi_{\tilde{\bar{\mathbf{c}}}_k\mathbf{x}}(\widetilde{G}_k))$$

Katagiri et al. [38] first formulated MOP3.16 as a multiobjective programming problem based on a probability maximization model. In Sections 3.2.2 and 3.2.3, we transform MOP3.16 into multiobjective programming problems through a probability maximization model and a fractile optimization model, respectively; in these cases, the decision maker need not specify permissible possibility levels or permissible probability levels.

3.2.2 A formulation using a probability model

For the objective functions of MOP3.16, if the decision maker specifies permissible possibility levels $h_i, i = 1, \cdots, k$ for a possibility measure $\Pi_{\tilde{\bar{\mathbf{c}}}_i(\omega)\mathbf{x}}(\widetilde{G}_i)$, then MOP3.16 can be formulated as the following multiobjective programming problem using a probability maximization model.

MOP 3.17 $(\mathbf{h})$

$$\max_{\mathbf{x}\in X}(\Pr(\omega \mid \Pi_{\tilde{\bar{\mathbf{c}}}_1(\omega)\mathbf{x}}(\widetilde{G}_1) \geq h_1), \cdots, \Pr(\omega \mid \Pi_{\tilde{\bar{\mathbf{c}}}_k(\omega)\mathbf{x}}(\widetilde{G}_k) \geq h_k)) \tag{3.78}$$

Here, $\Pr(\cdot)$ is a probability measure and $\mathbf{h} = (h_1, \cdots, h_k)$ is a k-dimensional vector of permissible possibility levels. In MOP3.17$(\mathbf{h})$, inequality $\Pi_{\tilde{\bar{\mathbf{c}}}_i(\omega)\mathbf{x}}(\widetilde{G}_i) \geq h_i$ can be equivalently transformed into

$$\begin{aligned} & \sup_y \min\{\mu_{\tilde{\bar{\mathbf{c}}}_i(\omega)\mathbf{x}}(y), \mu_{\widetilde{G}_i}(y)\} \geq h_i \\ \Leftrightarrow \quad & (\bar{\mathbf{d}}_i(\omega) - L^{-1}(h_i)\alpha_i)\mathbf{x} \leq \mu_{\widetilde{G}_i}^{-1}(h_i), \end{aligned}$$

where $L^{-1}(\cdot)$, $R^{-1}(\cdot)$ and $\mu_{\widetilde{G}_i}^{-1}(\cdot)$ are inverse functions of $L(\cdot)$, $R(\cdot)$, and $\mu_{\widetilde{G}_i}(\cdot)$, respectively. Therefore, using distribution function $\Phi(\cdot)$ of the standard Gaussian random variable, the objective functions in MOP3.17$(\mathbf{h})$ can be expressed as

$$\begin{aligned} & \Pr(\omega \mid \Pi_{\tilde{\bar{\mathbf{c}}}_i(\omega)\mathbf{x}}(\widetilde{G}_i) \geq h_i) \\ = \quad & \Phi\left(\frac{\mu_{\widetilde{G}_i}^{-1}(h_i) - (\mathbf{E}[\bar{\mathbf{d}}_i]\mathbf{x} - L^{-1}(h_i)\alpha_i\mathbf{x})}{\sqrt{\mathbf{x}^T V_i \mathbf{x}}}\right) \\ \stackrel{\text{def}}{=} \quad & p_i(\mathbf{x}, h_i), i = 1, \cdots, k. \end{aligned} \tag{3.79}$$

As a result, using $p_i(\mathbf{x}, h_i), i = 1, \cdots, k$, MOP3.17$(\mathbf{h})$ can be transformed into the following simple form [38].

MOP 3.18 $(\mathbf{h})$

$$\max_{\mathbf{x}\in X}(p_1(\mathbf{x}, h_1), \cdots, p_k(\mathbf{x}, h_k))$$

In MOP3.18$(\mathbf{h})$, the decision maker prefers not only larger values for permissible possibility level h_i but also larger values for corresponding distribution function $p_i(\mathbf{x}, h_i)$. Since these values conflict with one another, larger values of permissible possibility level h_i result in smaller values of corresponding distribution function $p_i(\mathbf{x}, h_i)$. From such a perspective, we consider the following multiobjective programming problem, which can be viewed as a natural extension of MOP3.18$(\mathbf{h})$.

MOP 3.19

$$\max_{\mathbf{x}\in X, h_i\in(0,1), i=1,\cdots,k}(p_1(\mathbf{x}, h_1), \cdots, p_k(\mathbf{x}, h_k), h_1, \cdots, h_k)$$

Given the imprecise nature of the decision maker's judgment, it is natural to assume that the decision maker has fuzzy goals for $p_i(\mathbf{x}, h_i), i = 1, \cdots, k$. In this subsection, we assume that such fuzzy goals can be quantified by eliciting the corresponding membership functions. Let us denote the membership function of probability function $p_i(\mathbf{x}, h_i)$ as $\mu_{p_i}(p_i(\mathbf{x}, h_i))$. Then, we can transform MOP3.19 into the following multiobjective programming problem.

MOP 3.20

$$\max_{\mathbf{x}\in X, h_i\in(0,1), i=1,\cdots,k} (\mu_{p_1}(p_1(\mathbf{x},h_1)),\cdots,\mu_{p_k}(p_k(\mathbf{x},h_k)),h_1,\cdots,h_k) \quad (3.80)$$

To appropriately elicit membership functions $\mu_{p_i}(p_i(\mathbf{x},h_i))$, we propose the following procedures. First, the decision maker sets the intervals for permissible possibility levels $h_i, i=1,\cdots,k$ as

$$H_i \stackrel{\text{def}}{=} [h_{i\min}, h_{i\max}], i=1,\cdots,k,$$

where $h_{i\min}(>0)$ is the maximum value of unacceptable levels and $h_{i\max}(<1)$ is the minimum value of sufficiently satisfactory levels. Corresponding to interval H_i, it seems natural to define the interval of $p_i(\mathbf{x},h_i)$ as

$$P_i(H_i) \stackrel{\text{def}}{=} [p_{i\min}, p_{i\max}] = \{p_i(\mathbf{x},h_i) \mid \mathbf{x}\in X, h_i\in H_i\}, i=1,\cdots,k.$$

We can obtain $p_{i\min}$ and $p_{i\max}$ as follows. First, $p_{i\max}$ can be obtained by solving optimization problems.

$$p_{i\max} \stackrel{\text{def}}{=} \max_{\mathbf{x}\in X} p_i(\mathbf{x},h_{i\min}), i=1,\cdots,k, \quad (3.81)$$

Next, to obtain $p_{i\min}, i=1,\cdots,k$, we first solve optimization problems,

$$\max_{\mathbf{x}\in X} p_i(\mathbf{x},h_{i\max}), i=1,\cdots,k, \quad (3.82)$$

denoting corresponding optimal solutions as $\mathbf{x}_i, i=1,\cdots,k$. Using optimal solution $\mathbf{x}_i, i=1,\cdots,k$, $p_{i\min}$ can be obtained as

$$p_{i\min} \stackrel{\text{def}}{=} \min_{\ell=1,\cdots,k,\ell\neq i} p_i(\mathbf{x}_\ell,h_{i\max}), i=1,\cdots,k. \quad (3.83)$$

Unfortunately, optimization problems (3.81) and (3.82) are nonlinear; however, since $\Phi(\cdot)$ is continuous and strictly increasing, we can easily solve (3.81) and (3.82) by applying Dinkelbach's algorithm [17]. After obtaining $p_{i\min}$ and $p_{i\max}, i=1,\cdots,k$, the decision maker can elicit his or her membership functions $\mu_{p_i}(p_i(\mathbf{x},h_i)), i=1,\cdots,k$ according to Assumption 3.3.

Assumption 3.3
$\mu_{p_i}(p_i(\mathbf{x},h_i)), i=1,\cdots,k$ are strictly increasing and continuous with respect to $p_i(\mathbf{x},h_i)\in P_i(H_i)$, and $\mu_{p_i}(p_{i\min})=0$, $\mu_{p_i}(p_{i\max})=1, i=1,\cdots,k$.

Note that $\mu_{p_i}(p_i(\mathbf{x},h_i))$ is strictly decreasing with respect to $h_i\in H_i$. If the decision maker adopts the fuzzy decision [6, 71, 140] to integrate $\mu_{p_i}(p_i(\mathbf{x},h_i))$ and h_i, MOP3.20 can be transformed into the following form.

MOP 3.21

$$\max_{\mathbf{x}\in X, h_i = H_i, i=1,\cdots,k} \left(\mu_{D_{p_1}}(\mathbf{x}, h_1), \cdots, \mu_{D_{p_k}}(\mathbf{x}, h_k)\right) \tag{3.84}$$

where

$$\mu_{D_{p_i}}(\mathbf{x}, h_i) \stackrel{\text{def}}{=} \min\{h_i, \mu_{p_i}(p_i(\mathbf{x}, h_i))\}. \tag{3.85}$$

To handle MOP3.21, we next introduce an MP-Pareto optimal solution.

Definition 3.3 $\mathbf{x}^* \in X, h_i^* \in H_i, i = 1, \cdots, k$ is an MP-Pareto optimal solution to MOP3.21 if and only if there does not exist another $\mathbf{x} \in X, h_i \in H_i, i = 1, \cdots, k$ such that $\mu_{D_{p_i}}(\mathbf{x}, h_i) \geq \mu_{D_{p_i}}(\mathbf{x}^*, h_i^*)$, $i = 1, \cdots, k$ with strict inequality holding for at least one i.

To generate a candidate for a satisfactory solution that is also MP-Pareto optimal, the decision maker specifies reference membership values [71]. Once reference membership values $\hat{\mu} = (\hat{\mu}_1, \cdots, \hat{\mu}_k)$ are specified, the corresponding MP-Pareto optimal solution is obtained by solving the following minmax problem.

MINMAX 3.5 ($\hat{\mu}$)

$$\min_{\mathbf{x}\in X, h_i \in H_i, i=1,\cdots,k, \lambda\in\Lambda} \lambda \tag{3.86}$$

subject to

$$\hat{\mu}_i - \mu_{p_i}(p_i(\mathbf{x}, h_i)) \leq \lambda, i = 1, \cdots, k \tag{3.87}$$

$$\hat{\mu}_i - h_i \leq \lambda, i = 1, \cdots, k \tag{3.88}$$

where

$$\Lambda \stackrel{\text{def}}{=} [\lambda_{\min}, \lambda_{\max}] = [\max_{i=1,\cdots,k} \hat{\mu}_i - 1, \max_{i=1,\cdots,k} \hat{\mu}_i]. \tag{3.89}$$

From Assumption 3.3, constraints (3.87) can be transformed into

$$\begin{aligned} & \hat{\mu}_i - \mu_{p_i}(p_i(\mathbf{x}, h_i)) \leq \lambda, \\ \Leftrightarrow \quad & p_i(\mathbf{x}, h_i) \geq \mu_{p_i}^{-1}(\hat{\mu}_i - \lambda), \\ \Leftrightarrow \quad & \mu_{\tilde{G}_i}^{-1}(h_i) \geq (\mathbf{E}[\bar{\mathbf{d}}_i]\mathbf{x} - L^{-1}(h_i)\alpha_i\mathbf{x}) + \Phi^{-1}(\mu_{p_i}^{-1}(\hat{\mu}_i - \lambda)) \cdot \sqrt{\mathbf{x}^T V_i \mathbf{x}}. \end{aligned} \tag{3.90}$$

Because $h_i \geq \hat{\mu}_i - \lambda$, it holds that $\mu_{\tilde{G}_i}^{-1}(h_i) \leq \mu_{\tilde{G}_i}^{-1}(\hat{\mu}_i - \lambda)$ and $L^{-1}(h_i) \leq L^{-1}(\hat{\mu}_i - \lambda)$. From these inequalities and Assumption 3.3, the following inequalities can be derived.

$$\begin{aligned}
\mu_{\tilde{G}_i}^{-1}(\hat{\mu}_i - \lambda) &\geq \mu_{\tilde{G}_i}^{-1}(h_i) \\
&\geq (\mathbf{E}[\bar{\mathbf{d}}_i]\mathbf{x} - L^{-1}(\lambda)\alpha_i\mathbf{x}) + \Phi^{-1}(\mu_{\bar{p}_i}^{-1}(h_i)) \cdot \sqrt{\mathbf{x}^T V_i \mathbf{x}} \\
&\geq (\mathbf{E}[\bar{\mathbf{d}}_i]\mathbf{x} - L^{-1}(\hat{\mu}_i - \lambda)\alpha_i\mathbf{x}) + \Phi^{-1}(\mu_{\bar{p}_i}^{-1}(\hat{\mu}_i - \lambda)) \cdot \sqrt{\mathbf{x}^T V_i \mathbf{x}}
\end{aligned} \tag{3.91}$$

Therefore, MINMAX3.5($\hat{\mu}$) can be reduced to the following minmax problem, where permissible possibility levels $h_i, i = 1, \cdots, k$ have disappeared.

MINMAX 3.6 ($\hat{\mu}$)

$$\min_{\mathbf{x} \in X, \lambda \in \Lambda} \lambda \tag{3.92}$$

subject to

$$\mu_{\tilde{G}_i}^{-1}(\hat{\mu}_i - \lambda) \geq (\mathbf{E}[\bar{\mathbf{d}}_i]\mathbf{x} - L^{-1}(\hat{\mu}_i - \lambda)\alpha_i\mathbf{x}) + \Phi^{-1}(\mu_{\bar{p}_i}^{-1}(\hat{\mu}_i - \lambda)) \cdot \sqrt{\mathbf{x}^T V_i \mathbf{x}}, \quad i = 1, \cdots, k \tag{3.93}$$

Since constraints (3.93) are nonlinear, it is difficult to solve MINMAX3.6($\hat{\mu}$) directly; however, we can easily obtain the optimal solution of MINMAX3.6($\hat{\mu}$) using the following additional assumption (i.e., in addition to Assumption 3.3).

Assumption 3.4

$$0.5 < p_{i\min} < p_{i\max} < 1, i = 1, \cdots, k$$

To solve MINMAX3.6($\hat{\mu}$), we first define the following functions corresponding to constraints (3.93).

$$\begin{aligned}
g_i(\mathbf{x}, \lambda) &\stackrel{\text{def}}{=} \mu_{\tilde{G}_i}^{-1}(\hat{\mu}_i - \lambda) - (\mathbf{E}[\bar{\mathbf{d}}_i]\mathbf{x} - L^{-1}(\hat{\mu}_i - \lambda)\alpha_i\mathbf{x}) \\
&\quad - \Phi^{-1}(\mu_{\bar{p}_i}^{-1}(\hat{\mu}_i - \lambda)) \cdot \sqrt{\mathbf{x}^T V_i \mathbf{x}}, i = 1, \cdots, k
\end{aligned}$$

From Assumption 3.4, it holds that $\Phi^{-1}(\mu_{\bar{p}_i}^{-1}(\hat{\mu}_i - \lambda)) > 0$ because $\mu_{\bar{p}_i}^{-1}(\hat{\mu}_i - \lambda) > 0.5$ for any $\lambda \in \Lambda$. This means that $g_i(\mathbf{x}, \lambda), i = 1, \cdots, k$ are concave with respect to $\mathbf{x} \in X$ for any fixed value $\lambda \in \Lambda$. Let us define the following feasible set $X(\lambda)$ of MINMAX3.6($\hat{\mu}$) for some fixed value $\lambda \in \Lambda$.

$$G(\lambda) \stackrel{\text{def}}{=} \{\mathbf{x} \in X \mid g_i(\mathbf{x}, \lambda) \geq 0, i = 1, \cdots, k\} \tag{3.94}$$

Clearly, $G(\lambda)$ is a convex set; $G(\lambda)$ satisfies the following property.

Property 3.1
If $0 < \lambda_1 < \lambda_2 < 1$, then it holds that $G(\lambda_1) \subset G(\lambda_2)$.

(Proof)
From Assumption 3.3 and $0 < \lambda_1 < \lambda_2 < 1$, it holds that $\mu_{\tilde{G}_i}^{-1}(\hat{\mu}_i - \lambda_1) < \mu_{\tilde{G}_i}^{-1}(\hat{\mu}_i - \lambda_2)$ and $\Phi^{-1}(\mu_{\bar{p}_i}^{-1}(\hat{\mu}_i - \lambda_1)) > \Phi^{-1}(\mu_{\bar{p}_i}^{-1}(\hat{\mu}_i - \lambda_2))$. This means that $g_i(\mathbf{x}, \lambda_1) < g_i(\mathbf{x}, \lambda_2)$ for any $\mathbf{x} \in X$. As a result, it holds that $G(\lambda_1) \subset G(\lambda_2)$ for any $0 < \lambda_1 < \lambda_2 < 1$.

From Property 3.1, we obtain optimal solution $(\mathbf{x}^*, \lambda^*)$ of MINMAX3.6($\hat{\mu}$) using the following simple algorithm based on the bisection method and convex programming techniques.

Algorithm 3.2

Step 1: *Set $\lambda_0 = \lambda_{\min}, \lambda_1 = \lambda_{\max}$, $\lambda \leftarrow (\lambda_0 + \lambda_1)/2$.*
Step 2: *Solve the following convex programming problem for a given value λ, denoting the optimal solution as $\mathbf{x}(\lambda)$.*

$$\max_{\mathbf{x} \in X} g_j(\mathbf{x}, \lambda) \tag{3.95}$$

subject to

$$g_i(\mathbf{x}, \lambda) \geq 0, i = 1, \cdots, k, i \neq j \tag{3.96}$$

Step 3: *If $\mid \lambda_1 - \lambda_0 \mid < \varepsilon$, then go to Step 4, where ε is a sufficiently small positive constant. If $g_j(\mathbf{x}(\lambda), \lambda) \geq 0$ and $g_i(\mathbf{x}(\lambda), \lambda) \geq 0$ for any $i = 1, \cdots, k, i \neq j$, then set $\lambda_1 \leftarrow \lambda, \lambda \leftarrow (\lambda_0 + \lambda_1)/2$, and go to Step 2. Otherwise, $\lambda_0 \leftarrow \lambda, \lambda \leftarrow (\lambda_0 + \lambda_1)/2$, and go to Step 2.*
Step 4: *Set $\lambda^* \leftarrow \lambda$ and $\mathbf{x}^* \leftarrow \mathbf{x}(\lambda)$. The optimal solution $(\mathbf{x}^*, \lambda^*)$ of MINMAX3.6($\hat{\mu}$) is obtained.*

The relationships between optimal solution $(\mathbf{x}^*, \lambda^*)$ of MINMAX3.6($\hat{\mu}$) and MP-Pareto optimal solutions to MOP3.21 are characterized by the following theorem.

Theorem 3.4
(1) If $\mathbf{x}^* \in X, \lambda^* \in \Lambda$ is a unique optimal solution of MINMAX3.6($\hat{\mu}$), then $\mathbf{x}^* \in X, h_i^* \stackrel{\text{def}}{=} \hat{\mu}_i - \lambda^* \in H_i, i = 1, \cdots, k$ is an MP-Pareto optimal solution to MOP3.21.

(2) If $\mathbf{x}^* \in X, h_i^* \in H_i, i = 1, \cdots, k$ is an MP-Pareto optimal solution to MOP3.21, then $\mathbf{x}^* \in X$, $\lambda^* \stackrel{\text{def}}{=} \hat{\mu}_i - h_i^* = \hat{\mu}_i - \mu_{p_i}(p_i(\mathbf{x}^*, h_i^*)), i = 1, \cdots, k$ is an optimal solution of MINMAX3.6($\hat{\mu}$) for some reference membership values $\hat{\mu} = (\hat{\mu}_1, \cdots, \hat{\mu}_k)$.

(Proof)
(1) From (3.93), it holds that $\hat{\mu}_i-\lambda^* \le \mu_{p_i}(p_i(\mathbf{x}^*,\hat{\mu}_i-\lambda^*)), i=1,\cdots,k$. Assume that $\mathbf{x}^* \in X, \hat{\mu}_i-\lambda^* \in H_i, i=1,\cdots,k$ is not an MP-Pareto optimal solution to MOP3.21. Then, there exists $\mathbf{x} \in X, h_i \in H_i, i=1,\cdots,k$ such that

$$\begin{aligned}\mu_{D_{p_i}}(\mathbf{x},h_i) &= \min\{h_i, \mu_{p_i}(p_i(\mathbf{x},h_i))\} \\ &\ge \mu_{D_{p_i}}(\mathbf{x}^*,\hat{\mu}_i-\lambda^*) \\ &= \hat{\mu}_i-\lambda^*, i=1,\cdots,k,\end{aligned}$$

with strict inequality holding for at least one i. Then it holds that

$$h_i \ge \hat{\mu}_i-\lambda^*, i=1,\cdots,k \tag{3.97}$$

$$\mu_{p_i}(p_i(\mathbf{x},h_i)) \ge \hat{\mu}_i-\lambda^*, i=1,\cdots,k. \tag{3.98}$$

From Assumption 3.3, (3.79), and $L^{-1}(h_i) \le L^{-1}(\hat{\mu}_i-\lambda^*)$, (3.97) and (3.98) can be transformed as follows.

$$\begin{aligned}\mu_{\tilde{G}_i}^{-1}(h_i) &\le \mu_{\tilde{G}_i}^{-1}(\hat{\mu}_i-\lambda^*), i=1,\cdots,k \\ \mu_{\tilde{G}_i}^{-1}(h_i) &\ge (\mathbf{E}[\bar{\mathbf{c}}_i]\mathbf{x}-L^{-1}(\hat{\mu}_i-\lambda^*)\alpha_i\mathbf{x})+\Phi^{-1}(\mu_{p_i}^{-1}(\hat{\mu}_i-\lambda^*))\cdot\sqrt{\mathbf{x}^T V_i \mathbf{x}}, \\ & \quad i=1,\cdots,k\end{aligned}$$

As a result, there exists $\mathbf{x} \in X$ such that

$$\begin{aligned}&\mu_{\tilde{G}_i}^{-1}(\hat{\mu}_i-\lambda^*)-(\mathbf{E}[\bar{\mathbf{c}}_i]\mathbf{x}-L^{-1}(\hat{\mu}_i-\lambda^*)\alpha_i\mathbf{x}) \\ \ge\ &\Phi^{-1}(\mu_{p_i}^{-1}(\hat{\mu}_i-\lambda^*))\cdot\sqrt{\mathbf{x}^T V_i \mathbf{x}}, i=1,\cdots,k,\end{aligned}$$

with strict inequality holding for at least one i, which contradicts the fact that $\mathbf{x}^* \in X, \lambda^* \in \Lambda$ is a unique optimal solution to MINMAX3.6($\hat{\mu}$).

(2) Assume that $\mathbf{x}^* \in X, \lambda^* \in \Lambda$ is not an optimal solution of MINMAX3.6($\hat{\mu}$) for any reference membership values $\hat{\mu}=(\hat{\mu}_1,\cdots,\hat{\mu}_k)$, which satisfies

$$\hat{\mu}_i-\lambda^* = h_i^* = \mu_{p_i}(p_i(\mathbf{x}^*,h_i^*)), i=1,\cdots,k. \tag{3.99}$$

Then, there exists some $\mathbf{x} \in X, \lambda < \lambda^*$ such that

$$\begin{aligned}&\mu_{\tilde{G}_i}^{-1}(\hat{\mu}_i-\lambda)-(\mathbf{E}[\bar{\mathbf{c}}_i]\mathbf{x}-L^{-1}(\hat{\mu}_i-\lambda)\alpha_i\mathbf{x}) \\ &\ge \Phi^{-1}(\mu_{p_i}^{-1}(\hat{\mu}_i-\lambda^*))\cdot\sqrt{\mathbf{x}^T V_i \mathbf{x}}, \\ \Leftrightarrow\ &\mu_{p_i}(p_i(\mathbf{x},\hat{\mu}_i-\lambda)) \ge \hat{\mu}_i-\lambda, i=1,\cdots,k.\end{aligned} \tag{3.100}$$

Because of (3.99), (3.100), and $\hat{\mu}_i-\lambda > \hat{\mu}_i-\lambda^*, i=1,\cdots,k$, the following inequalities hold.

$$\mu_{p_i}(p_i(\mathbf{x},h_i)) > \mu_{p_i}(p_i(\mathbf{x}^*,h_i^*)), i=1,\cdots,k$$

Here, $h_i \stackrel{\text{def}}{=} \hat{\mu}_i - \lambda \in H_i$. Then, because $h_i > h_i^*$, there exists $\mathbf{x} \in X, h_i \in H_i, i = 1, \cdots, k$ such that

$$\mu_{D_{p_i}}(\mathbf{x}, h_i) > \mu_{D_{p_i}}(\mathbf{x}^*, h_i^*), i = 1, \cdots, k.$$

This contradicts the fact that $\mathbf{x}^* \in X, h_i^* \in H_i, i = 1, \cdots, k$ is an MP-Pareto optimal solution to MOP3.21. □

3.2.3 A formulation using a fractile model

For the objective function of MOP3.16, if the decision maker specifies permissible probability levels $\hat{p}_i, i = 1, \cdots, k$, MOP3.16 can be formulated as the following multiobjective programming problem using a fractile optimization model.

MOP 3.22 $(\hat{\mathbf{p}})$

$$\max_{\mathbf{x} \in X, h_i \in [0,1], i=1,\cdots,k} (h_1, \cdots, h_k)$$

subject to

$$\Pr(\omega \mid \Pi_{\tilde{\bar{\mathbf{c}}}_i(\omega)\mathbf{x}}(\widetilde{G}_i) \geq h_i) \geq \hat{p}_i, i = 1, \cdots, k$$

Here, $\Pr(\cdot)$ is a probability measure and $\hat{\mathbf{p}} = (\hat{p}_1, \cdots, \hat{p}_k)$ is a k-dimensional vector of permissible probability levels. In MOP3.22($\hat{\mathbf{p}}$), $\Pr(\omega \mid \Pi_{\tilde{\bar{\mathbf{c}}}_i(\omega)\mathbf{x}}(\widetilde{G}_i) \geq h_i)$ can be equivalently transformed into

$$\begin{aligned} & \Pr(\omega \mid \Pi_{\tilde{\bar{\mathbf{c}}}_i(\omega)\mathbf{x}}(\widetilde{G}_i) \geq h_i) \\ = \; & \Phi\left(\frac{\mu_{\widetilde{G}_i}^{-1}(h_i) - (\mathbf{E}[\bar{\mathbf{d}}_i]\mathbf{x} - L^{-1}(h_i)\alpha_i\mathbf{x})}{\sqrt{\mathbf{x}^T V_i \mathbf{x}}} \right) \\ \stackrel{\text{def}}{=} \; & p_i(\mathbf{x}, h_i), i = 1, \cdots, k, \end{aligned} \tag{3.101}$$

where $L^{-1}(\cdot)$ and $R^{-1}(\cdot)$ are inverse functions of $L(\cdot)$ and $R(\cdot)$; further, $\Phi(\cdot)$ is a distribution function of a standard Gaussian random variable. Using $p_i(\mathbf{x}, h_i), i = 1, \cdots, k$, MOP3.22($\hat{\mathbf{p}}$) can be transformed into the following form [38].

MOP 3.23 $(\hat{\mathbf{p}})$

$$\max_{\mathbf{x} \in X, h_i \in [0,1], i=1,\cdots,k} (h_1, \cdots, h_k)$$

subject to

$$p_i(\mathbf{x}, h_i) \geq \hat{p}_i, i = 1, \cdots, k \tag{3.102}$$

Because distribution function $\Phi(\cdot)$ is continuous and strictly increasing, constraints (3.102) can be transformed into

$$\begin{aligned} & \hat{p}_i \leq p_i(\mathbf{x}, h_i) \\ \Leftrightarrow \quad & \mu_{\tilde{G}_i}^{-1}(h_i) \geq (\mathbf{E}[\bar{\mathbf{d}}_i]\mathbf{x} - L^{-1}(h_i)\alpha_i\mathbf{x}) + \Phi^{-1}(\hat{p}_i) \cdot \sqrt{\mathbf{x}^T V_i \mathbf{x}}. \end{aligned} \tag{3.103}$$

The right-hand-side of inequality (3.103) can be defined as follows.

$$f_i(\mathbf{x}, h_i, \hat{p}_i) \stackrel{\text{def}}{=} (\mathbf{E}[\bar{\mathbf{d}}_i]\mathbf{x} - L^{-1}(h_i)\alpha_i\mathbf{x}) + \Phi^{-1}(\hat{p}_i) \cdot \sqrt{\mathbf{x}^T V_i \mathbf{x}} \tag{3.104}$$

Then, MOP3.23($\hat{\mathbf{p}}$) can be equivalently transformed into the following form, because $\mu_{\tilde{G}_i}(\cdot), i = 1, \cdots, k$ is continuous and strictly decreasing.

MOP 3.24 ($\hat{\mathbf{p}}$)

$$\max_{\mathbf{x} \in X, h_i \in [0,1], i=1,\cdots,k} (h_1, \cdots, h_k)$$

subject to

$$\mu_{\tilde{G}_i}(f_i(\mathbf{x}, h_i, \hat{p}_i)) \geq h_i, i = 1, \cdots, k \tag{3.105}$$

In MOP3.24($\hat{\mathbf{p}}$), we focus here on inequality (3.105); $f_i(\mathbf{x}, h_i, \hat{p}_i)$ is continuous and strictly increasing with respect to h_i for any $\mathbf{x} \in X$, because $L^{-1}(h_i)$ is continuous and strictly decreasing with respect to h_i. This means that the left-hand-side of inequality (3.105) is continuous and strictly decreasing with respect to h_i for any $\mathbf{x} \in X$. Because the right-hand-side of inequality (3.105) is continuous and strictly increasing with respect to h_i, inequality (3.105) must always satisfy the active condition, i.e., $\mu_{\tilde{G}_i}(f_i(\mathbf{x}, h_i, \hat{p}_i)) = h_i$, $i = 1, \cdots, k$ at the optimal solution of MOP3.24($\hat{\mathbf{p}}$). From this perspective, MOP3.24($\hat{\mathbf{p}}$) can be equivalently expressed in the following form.

MOP 3.25 ($\hat{\mathbf{p}}$)

$$\max_{\mathbf{x} \in X, h_i \in [0,1], i=1,\cdots,k} (\mu_{\tilde{G}_1}(f_1(\mathbf{x}, h_1, \hat{p}_1)), \cdots, \mu_{\tilde{G}_k}(f_k(\mathbf{x}, h_k, \hat{p}_k)))$$

subject to

$$\mu_{\tilde{G}_i}(f_i(\mathbf{x}, h_i, \hat{p}_i)) = h_i, i = 1, \cdots, k$$

In MOP3.25($\hat{\mathbf{p}}$), the decision maker prefers not only larger values of permissible probability level $\hat{p}_i$ but also larger values of corresponding membership function $\mu_{\tilde{G}_i}(f_i(\mathbf{x}, h_i, \hat{p}_i))$. From such a perspective, we consider the following multiobjective programming problem, which can be interpreted as a natural extension of MOP3.25($\hat{\mathbf{p}}$).

MOP 3.26

$$\max_{\mathbf{x}\in X, h_i\in[0,1], \hat{p}_i\in(0,1), i=1,\cdots,k} (\mu_{\tilde{G}_1}(f_1(\mathbf{x},h_1,\hat{p}_1)),\cdots,\mu_{\tilde{G}_k}(f_k(\mathbf{x},h_k,\hat{p}_k)),\hat{p}_1,\cdots,\hat{p}_k)$$

subject to

$$\mu_{\tilde{G}_i}(f_i(\mathbf{x},h_i,\hat{p}_i)) = h_i, i=1,\cdots,k$$

Note that in MOP3.26, permissible probability levels $(\hat{p}_1,\cdots,\hat{p}_k)$ are decision variables, not fixed values.

Considering the imprecise nature of the decision maker's judgment, we assume the decision maker has a fuzzy goal for each permissible probability level $\hat{p}_i$. Such a fuzzy goal can be quantified by eliciting the corresponding membership function. Let us denote a membership function of permissible probability level $\hat{p}_i$ as $\mu_{\hat{p}_i}(\hat{p}_i)$. Then, MOP3.26 can be transformed into the following multiobjective programming problem.

MOP 3.27

$$\max_{\mathbf{x}\in X, h_i\in[0,1], \hat{p}_i\in(0,1), i=1,\cdots,k} (\mu_{\tilde{G}_1}(f_1(\mathbf{x},h_1,\hat{p}_1)),\cdots, \mu_{\tilde{G}_k}(f_k(\mathbf{x},h_k,\hat{p}_k)),\mu_{\hat{p}_1}(\hat{p}_1),\cdots,\mu_{\hat{p}_k}(\hat{p}_k))$$

subject to

$$\mu_{\tilde{G}_i}(f_i(\mathbf{x},h_i,\hat{p}_i)) = h_i, i=1,\cdots,k$$

To appropriately elicit membership functions $\mu_{\hat{p}_i}(\hat{p}_i), i=1,\cdots,k$, we propose the following procedures. First, the decision maker sets the intervals for permissible probability levels $\hat{p}_i$ as

$$P_i \stackrel{\text{def}}{=} [p_{i\min}, p_{i\max}], i=1,\cdots,k,$$

where $p_{i\min}$ is the maximum unacceptable value of $\hat{p}_i$ and $p_{i\max}$ is the sufficiently satisfactory minimum value of $\hat{p}_i$. Throughout this subsection, we make the following assumption for $\mu_{\hat{p}_i}(\hat{p}_i)$.

Assumption 3.5
$\mu_{\hat{p}_i}(\hat{p}_i), i=1,\cdots,k$ are strictly increasing and continuous with respect to $\hat{p}_i \in P_i$, and $\mu_{\hat{p}_i}(p_{i\min}) = 0$, $\mu_{\hat{p}_i}(p_{i\max}) = 1$, and $0.5 < p_{i\min} < p_{i\max} < 1, i=1,\cdots,k$.

Corresponding to interval P_i, the interval of a permissible possibility level h_i can be computed as

$$\begin{aligned} H_i(P_i) &\stackrel{\text{def}}{=} [h_{i\min}, h_{i\max}] \\ &= \{h_i = \mu_{\tilde{G}_i}(f_i(\mathbf{x},h_i,\hat{p}_i)) \mid \mathbf{x}\in X, \hat{p}_i \in P_i\}, i=1,\cdots,k. \end{aligned}$$

The maximum value $h_{i\max}$ can be obtained by solving the following problem.

$$h_{i\max} \stackrel{\text{def}}{=} \max_{\mathbf{x}\in X, h_i\in[0,1]} \mu_{\tilde{G}_i}(f_i(\mathbf{x}, h_i, p_{i\min}))$$

$$\text{subject to } h_i = \mu_{\tilde{G}_i}(f_i(\mathbf{x}, h_i, p_{i\min}))$$

This is equivalent to the following problem.

$$\max_{\mathbf{x}\in X, h_i\in[0,1]} h_i$$

subject to

$$\mu_{\tilde{G}_i}^{-1}(h_i) = (\mathbf{E}[\bar{\mathbf{d}}_i]\mathbf{x} - L^{-1}(h_i)\alpha_i\mathbf{x}) + \Phi^{-1}(p_{i\min}) \cdot \sqrt{\mathbf{x}^T V_i \mathbf{x}}$$

The minimum value $h_{i\min}$ is obtained by the following procedure. First, we solve the following optimization problem for each $i = 1, \cdots, k$.

$$\max_{\mathbf{x}\in X, h_i\in[0,1]} h_i$$

$$\text{subject to } h_i = \mu_{\tilde{G}_i}(f_i(\mathbf{x}, h_i, p_{i\max}))$$

Using optimal solutions $\mathbf{x}_i^* \in X, h_i^* \in [0,1], i = 1, \cdots, k$ of the above problems, we can obtain $h_{i\min}$ as follows.

$$h_{i\min} \stackrel{\text{def}}{=} \min_{\ell=1,\cdots,k,\ell\neq i} \mu_{\tilde{G}_i}(f_i(\mathbf{x}_\ell^*, h_\ell^*, p_{i\max}))$$

If the decision maker adopts the fuzzy decision [6, 71, 140] to integrate $\mu_{\tilde{G}_i}(f_i(\mathbf{x}, h_i, \hat{p}_i))$ and $\mu_{\hat{p}_i}(\hat{p}_i)$, MOP3.27 can be transformed into the following form.

MOP 3.28

$$\max_{\mathbf{x}\in X, \hat{p}_i\in P_i, h_i=H_i(P_i), i=1,\cdots,k} \left(\mu_{D_{f_1}}(\mathbf{x}, h_1, \hat{p}_1), \cdots, \mu_{D_{f_k}}(\mathbf{x}, h_k, \hat{p}_k)\right) \tag{3.106}$$

subject to

$$\mu_{\tilde{G}_i}(f_i(\mathbf{x}, h_i, \hat{p}_i)) = h_i, i = 1, \cdots, k$$

where

$$\mu_{D_{f_i}}(\mathbf{x}, h_i, \hat{p}_i) \stackrel{\text{def}}{=} \min\{\mu_{\tilde{G}_i}(f_i(\mathbf{x}, h_i, \hat{p}_i)), \mu_{\hat{p}_i}(\hat{p}_i)\}, i = 1, \cdots, k \tag{3.107}$$

are integrated membership functions.

To handle MOP3.28, we introduce an MF-Pareto optimal solution below.

Definition 3.4 $\mathbf{x}^* \in X, \hat{p}_i^* \in P_i, h_i^* \in H_i(P_i), i = 1, \cdots, k$ is an MF-Pareto optimal solution to MOP3.28 if and only if there does not exist another $\mathbf{x} \in X, \hat{p}_i \in P_i, h_i \in H_i(P_i), i = 1, \cdots, k$ such that $\mu_{D_{f_i}}(\mathbf{x}, h_i, \hat{p}_i) \geq \mu_{D_{f_i}}(\mathbf{x}^*, h_i^*, \hat{p}_i^*)$ $i =$

$1, \cdots, k$ with strict inequality holding for at least one i, where $\mu_{\tilde{G}_i}(f_i(\mathbf{x}^*, h_i^*, \hat{p}_i^*)) = h_i^*, \mu_{\tilde{G}_i}(f_i(\mathbf{x}, h_i, \hat{p}_i)) = h_i, i = 1, \cdots, k$.

To generate a candidate for a satisfactory solution that is also MF-Pareto optimal, the decision maker specifies reference membership values [71]. Once reference membership values $\hat{\mu} = (\hat{\mu}_1, \cdots, \hat{\mu}_k)$ are specified, the corresponding MF-Pareto optimal solution is obtained by solving the following minmax problem.

MINMAX 3.7 ($\hat{\mu}$)

$$\min_{\mathbf{x}\in X, \hat{p}_i \in P_i, h_i \in H_i(P_i), i=1,\cdots,k, \lambda \in \Lambda} \lambda$$

subject to

$$\hat{\mu}_i - \mu_{\hat{p}_i}(\hat{p}_i) \le \lambda, i = 1, \ldots, k, \tag{3.108}$$

$$\hat{\mu}_i - h_i \le \lambda, i = 1, \ldots, k, \tag{3.109}$$

$$\mu_{\tilde{G}_i}(f_i(\mathbf{x}, h_i, \hat{p}_i)) = h_i, i = 1, \ldots, k \tag{3.110}$$

where

$$\Lambda \stackrel{\text{def}}{=} [\lambda_{\min}, \lambda_{\max}] = [\max_{i=1,\ldots,k} \hat{\mu}_i - 1, \max_{i=1,\ldots,k} \hat{\mu}_i]. \tag{3.111}$$

In MINMAX3.7($\hat{\mu}$), constraints (3.109) and (3.110) can be transformed as follows.

$$\begin{aligned} & h_i = \mu_{\tilde{G}_i}(f_i(\mathbf{x}, h_i, \hat{p}_i)) \ge \hat{\mu}_i - \lambda, \\ \Leftrightarrow \quad & \mu_{\tilde{G}_i}^{-1}(h_i) = f_i(\mathbf{x}, h_i, \hat{p}_i) \le \mu_{\tilde{G}_i}^{-1}(\hat{\mu}_i - \lambda) \\ \Leftrightarrow \quad & \mu_{\tilde{G}_i}^{-1}(h_i) = (\mathbf{E}[\bar{\mathbf{d}}_i]\mathbf{x} - L^{-1}(h_i)\alpha_i\mathbf{x}) + \Phi^{-1}(\hat{p}_i) \cdot \sqrt{\mathbf{x}^T V_i \mathbf{x}} \\ & \le \mu_{\tilde{G}_i}^{-1}(\hat{\mu}_i - \lambda) \end{aligned} \tag{3.112}$$

Because of $L^{-1}(h_i) \le L^{-1}(\hat{\mu}_i - \lambda)$, the following relation holds for the left-hand-side of inequality (3.112).

$$\begin{aligned} & (\mathbf{E}[\bar{\mathbf{d}}_i]\mathbf{x} - L^{-1}(h_i)\alpha_i\mathbf{x}) + \Phi^{-1}(\hat{p}_i) \cdot \sqrt{\mathbf{x}^T V_i \mathbf{x}} \\ \ge \quad & (\mathbf{E}[\bar{\mathbf{d}}_i]\mathbf{x} - L^{-1}(\hat{\mu}_i - \lambda)\alpha_i\mathbf{x}) + \Phi^{-1}(\hat{p}_i) \cdot \sqrt{\mathbf{x}^T V_i \mathbf{x}} \end{aligned} \tag{3.113}$$

From inequalities (3.112) and (3.113), the following inequality holds.

$$\mu_{\tilde{G}_i}^{-1}(\hat{\mu}_i - \lambda) \ge (\mathbf{E}[\bar{\mathbf{d}}_i]\mathbf{x} - L^{-1}(\hat{\mu}_i - \lambda)\alpha_i\mathbf{x}) + \Phi^{-1}(\hat{p}_i) \cdot \sqrt{\mathbf{x}^T V_i \mathbf{x}} \tag{3.114}$$

On the other hand, from Assumption 3.5 and inequalities (3.108), it holds that

$\hat{p}_i \geq \mu_{\tilde{p}_i}^{-1}(\hat{\mu}_i - \lambda), i = 1, \cdots, k$. As a result, inequality (3.114) can be transformed as

$$\begin{aligned} \mu_{\tilde{G}_i}^{-1}(\hat{\mu}_i - \lambda) &\geq (\mathbf{E}[\bar{\mathbf{d}}_i]\mathbf{x} - L^{-1}(\hat{\mu}_i - \lambda)\alpha_i\mathbf{x}) + \Phi^{-1}(\hat{p}_i) \cdot \sqrt{\mathbf{x}^T V_i \mathbf{x}} \\ &\geq (\mathbf{E}[\bar{\mathbf{d}}_i]\mathbf{x} - L^{-1}(\hat{\mu}_i - \lambda)\alpha_i\mathbf{x}) + \Phi^{-1}(\mu_{\tilde{p}_i}^{-1}(\hat{\mu}_i - \lambda)) \cdot \sqrt{\mathbf{x}^T V_i \mathbf{x}}. \end{aligned}$$

Therefore, MINMAX3.7($\hat{\mu}$) can be reduced to the following simple minmax problem, where both decision variables $\hat{p}_i$, $i = 1, \cdots, k$ and $h_i, i = 1, \cdots, k$ have disappeared.

MINMAX 3.8 ($\hat{\mu}$)

$$\min_{\mathbf{x} \in X, \lambda \in \Lambda} \lambda$$

subject to

$$\mu_{\tilde{G}_i}^{-1}(\hat{\mu}_i - \lambda) \geq (\mathbf{E}[\bar{\mathbf{d}}_i]\mathbf{x} - L^{-1}(\hat{\mu}_i - \lambda)\alpha_i\mathbf{x}) + \Phi^{-1}(\mu_{\tilde{p}_i}^{-1}(\hat{\mu}_i - \lambda)) \cdot \sqrt{\mathbf{x}^T V_i \mathbf{x}},$$
$$i = 1, \cdots, k \tag{3.115}$$

Note that MINMAX3.8($\hat{\mu}$) can be solved by applying Algorithm 3.2.

The relationships between optimal solution $\mathbf{x}^* \in X, \lambda^* \in \Lambda$ of MINMAX3.8 ($\hat{\mu}$) and MF-Pareto optimal solutions to MOP3.28 are characterized by the following theorem.

Theorem 3.5

(1) If $\mathbf{x}^* \in X, \lambda^* \in \Lambda$ is a unique optimal solution of MINMAX3.8($\hat{\mu}$), then $\mathbf{x}^* \in X, \hat{p}_i^* \stackrel{\text{def}}{=} \mu_{\tilde{p}_i}^{-1}(\hat{\mu}_i - \lambda^*), h_i^* \stackrel{\text{def}}{=} \hat{\mu}_i - \lambda^*, i = 1, \ldots, k$ is an MF-Pareto optimal solution to MOP3.28.

(2) If $\mathbf{x}^* \in X, \hat{p}_i^* \in P_i, h_i^* \in H_i(P_i), i = 1, \ldots, k$ is an MF-Pareto optimal solution to MOP3.28, then $\mathbf{x}^* \in X$, $\lambda^* \stackrel{\text{def}}{=} \hat{\mu}_i - \mu_{\hat{p}_i}(\hat{p}_i^*) = \hat{\mu}_i - \mu_{\tilde{G}_i}(f_i(\mathbf{x}^*, h_i^*, \hat{p}_i^*)), i = 1, \ldots, k$ is an optimal solution of MINMAX3.8($\hat{\mu}$) for some reference membership values $\hat{\mu} = (\hat{\mu}_1, \cdots, \hat{\mu}_k)$.

(Proof)
(1) From (3.115), it holds that

$$\hat{\mu}_i - \lambda^* \leq \mu_{\tilde{G}_i}(f_i(\mathbf{x}^*, \hat{\mu}_i - \lambda^*, \mu_{\tilde{p}_i}^{-1}(\hat{\mu}_i - \lambda^*))).$$

Assume that $\mathbf{x}^* \in X, \hat{\mu}_i - \lambda^* \in H_i(P_i), \mu_{\tilde{p}_i}^{-1}(\hat{\mu}_i - \lambda^*) \in P_i, i = 1, \ldots, k$ is not an MF-Pareto optimal solution to MOP3.28. Then, there exists $\mathbf{x} \in X, \hat{p}_i \in P_i, h_i \in H_i(P_i), i = 1, \ldots, k$, such that

$$\begin{aligned} \mu_{D_{f_i}}(\mathbf{x}, h_i, \hat{p}_i) &= \min\{\mu_{\hat{p}_i}(\hat{p}_i), \mu_{\tilde{G}_i}(f_i(\mathbf{x}, h_i, \hat{p}_i))\} \\ &\geq \mu_{D_{f_i}}(\mathbf{x}^*, \hat{\mu}_i - \lambda^*, \mu_{\tilde{p}_i}^{-1}(\hat{\mu}_i - \lambda^*)) \\ &= \hat{\mu}_i - \lambda^*, i = 1, \ldots, k, \end{aligned}$$

with strict inequality holding for at least one i, and $\mu_{\tilde{G}_i}(f_i(\mathbf{x},h_i,\hat{p}_i)) = h_i, i = 1,\cdots,k$. Then it holds that

$$\begin{aligned} \mu_{\hat{p}_i}(\hat{p}_i) &\geq \hat{\mu}_i - \lambda^*, i = 1,\cdots,k \quad (3.116)\\ \mu_{\tilde{G}_i}(f_i(\mathbf{x},h_i,\hat{p}_i)) &\geq \hat{\mu}_i - \lambda^*, i = 1,\cdots,k. \quad (3.117)\end{aligned}$$

From Assumption 3.5, (3.104), and $L^{-1}(h_i) \leq L^{-1}(\hat{\mu}_i - \lambda^*)$, (3.116) and (3.117) can be transformed as follows.

$$\begin{aligned} \hat{p}_i &\geq \mu_{\hat{p}_i}^{-1}(\hat{\mu}_i - \lambda^*), i = 1,\ldots,k\\ \hat{p}_i &\leq \Phi\left(\frac{\mu_{\tilde{G}_i}^{-1}(\hat{\mu}_i - \lambda^*) - (\mathbf{E}[\bar{\mathbf{d}}_i]\mathbf{x} - L^{-1}(\hat{\mu}_i - \lambda^*)\alpha_i\mathbf{x})}{\sqrt{\mathbf{x}^T V_i \mathbf{x}}}\right), i = 1,\ldots,k\end{aligned}$$

As a result, there exists $\mathbf{x} \in X$ such that

$$\begin{aligned} &\mu_{\tilde{G}_i}^{-1}(\hat{\mu}_i - \lambda^*) - (\mathbf{E}[\bar{\mathbf{d}}_i]\mathbf{x} - L^{-1}(\hat{\mu}_i - \lambda^*)\alpha_i\mathbf{x})\\ \geq\ &\Phi^{-1}(\mu_{p_i}^{-1}(\hat{\mu}_i - \lambda^*)) \cdot \sqrt{\mathbf{x}^T V_i \mathbf{x}}, i = 1,\cdots,k,\end{aligned}$$

with strict inequality holding for at least one i, which contradicts the fact that $\mathbf{x}^* \in X, \lambda^* \in \Lambda$ is a unique optimal solution of MINMAX3.8($\hat{\mu}$).

(2) Assume that $\mathbf{x}^* \in X, \lambda^* \in \Lambda$ is not an optimal solution of MINMAX3.8($\hat{\mu}$) for any reference membership values $\hat{\mu} = (\hat{\mu}_1,\ldots,\hat{\mu}_k)$ that satisfy

$$\hat{\mu}_i - \lambda^* = \mu_{\hat{p}_i}(\hat{p}_i^*) = \mu_{\tilde{G}_i}(f_i(\mathbf{x}^*,h_i^*,\hat{p}_i^*)), i = 1,\ldots,k. \quad (3.118)$$

Then, there exists some $\mathbf{x} \in X, \lambda < \lambda^*$ such that

$$\begin{aligned} &\mu_{\tilde{G}_i}^{-1}(\hat{\mu}_i - \lambda) - (\mathbf{E}[\bar{\mathbf{d}}_i]\mathbf{x} - L^{-1}(\hat{\mu}_i - \lambda)\alpha_i\mathbf{x})\\ &\geq \Phi^{-1}(\mu_{p_i}^{-1}(\hat{\mu}_i - \lambda)) \cdot \sqrt{\mathbf{x}^T V_i \mathbf{x}}\\ \Leftrightarrow\ &\mu_{\tilde{G}_i}(f_i(\mathbf{x},\hat{\mu}_i - \lambda, \mu_{\hat{p}_i}^{-1}(\hat{\mu}_i - \lambda)) \geq \hat{\mu}_i - \lambda, i = 1,\ldots,k. \quad (3.119)\end{aligned}$$

Because of (3.118), (3.119) and $\hat{\mu}_i - \lambda > \hat{\mu}_i - \lambda^*, i = 1,\ldots,k$, the following inequalities hold.

$$\begin{aligned} \mu_{\hat{p}_i}(\hat{p}_i) &> \mu_{\hat{p}_i}(\hat{p}_i^*), i = 1,\ldots,k\\ \mu_{\tilde{G}_i}(f_i(\mathbf{x},\hat{h}_i,\hat{p}_i)) &> \mu_{\tilde{G}_i}(f_i(\mathbf{x}^*,h_i^*,\hat{p}_i^*)), i = 1,\ldots,k\end{aligned}$$

Here, $\hat{p}_i \stackrel{\text{def}}{=} \mu_{\hat{p}_i}^{-1}(\mu_i - \lambda) \in P_i$, $\hat{h}_i \stackrel{\text{def}}{=} \hat{\mu}_i - \lambda \in H_i(P_i), i = 1,\ldots,k$, indicating that there exists some $\mathbf{x} \in X, \hat{p}_i \in P_i, \hat{h}_i \in H_i(P_i), i = 1,\ldots,k$ such that $\mu_{D_{f_i}}(\mathbf{x},\hat{h}_i,\hat{p}_i) > \mu_{D_{f_i}}(\mathbf{x}^*,h_i^*,\hat{p}_i^*), i = 1,\ldots,k$. This contradicts the fact that $\mathbf{x}^* \in X, \hat{p}_i^* \in P_i, h_i^* \in H_i(P_i), i = 1,\ldots,k$ is an MF-Pareto optimal solution to MOP3.28. □

3.2.4 An interactive convex programming algorithm

In this subsection, we propose an interactive algorithm to obtain a satisfactory solution from an MP-Pareto optimal solution set. Unfortunately, there is no guarantee that optimal solution $\mathbf{x}^* \in X, \lambda^* \in \Lambda$ to MINMAX3.6($\hat{\mu}$) is MP-Pareto optimal, if $\mathbf{x}^* \in X, \lambda^* \in \Lambda$ is not unique. To guarantee MP-Pareto optimality, we first assume that k constraints (3.93) of MINMAX3.6($\hat{\mu}$) are active at optimal solution $\mathbf{x}^* \in X, \lambda^* \in \Lambda$, i.e.,

$$\begin{aligned}\mu_{\tilde{G}_i}^{-1}(\hat{\mu}_i - \lambda^*) &= (\mathbf{E}[\bar{\mathbf{d}}_i]\mathbf{x}^* - L^{-1}(\hat{\mu}_i - \lambda^*)\alpha_i \mathbf{x}^*) \\ &\quad + \Phi^{-1}(\mu_{p_i}^{-1}(\hat{\mu}_i - \lambda^*)) \cdot \sqrt{\mathbf{x}^{*T} V_i \mathbf{x}^*}, i = 1, \cdots, k.\end{aligned}$$

If the j-th constraint of (3.93) is inactive, i.e., $\mu_{p_j}^{-1}(\hat{\mu}_j - \lambda^*) < p_j(\mathbf{x}^*, \hat{\mu}_j - \lambda^*)$, we can convert the inactive constraint into the active one by applying the following algorithm, where

$$\begin{aligned}h_j(\hat{\mu}_j) &\stackrel{\text{def}}{=} \mu_{\tilde{G}_i}^{-1}(\hat{\mu}_j - \lambda^*) - (\mathbf{E}[\bar{\mathbf{d}}_j]\mathbf{x}^* - L^{-1}(\hat{\mu}_j - \lambda^*)\alpha_i \mathbf{x}^*) \\ &\quad - \Phi^{-1}(\mu_{p_j}^{-1}(\hat{\mu}_j - \lambda^*)) \cdot \sqrt{\mathbf{x}^{*T} V_j \mathbf{x}^*}. \end{aligned} \tag{3.120}$$

Algorithm 3.3

Step 1: *Set* $\mu_j^L \leftarrow \lambda^*, \mu_j^R \leftarrow \lambda^* + 1$.
Step 2: *Set* $\mu_j \leftarrow (\mu_j^L + \mu_j^R)/2$.
Step 3: *If* $h_j(\mu_j) > 0$, *then* $\mu_j^L \leftarrow \mu_j$ *and go to Step 2, else if* $h_j(\mu_j) < 0$, *then* $\mu_j^R \leftarrow \mu_j$ *and go to Step 2, else update the reference membership value as* $\hat{\mu}_j \leftarrow \mu_j$ *and stop.*

For the optimal solution $\mathbf{x}^* \in X, \lambda^* \in \Lambda$ of MINMAX3.6($\hat{\mu}$), where the active conditions for (3.93) are satisfied, we solve MP-Pareto optimality test problems for each $j = 1, \cdots, k$.

Test 3.2

$$w_j \stackrel{\text{def}}{=} \min_{\mathbf{x} \in X} f_j(\mathbf{x}, \mu_{p_j}^{-1}(\hat{\mu}_j - \lambda^*)), j = 1, \cdots, k$$

subject to

$$f_i(\mathbf{x}, \mu_{p_i}^{-1}(\hat{\mu}_i - \lambda^*)) \leq f_i(\mathbf{x}^*, \mu_{p_i}^{-1}(\hat{\mu}_i - \lambda^*)), i = 1, \cdots, k$$

where

$$\begin{aligned}&f_j(\mathbf{x}, \mu_{p_j}^{-1}(\hat{\mu}_j - \lambda^*)) \\ \stackrel{\text{def}}{=}\ &(\mathbf{E}[\bar{\mathbf{d}}_j]\mathbf{x} - L^{-1}(\hat{\mu}_j - \lambda^*)\alpha_i \mathbf{x}) + \Phi^{-1}(\mu_{p_j}^{-1}(\hat{\mu}_j - \lambda^*)) \cdot \sqrt{\mathbf{x}^T V_j \mathbf{x}}.\end{aligned}$$

Note that TEST3.2 is a convex programming problem. For the optimal solution of TEST3.2, the following theorem holds.

Theorem 3.6
If $w_i = f_i(\mathbf{x}^*, \mu_{\tilde{p}_i}^{-1}(\hat{\mu}_i - \lambda^*)), i = 1, \cdots, k$, then $\mathbf{x}^* \in X, \hat{\mu}_i - \lambda^* \in H_i, i = 1, \cdots, k$ is an MP-Pareto optimal solution to MOP3.21.

Given the above, we next present the interactive algorithm for deriving a satisfactory solution from an MP-Pareto optimal solution set.

Algorithm 3.4

Step 1: *The decision maker sets membership functions $\mu_{\tilde{G}_i}(y), i = 1, \cdots, k$ for the objective functions in MOP3.15.*
Step 2: *The decision maker sets his or her membership function $\mu_{p_i}(p_i(\mathbf{x}, h_i))$, according to Assumptions 3.3 and 3.4.*
Step 3: *Set initial reference membership values as $\hat{\mu}_i = 1, i = 1, \cdots, k$.*
Step 4: *Solve MINMAX3.6($\hat{\mu}$) and obtain optimal solution $(\mathbf{x}^*, \lambda^*)$. Corresponding to optimal solution $(\mathbf{x}^*, \lambda^*)$, TEST3.2 is formulated and solved.*
Step 5: *If the decision maker is satisfied with the current values of MP-Pareto optimal solution $\mu_{D_{p_i}}(\mathbf{x}^*, h_i^*), i = 1, \cdots, k$ where $h_i^* \stackrel{\text{def}}{=} \hat{\mu}_i - \lambda^*, i = 1, \cdots, k$, then stop. Otherwise, the decision maker updates his or her reference membership values $\hat{\mu}_i, i = 1, \cdots, k$, and returns to Step 4.*

3.2.5 A numerical example

To demonstrate our proposed algorithm from Section 3.2.4 above, we consider the following fuzzy random three objective programming problem.

MOP 3.29

$$\min \tilde{\bar{\mathbf{C}}}\mathbf{x} = (\tilde{\bar{\mathbf{c}}}_1\mathbf{x}, \tilde{\bar{\mathbf{c}}}_2\mathbf{x}, \tilde{\bar{\mathbf{c}}}_3\mathbf{x}) \tag{3.121}$$

subject to

$$\mathbf{x} \in X \stackrel{\text{def}}{=} \{\mathbf{x} \in \mathrm{R}^4 \mid \mathbf{a}_j\mathbf{x} \leq b_j, j = 1, 2, \mathbf{x} \geq \mathbf{0}\} \tag{3.122}$$

Here, $\mathbf{x} = (x_1, x_2, x_3, x_4)^T$ is a four-dimensional decision column vector, $\mathbf{a}_j, j = 1, 2$ are four-dimensional coefficient row vectors defined as $\mathbf{a}_1 = (7, 3, 4, 6), \mathbf{a}_2 = (-5, -6, -7, -9)$, and $(b_1, b_2) = (27, -15)$, and $\tilde{\bar{\mathbf{c}}}_i = (\tilde{\bar{c}}_{i1}, \tilde{\bar{c}}_{i2}, \tilde{\bar{c}}_{i3}, \tilde{\bar{c}}_{i4}), i = 1, 2, 3$ are four-dimensional coefficient row vectors of fuzzy random variables. Each coefficient $\tilde{\bar{c}}_{ij}$ is an LR-type fuzzy random variable. More specifically, for given elementary event ω, realized value $\tilde{\bar{c}}_{ij}(\omega)$ of each coefficient $\tilde{\bar{c}}_{ij}$ is an LR fuzzy number characterized by membership function

$$\mu_{\widetilde{c}_{ij}(\omega)}(s) = \begin{cases} L\left(\frac{d_{ij}(\omega)-s}{\alpha_{ij}}\right), & s \leq d_{ij}(\omega) \\ R\left(\frac{s-d_{ij}(\omega)}{\beta_{ij}}\right), & s > d_{ij}(\omega), \end{cases}$$

where L and R are defined as $L(x) = R(x) = \max\{0, 1- \mid x \mid\}$, and left spread parameters $\alpha_{ij}, i = 1,2,3, j = 1,\cdots,4$ are set as

$$\begin{aligned} (\alpha_{11}, \alpha_{12}, \alpha_{13}, \alpha_{14}) &= (0.4, 0.5, 0.4, 0.6) \\ (\alpha_{21}, \alpha_{22}, \alpha_{23}, \alpha_{24}) &= (2.5, 1.5, 0.2, 0.4) \\ (\alpha_{31}, \alpha_{32}, \alpha_{33}, \alpha_{34}) &= (2.0, 1.3, 1.8, 2.3), \end{aligned}$$

and each right spread parameters β_{ij} are the same as α_{ij}. The mean value of LR-type fuzzy random variable $\widetilde{\overline{c}}_{ij}$ is denoted as $\overline{d}_{ij}$ which is the Gaussian random variable. For vectors of the Gaussian random variable $\overline{\mathbf{d}}_i = (\overline{d}_{i1}, \cdots, \overline{d}_{i4}), i = 1,2,3$, the corresponding mean vectors are set as

$$\begin{aligned} \mathbf{E}[\overline{\mathbf{d}}_1] &= (E[\overline{d}_{11}], E[\overline{d}_{12}], E[\overline{d}_{13}], E[\overline{d}_{14}]) = (2,3,2,4), \\ \mathbf{E}[\overline{\mathbf{d}}_2] &= (E[\overline{d}_{21}], E[\overline{d}_{22}], E[\overline{d}_{23}], E[\overline{d}_{24}]) = (10,-7,1,-2), \\ \mathbf{E}[\overline{\mathbf{d}}_3] &= (E[\overline{d}_{31}], E[\overline{d}_{32}], E[\overline{d}_{33}], E[\overline{d}_{34}]) = (-8,-5,-7,-14), \end{aligned}$$

and the variance covariance matrices $V_i, i = 1,2,3$ for $\overline{\mathbf{d}}_i, i = 1,2,3$ are set as follows.

$$V_1 = \begin{pmatrix} 25 & -1 & 0.8 & -2 \\ -1 & 4 & -1.2 & 1.2 \\ 0.8 & -1.2 & 4 & 2 \\ -2 & 1.2 & 2 & 9 \end{pmatrix}$$

$$V_2 = \begin{pmatrix} 16 & 1.4 & -1.2 & 1.4 \\ 1.4 & 1 & 1.5 & -0.8 \\ -1.2 & 1.5 & 25 & -0.6 \\ 1.4 & -0.8 & -0.6 & 4 \end{pmatrix}$$

$$V_3 = \begin{pmatrix} 4 & -1.9 & 1.5 & 1.8 \\ -1.9 & 25 & 0.8 & -0.4 \\ 1.5 & 0.8 & 9 & 2.5 \\ 1.8 & -0.4 & 2.5 & 36 \end{pmatrix}$$

We assume that the hypothetical decision maker sets his or her membership functions for the fuzzy goals of $\widetilde{\overline{\mathbf{c}}}_i\mathbf{x}, i = 1,2,3$ as follows (i.e., Step 1).

$$\begin{aligned} \mu_{\widetilde{G}_1}(y_1) &= \frac{y_1 - 57.0}{57.0 - 3.0} \\ \mu_{\widetilde{G}_2}(y_2) &= \frac{y_2 - 15.0}{15.0 - (-70.0)} \\ \mu_{\widetilde{G}_3}(y_3) &= \frac{y_3 - 3.0}{3.0 - (-65.0)} \end{aligned}$$

According to Assumptions 3.3 and 3.4, we also assume that the hypothetical decision maker sets his or her membership functions for $p_i(\mathbf{x}, h_i), i = 1,2,3$ as follows (i.e., Step 2).

$$\begin{aligned}
\mu_{p_1}(p_1(\mathbf{x}, h_1)) &= \frac{0.55 - \hat{p}_1}{0.55 - 0.9} \\
\mu_{\hat{p}_2}(p_2(\mathbf{x}, h_2)) &= \frac{0.6 - \hat{p}_2}{0.6 - 0.95} \\
\mu_{\hat{p}_3}(p_3(\mathbf{x}, h_3)) &= \frac{0.55 - \hat{p}_3}{0.55 - 0.88}
\end{aligned}$$

Next, we set the initial reference membership values as $(\hat{\mu}_1, \hat{\mu}_2) = (1,1)$ (i.e., Step 3). By solving MINMAX3.6($\hat{\mu}$), the corresponding MP-Pareto optimal solution is obtained as (i.e., Step 4)

$$\mu_{D_{p_i}}(p_i(\mathbf{x}^*, \hat{\mu}_i - \lambda^*)) = 0.56929, i = 1,2,3.$$

The hypothetical decision maker is not satisfied with the current value of the MP-Pareto optimal solution. To improve $\mu_{D_{p_1}}(\cdot)$ at the expense of $\mu_{D_{p_2}}(\cdot)$ and $\mu_{D_{p_3}}(\cdot)$, he or she updates his or her reference membership values as (i.e., Step 5)

$$(\hat{\mu}_1, \hat{\mu}_2, \hat{\mu}_3) = (0.65, 0.55, 0.53).$$

Then, the corresponding MP-Pareto optimal solution is obtained as (i.e., Step 4)

$$\begin{aligned}
& (\mu_{D_{p_1}}(p_1(\mathbf{x}^*, \hat{\mu}_1 - \lambda^*)), \mu_{D_{p_2}}(p_2(\mathbf{x}^*, \hat{\mu}_2 - \lambda^*)), \mu_{D_{p_3}}(p_3(\mathbf{x}^*, \hat{\mu}_3 - \lambda^*))) \\
= \; & (0.62594, 0.52594, 0.50594).
\end{aligned}$$

For the current value of the MP-Pareto optimal solution, the hypothetical decision maker updates his or her reference membership values as

$$(\hat{\mu}_1, \hat{\mu}_2, \hat{\mu}_3) = (0.625, 0.526, 0.515),$$

to improve $\mu_{D_{p_3}}(\cdot)$ at the expense (albeit small) of $\mu_{D_{p_1}}(\cdot)$ (i.e., Step 5). The corresponding MP-Pareto optimal solution is then obtained as (i.e., Step 4)

$$\begin{aligned}
& (\mu_{D_{p_1}}(p_1(\mathbf{x}^*, \hat{\mu}_1 - \lambda^*)), \mu_{D_{p_2}}(p_2(\mathbf{x}^*, \hat{\mu}_2 - \lambda^*)), \mu_{D_{p_3}}(p_3(\mathbf{x}^*, \hat{\mu}_3 - \lambda^*))) \\
= \; & (0.62342, 0.52442, 0.51342).
\end{aligned}$$

Since the hypothetical decision maker is then satisfied with the current value of the MP-Pareto optimal solution, the interactive processes stop (i.e., Step 5). The interactive processes with this hypothetical decision maker are summarized in Table 3.3. In the table, note that the proper balance between membership functions $\mu_{p_i}(p_i(\mathbf{x}^*, \hat{\mu}_i - \lambda^*))$ and $\mu_{\tilde{G}_i}(y^*)(= h_i^* = \hat{\mu}_i - \lambda^*)$ is attained at the MP-Pareto optimal solution at each iteration.

Table 3.3: Interactive processes in MOP3.29

	1	2	3
$\hat{\mu}_1$	1	0.65	0.625
$\hat{\mu}_2$	1	0.55	0.526
$\hat{\mu}_3$	1	0.53	0.515
$\mu_{D_{p_1}}(\mathbf{x}^*, \hat{\mu}_1 - \lambda^*)$	0.56929	0.62594	0.62342
$\mu_{D_{p_2}}(\mathbf{x}^*, \hat{\mu}_2 - \lambda^*)$	0.56929	0.52594	0.52442
$\mu_{D_{p_2}}(\mathbf{x}^*, \hat{\mu}_3 - \lambda^*)$	0.56929	0.50594	0.51342
$p_1(\mathbf{x}^*, \hat{\mu}_1 - \lambda^*)$	0.74925	0.76908	0.76820
$p_2(\mathbf{x}^*, \hat{\mu}_2 - \lambda^*)$	0.79925	0.78408	0.78355
$p_3(\mathbf{x}^*, \hat{\mu}_3 - \lambda^*)$	0.73787	0.71696	0.71943
$f_1^* \stackrel{\text{def}}{=} \mu_{\tilde{G}_1}^{-1}(\hat{\mu}_1 - \lambda^*)$	26.258	23.199	23.335
$f_2^* \stackrel{\text{def}}{=} \mu_{\tilde{G}_2}^{-1}(\hat{\mu}_2 - \lambda^*)$	-33.390	-29.705	-29.575
$f_3^* \stackrel{\text{def}}{=} \mu_{\tilde{G}_3}^{-1}(\hat{\mu}_3 - \lambda^*)$	-35.712	-31.404	-31.912

3.3 Interactive decision making for MOFRPs using an expectation variance model

Katagiri et al. [37, 40] formulated two MOFRPs using an expectation model and a variance model, proposing interactive decision making methods to obtain satisfactory solutions; however, in their methods, the balance between expectations and variances of the objective functions in MOFRPs was not considered. In this section, we therefore formulate MOFRPs using both an expectation model and a variance model, thus proposing an interactive convex programming algorithm [127, 128] to obtain satisfactory solutions. In Section 3.3.1, using a possibility measure [18], we transform MOFRPs into multiobjective stochastic programming problems. In Section 3.3.2, we formulate transformed multiobjective stochastic programming problems using an expectation model and a variance model. Corresponding to such models, we define an E-Pareto optimal solution and a V-Pareto optimal solution. Next, in Section 3.3.3, given the assumption that the decision maker can specify membership functions for the coefficients of variation for the degree of possibility, we formulate MOFRPs using both an expectation model and a variance model simultaneously, thus defining an EV-Pareto optimal solution. In Section 3.3.4, we propose an interactive convex programming algorithm for obtaining a satisfactory solution from an EV-Pareto optimal solution set. Finally, in Section 3.3.5, we apply an interactive algorithm to a numerical example with a hypothetical decision maker.

3.3.1 MOFRPs and a possibility measure

In this subsection, we focus on MOFRPs of the following form.

MOP 3.30

$$\min_{\mathbf{x}\in X} \tilde{\bar{C}}\mathbf{x} = (\tilde{\bar{\mathbf{c}}}_1\mathbf{x}, \cdots, \tilde{\bar{\mathbf{c}}}_k\mathbf{x}) \tag{3.123}$$

Here, $\mathbf{x} = (x_1, \cdots, x_n)^T$ is an n-dimensional decision variable column vector. X is a linear constraint set with respect to $\mathbf{x}$, and $\tilde{\bar{\mathbf{c}}}_i = (\tilde{\bar{c}}_{i1}, \cdots, \tilde{\bar{c}}_{in}), i = 1, \cdots, k$ are coefficient vectors of objective function $\tilde{\bar{\mathbf{c}}}_i\mathbf{x}$, whose elements are fuzzy random variables. Note that symbols "–" and "~" indicate randomness and fuzziness, respectively.

In this subsection, we assume that given the occurrence of each scenario $\ell_i \in \{1, \cdots, L_i\}$, $\tilde{c}_{ij\ell_i}$ is a realization of fuzzy random variable $\tilde{\bar{c}}_{ij}$, which is a fuzzy number whose membership function is defined as [77]

$$\mu_{\tilde{c}_{ij\ell_i}}(t) = \begin{cases} \max\left\{1 - \frac{d_{ij\ell_i} - t}{\alpha_{ij}}, 0\right\}, & t \le d_{ij\ell_i} \\ \max\left\{1 - \frac{t - d_{ij\ell_i}}{\beta_{ij}}, 0\right\}, & t > d_{ij\ell_i}, \end{cases} \tag{3.124}$$

where parameters $\alpha_{ij} > 0$ and $\beta_{ij} > 0$ are constants and $d_{ij\ell_i}$ varies depending on which scenario ℓ_i occurs. Moreover, we assume that scenario ℓ_i occurs with probability $p_{i\ell_i}$, where $\sum_{\ell_i=1}^{L_i} p_{i\ell_i} = 1$ for $i = 1, \cdots, k$.

By Zadeh's extension principle [18], realization $\tilde{\mathbf{c}}_{i\ell_i}\mathbf{x}$ becomes a fuzzy number characterized by membership function.

$$\mu_{\tilde{\mathbf{c}}_{i\ell_i}\mathbf{x}}(y) = \begin{cases} \max\left\{1 - \frac{\mathbf{d}_{i\ell_i}\mathbf{x} - y}{\alpha_i\mathbf{x}}, 0\right\}, & y \le \mathbf{d}_{i\ell_i}\mathbf{x} \\ \max\left\{1 - \frac{y - \mathbf{d}_{i\ell_i}\mathbf{x}}{\beta_i\mathbf{x}}, 0\right\}, & y > \mathbf{d}_{i\ell_i}\mathbf{x}, \end{cases} \tag{3.125}$$

where $\mathbf{d}_{i\ell_i} \overset{\text{def}}{=} (d_{i1\ell_i}, \cdots, d_{in\ell_i})$, $\alpha_i \overset{\text{def}}{=} (\alpha_{i1}, \cdots, \alpha_{in}) \ge \mathbf{0}$, and $\beta_i \overset{\text{def}}{=} (\beta_{i1}, \cdots, \beta_{in}) \ge \mathbf{0}$.

Considering the imprecise nature of the decision maker's judgment, it is natural to assume that the decision maker has a fuzzy goal for each objective function in MOP3.30. In this subsection, we assume that such a fuzzy goal $\tilde{G}_i$ can be quantified by eliciting the corresponding linear membership function defined as

$$\mu_{\tilde{G}_i}(y_i) = \begin{cases} 1 & y_i < z_i^1 \\ \dfrac{y_i - z_i^0}{z_i^1 - z_i^0} & z_i^1 \le y_i \le z_i^0 \\ 0 & y_i > z_i^0, \end{cases} \tag{3.126}$$

where z_i^0 represents the minimum value of an unacceptable level of the objective function, and z_i^1 represents the maximum value of a sufficiently satisfactory level of the objective function. By using the concept of possibility measure [18], the degree of possibility that objective function value $\tilde{\tilde{\mathbf{c}}}_i\mathbf{x}$ satisfies fuzzy goal $\widetilde{G}_i$ is expressed as [38]

$$\Pi_{\tilde{\tilde{\mathbf{c}}}_i\mathbf{x}}(\widetilde{G}_i) \stackrel{\text{def}}{=} \sup_y \min\{\mu_{\tilde{\tilde{\mathbf{c}}}_i\mathbf{x}}(y), \mu_{\widetilde{G}_i}(y)\}. \tag{3.127}$$

Note that if scenario ℓ_i occurs with probability $p_{i\ell_i}$ then the value of the possibility measure can be represent as

$$\Pi_{\tilde{\mathbf{c}}_{i\ell_i}\mathbf{x}}(\widetilde{G}_i) \stackrel{\text{def}}{=} \sup_y \min\{\mu_{\tilde{\mathbf{c}}_{i\ell_i}\mathbf{x}}(y), \mu_{\widetilde{G}_i}(y)\}. \tag{3.128}$$

Using the above possibility measure, MOP3.30 can be transformed into the following multiobjective stochastic programming problem.

MOP 3.31

$$\max_{\mathbf{x}\in X}(\Pi_{\tilde{\tilde{\mathbf{c}}}_1\mathbf{x}}(\widetilde{G}_1), \cdots, \Pi_{\tilde{\tilde{\mathbf{c}}}_k\mathbf{x}}(\widetilde{G}_k)) \tag{3.129}$$

3.3.2 Formulations using an expectation model and a variance model

Katagiri et al. [37, 40] formulated MOP3.30 as the multiobjective programming problems using an expectation model (E-model) and a variance model (V-model). We first explain the E-model for MOP3.30 formulated as follows.

MOP 3.32

$$\max_{\mathbf{x}\in X}(E[\Pi_{\tilde{\tilde{\mathbf{c}}}_1\mathbf{x}}(\widetilde{G}_1)], \cdots, E[\Pi_{\tilde{\tilde{\mathbf{c}}}_k\mathbf{x}}(\widetilde{G}_k)]) \tag{3.130}$$

Here, $E[\cdot]$ denotes the expectation operator. To handle MOP3.32, we next introduce an E-Pareto optimal solution.

Definition 3.5 $\mathbf{x}^* \in X$ is an E-Pareto optimal solution to MOP3.32 if and only if there does not exist another $\mathbf{x} \in X$ such that $E[\Pi_{\tilde{\tilde{\mathbf{c}}}_i\mathbf{x}}(\widetilde{G}_i)] \geq E[\Pi_{\tilde{\tilde{\mathbf{c}}}_i\mathbf{x}^*}(\widetilde{G}_i)]$, $i = 1, \cdots, k$, with strict inequality holding for at least one i.

Note that using (3.125) and (3.126), (3.128) can be represented as [77]

$$\Pi_{\tilde{\mathbf{c}}_{i\ell_i}\mathbf{x}}(\widetilde{G}_i) = \frac{\sum_{j=1}^n(\alpha_{ij} - d_{ij\ell_i})x_j + z_i^0}{\sum_{j=1}^n \alpha_{ij}x_j - z_i^1 + z_i^0}, i = 1, \cdots, k, \ell_i = \{1, \cdots, L_i\}. \tag{3.131}$$

Since the probability that scenario ℓ_i occurs is $p_{i\ell_i}$, $E[\Pi_{\tilde{\bar{\mathbf{c}}}_i\mathbf{x}}(\widetilde{G}_i)], i=1,\cdots,k$ can be computed as

$$\begin{aligned} E[\Pi_{\tilde{\bar{\mathbf{c}}}_i\mathbf{x}}(\widetilde{G}_i)] &= \sum_{\ell_i=1}^{L_i} p_{i\ell_i}\Pi_{\tilde{\bar{\mathbf{c}}}_{i\ell_i}\mathbf{x}}(\widetilde{G}_i) \\ &= \frac{\sum_{j=1}^{n}(\alpha_{ij}-\sum_{\ell_i=1}^{L_i} p_{i\ell_i}d_{ij\ell_i})x_j+z_i^0}{\sum_{j=1}^{n}\alpha_{ij}x_j-z_i^1+z_i^0} \\ &\stackrel{\text{def}}{=} Z_i^E(\mathbf{x}), i=1,\cdots,k. \end{aligned} \tag{3.132}$$

Then, MOP3.32 can be transformed into MOP3.33.

MOP 3.33

$$\max_{\mathbf{x}\in X}(Z_1^E(\mathbf{x}),\cdots,Z_k^E(\mathbf{x})) \tag{3.133}$$

Next, consider the V-model for MOP3.30. The multiobjective programming problem based on the V-model can be formulated as follows, with each variance for $\Pi_{\tilde{\bar{\mathbf{c}}}_i\mathbf{x}}(\widetilde{G}_i)$ minimized under constraints $E[\Pi_{\tilde{\bar{\mathbf{c}}}_i\mathbf{x}}(\widetilde{G}_i)] \geq \xi_i, i=1,\cdots,k$.

MOP 3.34

$$\min_{\mathbf{x}\in X}(V[\Pi_{\tilde{\bar{\mathbf{c}}}_1\mathbf{x}}(\widetilde{G}_1)],\cdots,V[\Pi_{\tilde{\bar{\mathbf{c}}}_k\mathbf{x}}(\widetilde{G}_k)]) \tag{3.134}$$

subject to

$$E[\Pi_{\tilde{\bar{\mathbf{c}}}_i\mathbf{x}}(\widetilde{G}_i)] \geq \xi_i, i=1,\cdots,k \tag{3.135}$$

Here, $V[\cdot]$ denotes the variance operator, and ξ_i represents a permissible expectation level for $E[\Pi_{\tilde{\bar{\mathbf{c}}}_i\mathbf{x}}(\widetilde{G}_i)]$. Next, we denote the feasible set of MOP3.34 as

$$X(\xi) \stackrel{\text{def}}{=} \{\mathbf{x}\in X \mid E[\Pi_{\tilde{\bar{\mathbf{c}}}_i\mathbf{x}}(\widetilde{G}_i)] \geq \xi_i, i=1,\cdots,k\}. \tag{3.136}$$

Similar to the E-model, to handle MOP3.34, we define a V-Pareto optimal solution below.

Definition 3.6 $\mathbf{x}^* \in X(\xi)$ is a V-Pareto optimal solution to MOP3.34 if and only if there does not exist another $\mathbf{x}\in X(\xi)$ such that $V[\Pi_{\tilde{\bar{\mathbf{c}}}_i\mathbf{x}}(\widetilde{G}_i)] \leq V[\Pi_{\tilde{\bar{\mathbf{c}}}_i\mathbf{x}^*}(\widetilde{G}_i)]$, $i=1,\cdots,k$, with strict inequality holding for at least one i.

Note that $V[\Pi_{\widetilde{\bar{\mathbf{c}}}_i\mathbf{x}}(\widetilde{G}_i)]$ can be represented as [77]

$$\begin{aligned} V[\Pi_{\widetilde{\bar{\mathbf{c}}}_i\mathbf{x}}(\widetilde{G}_i)] &= \frac{1}{(\sum_{j=1}^{n} \alpha_{ij}x_j - z_i^1 + z_i^0)^2} V\left[\sum_{j=1}^{n} \bar{d}_{ij}x_j\right] \\ &= \frac{1}{(\sum_{j=1}^{n} \alpha_{ij}x_j - z_i^1 + z_i^0)^2} \mathbf{x}^T \mathbf{V}_i \mathbf{x} \\ &\stackrel{\text{def}}{=} Z_i^V(\mathbf{x}), \end{aligned} \tag{3.137}$$

where $\mathbf{V}_i, i = 1, \cdots, k$ are variance covariance matrices of $\bar{\mathbf{d}}_i$ expressed by

$$\mathbf{V}_i = \begin{pmatrix} v_{11}^i & v_{12}^i & \cdots & v_{1n}^i \\ v_{21}^i & v_{22}^i & \cdots & v_{2n}^i \\ \vdots & \vdots & \ddots & \vdots \\ v_{n1}^i & v_{n2}^i & \cdots & v_{nn}^i \end{pmatrix}, i = 1, \cdots, k, \tag{3.138}$$

and

$$\begin{aligned} v_{jj}^i &= V[\bar{d}_{jj}] \\ &= \sum_{\ell_i=1}^{L_i} p_{i\ell_i} d_{ij\ell_i}^2 - \left(\sum_{\ell_i=1}^{L_i} p_{i\ell_i} d_{ij\ell_i}\right)^2, j = 1, \cdots, n, \end{aligned} \tag{3.139}$$

$$\begin{aligned} v_{jr}^i &= \mathrm{Cov}[\bar{d}_{ij}, \bar{d}_{ir}] \\ &= E[\bar{d}_{ij} \cdot \bar{d}_{ir}] - E[\bar{d}_{ij}]E[\bar{d}_{ir}] \\ &= \sum_{\ell_i=1}^{L_i} p_{i\ell_i} d_{ij\ell_i} d_{ir\ell_i} - \sum_{\ell_i=1}^{L_i} p_{i\ell_i} d_{ij\ell_i} \sum_{\ell_i=1}^{L_i} p_{i\ell_i} d_{ir\ell_i}, \\ & \quad j, r = 1, \cdots, n, j \neq r. \end{aligned} \tag{3.140}$$

Furthermore, inequalities (3.135) can be expressed as

$$\sum_{j=1}^{n} \left(\sum_{\ell_i=1}^{L_i} p_{i\ell_i} d_{ij\ell_i} - (1 - \xi_i)\alpha_{ij}\right) x_j \leq z_i^0 - \xi_i(z_i^0 - z_i^1), i = 1, \cdots, k. \tag{3.141}$$

Then, MOP3.34 can be transformed into MOP3.35.

MOP 3.35

$$\min_{\mathbf{x}\in X} (Z_1^V(\mathbf{x}), \cdots, Z_k^V(\mathbf{x})) \tag{3.142}$$

subject to

$$\sum_{j=1}^{n} \left(\sum_{\ell_i=1}^{L_i} p_{i\ell_i} d_{ij\ell_i} - (1 - \xi_i)\alpha_{ij}\right) x_j \leq z_i^0 - \xi_i(z_i^0 - z_i^1), i = 1, \cdots, k$$

From $\mathbf{x}^T\mathbf{V}_i\mathbf{x} > 0$, $\sum_{j=1}^{n}\alpha_{ij}x_j - z_i^1 + z_i^0 > 0$ and the positive-semidefinite property of V_i, MOP3.35 can be equivalently transformed as follows, with $Z_i^{SD}(\mathbf{x}), i = 1,\cdots,k$ representing standard deviations for $\Pi_{\tilde{\bar{\mathbf{c}}}_i\mathbf{x}}(\widetilde{G}_i)$.

MOP 3.36

$$\min_{\mathbf{x}\in X}(Z_1^{SD}(\mathbf{x}),\cdots,Z_k^{SD}(\mathbf{x})) \tag{3.143}$$

subject to

$$\sum_{j=1}^{n}\left(\sum_{\ell_i=1}^{L_i} p_{i\ell_i}d_{ij\ell_i} - (1-\xi_i)\alpha_{ij}\right)x_j \leq z_i^0 - \xi_i(z_i^0 - z_i^1), i = 1,\cdots,k$$

where

$$Z_i^{SD}(\mathbf{x}) \stackrel{\text{def}}{=} \frac{\sqrt{\mathbf{x}^T\mathbf{V}_i\mathbf{x}}}{\sum_{j=1}^{n}\alpha_{ij}x_j - z_i^1 + z_i^0}. \tag{3.144}$$

Note that $Z_i^E(\mathbf{x})$ and $Z_i^{SD}(\mathbf{x})$ are the statistical values for the same random function $\Pi_{\tilde{\bar{\mathbf{c}}}_i\mathbf{x}}(\widetilde{G}_i)$. When solving MOP3.30, it is natural for the decision maker to simultaneously consider both $Z_i^E(\mathbf{x})$ and $Z_i^{SD}(\mathbf{x})$ for each objective function $\Pi_{\tilde{\bar{\mathbf{c}}}_i\mathbf{x}}(\widetilde{G}_i)$ of MOP3.30, rather than considering them individually. Moreover, in practice, it is difficult for the decision maker to express his or her preference for standard deviations $Z_i^{SD}(\mathbf{x})$, $i = 1,\cdots,k$. From such a perspective, in the following subsections, we propose a hybrid model for MOP3.30 in which the E-model and the V-model are incorporated together; thus, we define the concept of EV-Pareto optimality. To derive a satisfactory solution for the decision maker from an EV-Pareto optimal solution set, we propose an interactive algorithm.

3.3.3 A formulation using an expectation variance model

In this subsection, we consider the following hybrid model for MOP3.31, where both the E-model and the V-model are incorporated together, i.e., $Z_i^E(\mathbf{x}), i = 1,\cdots,k$ are maximized and $Z_i^{SD}(\mathbf{x}), i = 1,\cdots,k$ are minimized.

MOP 3.37

$$\max_{\mathbf{x}\in X}\left(Z_1^E(\mathbf{x}),\cdots,Z_k^E(\mathbf{x}),-Z_1^{SD}(\mathbf{x}),\cdots,-Z_k^{SD}(\mathbf{x})\right)$$

In MOP3.37, $Z_i^E(\mathbf{x})$ and $Z_i^{SD}(\mathbf{x})$ represent the expected value and standard deviation of objective function $\Pi_{\tilde{\bar{\mathbf{c}}}_i\mathbf{x}}(\widetilde{G}_i)$ in MOP3.31 respectively. Note that $Z_i^E(\mathbf{x})$ can be interpreted as an expected value of satisfactory degree for $\Pi_{\tilde{\bar{\mathbf{c}}}_i\mathbf{x}}(\widetilde{G}_i)$, but $Z_i^{SD}(\mathbf{x})$ does not represent the satisfactory degree itself. Here, instead of $Z_i^{SD}(\mathbf{x})$, let us consider the coefficient of variation defined as

$$Z_i^{CV}(\mathbf{x}) \stackrel{\text{def}}{=} \frac{Z_i^{SD}(\mathbf{x})}{Z_i^E(\mathbf{x})}$$
$$= \frac{\sqrt{\mathbf{x}^T \mathbf{V}_i \mathbf{x}}}{\sum_{j=1}^n (\alpha_{ij} - \sum_{\ell_i=1}^{L_i} p_{i\ell_i} d_{ij\ell_i}) x_j + z_i^0}. \tag{3.145}$$

By using the coefficient of variation $Z_i^{CV}(\mathbf{x})$, we can transform MOP3.37 as follows.

MOP 3.38

$$\max_{\mathbf{x}\in X} \left(Z_1^E(\mathbf{x}), \cdots, Z_k^E(\mathbf{x}), -Z_1^{CV}(\mathbf{x}), \cdots, -Z_k^{CV}(\mathbf{x}) \right) \tag{3.146}$$

In MOP3.38, we assume that the decision maker has fuzzy goals for $Z_i^{CV}(\mathbf{x}), i = 1, \cdots, k$, with corresponding membership functions defined as $\mu_i^{CV}(Z_i^{CV}(\mathbf{x})), i = 1, \cdots, k$. To elicit membership function $\mu_i^{CV}(Z_i^{CV}(\mathbf{x}))$ appropriately, we can compute the minimum value and the maximum value of $\mu_i^{CV}(Z_i^{CV}(\mathbf{x}))$ as

$$CV_{i\min} \stackrel{\text{def}}{=} \min_{\mathbf{x}\in X} \frac{\sqrt{\mathbf{x}^T \mathbf{V}_i \mathbf{x}}}{\sum_{j=1}^n (\alpha_{ij} - \sum_{\ell_i=1}^{L_i} p_{i\ell_i} d_{ij\ell_i}) x_j + z_i^0} \tag{3.147}$$

$$CV_{i\max} \stackrel{\text{def}}{=} \max_{\mathbf{x}\in X} \frac{\sqrt{\mathbf{x}^T \mathbf{V}_i \mathbf{x}}}{\sum_{j=1}^n (\alpha_{ij} - \sum_{\ell_i=1}^{L_i} p_{i\ell_i} d_{ij\ell_i}) x_j + z_i^0}. \tag{3.148}$$

Problem (3.147) for $CV_{i\min}$ is easily solved by applying a Dinkelbach-type algorithm [17] or a hybrid method combining the bisection method and a convex programming technique. Unfortunately, problem (3.148) becomes a non-convex optimization problem. On interval $[CV_{i\min}, CV_{i\max}]$, the decision maker sets his or her membership function $\mu_i^{CV}(Z_i^{CV}(\mathbf{x}))$, which is strictly decreasing and continuous.

Assumption 3.6
$\mu_i^{CV}(Z_i^{CV}(\mathbf{x})), i = 1, \cdots, k$ are strictly decreasing and continuous with respect to $Z_i^{CV}(\mathbf{x})$, and $\mu_i^{CV}(CV_{i\min}) = 1$, $\mu_i^{CV}(CV_{i\max}) = 0, i = 1, \cdots, k$.

From the perspective that both $Z_i^E(\mathbf{x})$ and $\mu_i^{CV}(Z_i^{CV}(\mathbf{x}))$ indicate satisfactory degrees for objective function $\Pi_{\tilde{\bar{\mathbf{c}}}_i\mathbf{x}}(\widetilde{G}_i)$ in MOP3.31, we introduce the integrated membership function in which both satisfactory levels $Z_i^E(\mathbf{x})$ and $\mu_i^{CV}(Z_i^{CV}(\mathbf{x}))$ are combined via the fuzzy decision [6, 71, 140], i.e.,

$$\mu_{D_i}(\mathbf{x}) \stackrel{\text{def}}{=} \min\{Z_i^E(\mathbf{x}), \mu_i^{CV}(Z_i^{CV}(\mathbf{x}))\}. \tag{3.149}$$

Then, MOP3.38 can be transformed into the following multiobjective programming problem.

MOP 3.39

$$\max_{\mathbf{x}\in X} \quad (\mu_{D_1}(\mathbf{x}),\cdots,\mu_{D_k}(\mathbf{x})) \tag{3.150}$$

Here, $\mu_{D_i}(\mathbf{x})$ can be interpreted as an overall satisfactory degree for fuzzy goal $\widetilde{G}_i$. To handle MOP3.39, we define an EV-Pareto optimal solution below.

Definition 3.7 $\mathbf{x}^* \in X$ is an EV-Pareto optimal solution to MOP3.39 if and only if there does not exist another $\mathbf{x} \in X$ such that $\mu_{D_i}(\mathbf{x}) \geq \mu_{D_i}(\mathbf{x}^*)$, $i = 1,\cdots,k$, with strict inequality holding for at least one i.

To generate a candidate for a satisfactory solution from an EV-Pareto optimal solution set, the decision maker specifies reference membership values [71]. For reference membership values $\hat{\mu} = (\hat{\mu}_1,\cdots,\hat{\mu}_k)$, the corresponding EV-Pareto optimal solution is obtained by solving the following minmax problem.

MINMAX 3.9 ($\hat{\mu}$)

$$\min_{\mathbf{x}\in X,\lambda\in\Lambda} \lambda \tag{3.151}$$

subject to

$$\hat{\mu}_i - Z_i^E(\mathbf{x}) \leq \lambda, i = 1,\cdots,k \tag{3.152}$$

$$\hat{\mu}_i - \mu_i^{CV}(Z_i^{CV}(\mathbf{x})) \leq \lambda, i = 1,\cdots,k \tag{3.153}$$

where

$$\Lambda \stackrel{\text{def}}{=} \left[\max_{i=1,\cdots,k} \hat{\mu}_i - 1, \max_{i=1,\cdots,k} \hat{\mu}_i\right] = [\lambda_{\min},\lambda_{\max}]. \tag{3.154}$$

From the definition of $Z_i^E(\mathbf{x})$ and $\mu_i^{CV}(Z_i^{CV}(\mathbf{x}))$, constraints (3.152) and (3.153) can be equivalently transformed into the following respective forms.

$$\sum_{j=1}^{n}(\alpha_{ij} - \sum_{\ell_i=1}^{L_i} p_{i\ell_i} d_{ij\ell_i})x_j + z_i^0 \geq \left(\sum_{j=1}^{n} \alpha_{ij}x_j - z_i^1 + z_i^0\right)\cdot(\hat{\mu}_i - \lambda),$$
$$i = 1,\cdots,k \tag{3.155}$$

$$\left(\sum_{j=1}^{n}\left(\alpha_{ij} - \sum_{\ell_i=1}^{L_i} p_{i\ell_i} d_{ij\ell_i}\right)x_j + z_i^0\right)\cdot \mu_i^{CV^{-1}}(\hat{\mu}_i - \lambda) \geq \sqrt{\mathbf{x}^T\mathbf{V}_i\mathbf{x}},$$
$$i = 1,\cdots,k \tag{3.156}$$

The relationships between optimal solution $\mathbf{x}^* \in X, \lambda^* \in \Lambda$ of MINMAX3.9($\hat{\mu}$) and EV-Pareto optimal solutions to MOP3.39 can be characterized by the following theorems.

Theorem 3.7

If $\mathbf{x}^* \in X$, $\lambda^* \in \Lambda$ *is a unique optimal solution of MINMAX3.9(*$\hat{\mu}$*), then* $\mathbf{x}^*$ *is an EV-Pareto optimal solution to MOP3.39.*

(Proof)

Let us assume that $\mathbf{x}^* \in X$ is not an EV-Pareto optimal solution to MOP3.39. Then, there exists $\mathbf{x} \in X$ such that $\mu_{D_i}(\mathbf{x}) \geq \mu_{D_i}(\mathbf{x}^*), i = 1, \cdots, k$, with strict inequality holding for at least one i. This implies that

$$
\begin{aligned}
& \mu_{D_i}(\mathbf{x}) \geq \mu_{D_i}(\mathbf{x}^*) \\
\Leftrightarrow \quad & \hat{\mu}_i - \min\{Z_i^E(\mathbf{x}), \mu_i^{CV}(Z_i^{CV}(\mathbf{x}))\} \leq \hat{\mu}_i - \min\{Z_i^E(\mathbf{x}^*), \mu_i^{CV}(Z_i^{CV}(\mathbf{x}^*))\} \\
\Leftrightarrow \quad & \max\{\hat{\mu}_i - Z_i^E(\mathbf{x}), \hat{\mu}_i - \mu_i^{CV}(Z_i^{CV}(\mathbf{x}))\} \\
& \leq \max\{\hat{\mu}_i - Z_i^E(\mathbf{x}^*), \hat{\mu}_i - \mu_i^{CV}(Z_i^{CV}(\mathbf{x}^*))\} \leq \lambda^*, i = 1, \cdots, k.
\end{aligned}
$$

The above contradicts the assumption that $\mathbf{x}^* \in X$, $\lambda^* \in \Lambda$ is a unique optimal solution of MINMAX3.9($\hat{\mu}$). □

Theorem 3.8

If $\mathbf{x}^* \in X$ *is an EV-Pareto optimal solution to MOP3.39, then there exist reference membership values* $\hat{\mu} = (\hat{\mu}_1, \cdots, \hat{\mu}_k)$ *such that* $\mathbf{x}^* \in X$, $\lambda^* \stackrel{\text{def}}{=} \hat{\mu}_i - \mu_{D_i}(\mathbf{x}^*), i = 1, \cdots, k$ *is an optimal solution of MINMAX3.9(*$\hat{\mu}$*).*

(Proof)

Let us assume that $\mathbf{x}^* \in X$, $\lambda^* = \hat{\mu}_i - \mu_{D_i}(\mathbf{x}^*) = \max\{\hat{\mu}_i - Z_i^E(\mathbf{x}^*), \hat{\mu}_i - \mu_i^{CV}(Z_i^{CV}(\mathbf{x}^*))\}$, $i = 1, \cdots, k$, is not an optimal solution of MINMAX3.9($\hat{\mu}$) for any reference membership values $\hat{\mu} = (\hat{\mu}_1, \cdots, \hat{\mu}_k)$. Then, there exists $\mathbf{x} \in X$ and $\lambda < \lambda^*$ such that

$$
\begin{aligned}
& \begin{cases} \hat{\mu}_i - Z_i^E(\mathbf{x}) \leq \lambda < \lambda^* \\ \hat{\mu}_i - \mu_i^{CV}(Z_i^{CV}(\mathbf{x})) \leq \lambda < \lambda^* \end{cases} \\
\Leftrightarrow \quad & \hat{\mu}_i - \mu_{D_i}(\mathbf{x}) \leq \lambda < \lambda^* \\
\Leftrightarrow \quad & \hat{\mu}_i - \mu_{D_i}(\mathbf{x}) < \hat{\mu}_i - \mu_{D_i}(\mathbf{x}^*) \\
\Leftrightarrow \quad & \mu_{D_i}(\mathbf{x}) > \mu_{D_i}(\mathbf{x}^*), i = 1, \cdots, k, i = 1, \cdots, k.
\end{aligned}
$$

The above contradicts the fact that $\mathbf{x}^* \in X$ is an EV-Pareto optimal solution to MOP3.39. □

Unfortunately, since MINMAX3.9($\hat{\mu}$) is a nonlinear programming problem, it is difficult to solve it directly. To overcome such difficulties, we consider the following function with respect to constraints (3.155) and (3.156).

$$g_i(\mathbf{x},\lambda) \stackrel{\text{def}}{=} \sqrt{\mathbf{x}^T\mathbf{V}_i\mathbf{x}} - {\mu_i^{CV}}^{-1}(\hat{\mu}_i-\lambda)\cdot\left(\sum_{j=1}^{n}\left(\alpha_{ij}-\sum_{\ell_i=1}^{L_i}p_{i\ell_i}d_{ij\ell_i}\right)x_j+z_i^0\right),$$
$$i=1,\cdots,k \tag{3.157}$$

$$h_i(\mathbf{x},\lambda) \stackrel{\text{def}}{=} \left(\sum_{j=1}^{n}\alpha_{ij}x_j - z_i^1 + z_i^0\right)\cdot(\hat{\mu}_i-\lambda)-\sum_{j=1}^{n}(\alpha_{ij}-\sum_{\ell_i=1}^{L_i}p_{i\ell_i}d_{ij\ell_i})x_j - z_i^0,$$
$$i=1,\cdots,k \tag{3.158}$$

Note that $g_i(\mathbf{x},\lambda)$, $h_i(\mathbf{x},\lambda)$, $i=1,\cdots,k$ are convex with respect to $\mathbf{x}\in X$ for any fixed $\lambda\in\Lambda$. Let us define the following feasible set $G(\lambda)$ for some fixed $\lambda\in\Lambda$.

$$G(\lambda) \stackrel{\text{def}}{=} \{\mathbf{x}\in X \mid g_i(\mathbf{x},\lambda)\le 0, h_i(\mathbf{x},\lambda)\le 0, i=1,\cdots,k\} \tag{3.159}$$

Then, it is clear that $G(\lambda)$ is a convex set and satisfies the following property.

Property 3.2
If $\lambda_1,\lambda_2\in\Lambda, \lambda_1\le\lambda_2$, then it holds that $G(\lambda_1)\subset G(\lambda_2)$.

In what follows, we assume that $G(\lambda_{\min})=\phi$ and $G(\lambda_{\max})\ne\phi$. From Property 3.2, we obtain optimal solution $(\mathbf{x}^*,\lambda^*)$ of MINMAX3.9($\hat{\mu}$) using the following algorithm, which is based on the bisection method and a convex programming technique.

Algorithm 3.5

Step 1: *Set* $\lambda_0\leftarrow\lambda_{\min}$, $\lambda_1\leftarrow\lambda_{\max}$, $\lambda\leftarrow(\lambda_0+\lambda_1)/2$.
Step 2: *Solve the convex programming problem for fixed* $\lambda\in\Lambda$.

$$\min_{\mathbf{x}\in X} h_j(\mathbf{x},\lambda)$$

subject to

$$\begin{aligned} g_i(\mathbf{x},\lambda) &\le 0, i=1,\cdots,k \\ h_i(\mathbf{x},\lambda) &\le 0, i=1,\cdots,k \end{aligned}$$

where index j is one of $\{1,2,\cdots,k\}$, *and we denote the optimal solution as* $\mathbf{x}(\lambda)\in X$.
Step 3: *If* $|\lambda_0-\lambda_1|<\delta$, *go to Step 4, where* δ *is a sufficiently small positive number. If* $g_i(\mathbf{x}(\lambda),\lambda)\le 0$ *and* $h_i(\mathbf{x}(\lambda),\lambda)\le 0$, *for any* $i=1,\cdots,k$, *set* $\lambda_1\leftarrow\lambda$, $\lambda\leftarrow(\lambda_0+\lambda_1)/2$. *Otherwise, set* $\lambda_0\leftarrow\lambda$, $\lambda\leftarrow(\lambda_0+\lambda_1)/2$, *and return to Step 2.*
Step 4: *Adopt* $\mathbf{x}^*\leftarrow\mathbf{x}(\lambda)$, $\lambda^*\leftarrow\lambda$ *as an optimal solution of MINMAX3.9(*$\hat{\mu}$*).*

3.3.4 An interactive convex programming algorithm

In Theorem 3.7, if optimal solution $\mathbf{x}^* \in X$, $\lambda^* \in \Lambda$ of MINMAX3.9($\hat{\mu}$) is not unique, EV-Pareto optimality cannot be guaranteed. To guarantee EV-Pareto optimality for $(\mathbf{x}^*, \lambda^*)$, we must formulate an EV-Pareto optimality test problem. Before formulating such a test problem, without loss of generality, we assume the following inequalities hold at optimal solution $\mathbf{x}^* \in X, \lambda^* \in \Lambda$.

$$Z_i^E(\mathbf{x}^*) \leq \mu_i^{CV}(Z_i^{CV}(\mathbf{x}^*)), i \in I_1 \tag{3.160}$$

$$Z_i^E(\mathbf{x}^*) > \mu_i^{CV}(Z_i^{CV}(\mathbf{x}^*)), i \in I_2 \tag{3.161}$$

$$I_1 \cup I_2 = \{1, \cdots, k\}, I_1 \cap I_2 \neq \phi \tag{3.162}$$

Given the above conditions, we formulate the following EV-Pareto optimality test problem.

Test 3.3

$$\max_{\mathbf{x} \in X, \varepsilon_i \geq 0, i=1,\cdots,k} \sum_{i=1}^{k} \varepsilon_i$$

subject to

$$\begin{aligned}
&Z_i^E(\mathbf{x}) \geq Z_i^E(\mathbf{x}^*) + \varepsilon_i, i \in I_1 \\
&\mu_i^{CV}(Z_i^{CV}(\mathbf{x})) \geq Z_i^E(\mathbf{x}^*) + \varepsilon_i, i \in I_1 \\
&Z_i^E(\mathbf{x}) \geq \mu_i^{CV}(Z_i^{CV}(\mathbf{x}^*)) + \varepsilon_i, i \in I_2 \\
&\mu_i^{CV}(Z_i^{CV}(\mathbf{x})) \geq \mu_i^{CV}(Z_i^{CV}(\mathbf{x}^*)) + \varepsilon_i, \in I_2
\end{aligned}$$

The following theorem shows the relationships between the optimal solution of TEST3.3 and the EV-Pareto optimal solution to MOP3.39.

Theorem 3.9
Let $\check{\mathbf{x}} \in X$, $\check{\varepsilon}_i \geq 0$, $i = 1, \cdots, k$ be an optimal solution of TEST3.3 for optimal solution $\mathbf{x}^* \in X$, $\lambda^* \in \Lambda$ of MINMAX3.9($\hat{\mu}$). If $\sum_{i=1}^{k} \check{\varepsilon}_i = 0$, then $\mathbf{x}^* \in X$ is an EV-Pareto optimal solution to MOP3.39.

(Proof)
If $\mathbf{x}^* \in X$ is not an EV-Pareto optimal solution to MOP3.39, there exists some $\mathbf{x} \in X$ such that $\mu_{D_i}(\mathbf{x}) \geq \mu_{D_i}(\mathbf{x}^*), i = 1, \cdots, k$, with strict inequality holding for at least one i. From inequalities (3.160) and (3.161), this is equivalent to

$$\begin{aligned}
\min\{Z_i^E(\mathbf{x}), \mu_i^{CV}(Z_i^{CV}(\mathbf{x}))\} &\geq \min\{Z_i^E(\mathbf{x}^*), \mu_i^{CV}(Z_i^{CV}(\mathbf{x}^*))\} \\
&= \begin{cases} Z_i^E(\mathbf{x}^*), & i \in I_1 \\ \mu_i^{CV}(Z_i^{CV}(\mathbf{x}^*)), & i \in I_2. \end{cases}
\end{aligned}$$

As a result, the following four inequalities hold.

$$\begin{cases} Z_i^E(\mathbf{x}) \geq Z_i^E(\mathbf{x}^*), & i \in I_1 \\ \mu_i^{CV}(Z_i^{CV}(\mathbf{x})) \geq Z_i^E(\mathbf{x}^*), & i \in I_1 \\ Z_i^E(\mathbf{x}) \geq \mu_i^{CV}(Z_i^{CV}(\mathbf{x}^*)), & i \in I_2 \\ \mu_i^{CV}(Z_i^{CV}(\mathbf{x})) \geq \mu_i^{CV}(Z_i^{CV}(\mathbf{x}^*)), & i \in I_2, \end{cases} \tag{3.163}$$

with strict inequality holding for at least one $i \in I_1 \cup I_2$. Hence, there must exist at least one i such that $\check{\varepsilon}_i > 0$. This contradicts the assumption that $\check{\varepsilon}_i = 0$, $i = 1, \cdots, k$. □

Given the above, we can now construct an interactive algorithm for deriving a satisfactory solution from an EV-Pareto optimal solution set.

Algorithm 3.6

Step 1: *The decision maker sets membership function* $\mu_{\tilde{G}_i}(y)$, $i = 1, \cdots, k$ *for the fuzzy goals of the objective functions in MOP3.30.*
Step 2: *Considering intervals* $[CV_{i\min}, CV_{i\max}], i = 1, \cdots, k$, *the decision maker sets membership functions* $\mu_i^{CV}(Z_i^{CV}(\mathbf{x}))$, $i = 1, \cdots, k$ *according Assumption 3.6.*
Step 3: *Set initial reference membership values as* $\hat{\mu}_i = 1$, $i = 1, \cdots, k$.
Step 4: *Solve MINMAX3.9(*$\hat{\mu}$*) by applying Algorithm 3.5, and obtain optimal solution* $\mathbf{x}^* \in X, \lambda^* \in \Lambda$. *To guarantee EV-Pareto optimality for* $\mathbf{x}^* \in X$, *solve TEST3.3.*
Step 5: *If the decision maker is satisfied with the current value of the EV-Pareto optimal solution* $\mathbf{x}^* \in X$, *then stop. Otherwise, the decision maker updates his or her reference membership values* $\hat{\mu}_i$, $i = 1, \cdots, k$ *and returns to Step 4.*

3.3.5 A numerical example

To demonstrate the efficiency of Algorithm 3.6 and the corresponding interactive processes via a hypothetical decision maker, we consider the following three objective programming problem with fuzzy random variable coefficients [77].

MOP 3.40

$$\begin{aligned} \min \tilde{\bar{\mathbf{c}}}_1\mathbf{x} &= \tilde{\bar{c}}_{11}x_1 + \tilde{\bar{c}}_{12}x_2 + \tilde{\bar{c}}_{13}x_3 \\ \min \tilde{\bar{\mathbf{c}}}_2\mathbf{x} &= \tilde{\bar{c}}_{21}x_1 + \tilde{\bar{c}}_{22}x_2 + \tilde{\bar{c}}_{23}x_3 \\ \min \tilde{\bar{\mathbf{c}}}_3\mathbf{x} &= \tilde{\bar{c}}_{31}x_1 + \tilde{\bar{c}}_{32}x_2 + \tilde{\bar{c}}_{33}x_3 \end{aligned}$$

subject to

$$\begin{aligned} \mathbf{x} \in X \quad = \quad & \{(x_1, x_2, x_3) \geq \mathbf{0} \mid 3x_1 + 2x_2 + x_3 \leq 18, 2x_1 + x_2 + 2x_3 \leq 13, \\ & 3x_1 + 4x_2 + 3x_3 \geq 15, x_1 + 3x_2 + 2x_3 \leq 17\} \end{aligned}$$

Table 3.4: Parameters of fuzzy random variables in MOP3.40

scenario	$\ell_1=1$	$\ell_1=2$	$\ell_1=3$	α_{1j},β_{1j}
$d_{11\ell_1}$	–2.5	–2.0	–1.5	0.4
$d_{12\ell_1}$	–3.5	–3.0	–2.5	0.5
$d_{13\ell_1}$	–2.25	–2.0	–1.75	0.4
$p_{1\ell_1}$	0.25	0.4	0.35	
scenario	$\ell_2=1$	$\ell_2=2$	$\ell_2=3$	α_{2j},β_{2j}
$d_{21\ell_2}$	–2.5	–2.0	–1.5	0.3
$d_{22\ell_2}$	–0.75	–0.5	–0.25	0.4
$d_{23\ell_2}$	–2.5	–2.25	–2.0	0.3
$p_{2\ell_2}$	0.3	0.5	0.2	
scenario	$\ell_3=1$	$\ell_3=2$	$\ell_3=3$	α_{3j},β_{3j}
$d_{31\ell_3}$	3.0	3.25	3.5	0.4
$d_{32\ell_3}$	2.5	2.75	3.0	0.5
$d_{33\ell_3}$	4.5	4.75	5.0	0.4
$p_{3\ell_3}$	0.2	0.45	0.35	

We assume that realization $\tilde{c}_{ij\ell_i}$ of fuzzy random variable $\tilde{\bar{c}}_{ij}$ in MOP3.40 is an LR fuzzy number whose membership function is defined as

$$\mu_{\tilde{c}_{ij\ell_i}}(t) = \begin{cases} \max\left\{1 - \frac{d_{ij\ell_i}-t}{\alpha_{ij}}, 0\right\}, & t \le d_{ij\ell_i} \\ \max\left\{1 - \frac{t-d_{ij\ell_i}}{\beta_{ij}}, 0\right\}, & t > d_{ij\ell_i}, \end{cases}$$

where parameters $d_{ij\ell_i}$, α_{ij}, and β_{ij} are given in Table 3.4, and $p_{i\ell_i}$ represents the probability that scenario ℓ_i for discrete-type random variable $d_{ij\ell_i}$ occurs. According to (3.139) and (3.140), variance covariance matrices $V_i, i = 1,2,3$ are computed as

$$V_1 = \begin{pmatrix} 0.1475 & 0.1475 & 0.07375 \\ 0.1475 & 0.1475 & 0.07375 \\ 0.07375 & 0.07375 & 0.036875 \end{pmatrix}$$

$$V_2 = \begin{pmatrix} 0.1225 & 0.06125 & 0.06125 \\ 0.06125 & 0.030625 & 0.030625 \\ 0.06125 & 0.030625 & 0.030625 \end{pmatrix}$$

$$V_3 = \begin{pmatrix} 0.032969 & 0.032969 & 0.032969 \\ 0.032969 & 0.032969 & 0.032969 \\ 0.032969 & 0.032969 & 0.032969 \end{pmatrix}.$$

Let us assume that a hypothetical decision maker sets the membership func-

Table 3.5: Parameters of the membership functions in MOP3.40

	z_i^0	z_i^1	q_i^0	q_i^1
$i=1$	–7.5	–23.8181	0.45	0.25
$i=2$	–0.9375	–16.25	0.25	0.08
$i=3$	33.5	9.375	0.3	0.03

Table 3.6: Interactive processes in MOP3.40

Iteration	1	2	3
$\hat{\mu}_1$	1	0.85	0.85
$\hat{\mu}_2$	1	1	1
$\hat{\mu}_3$	1	0.85	0.75
x_1^*	0.5683	1.0762	0.8326
x_2^*	3.1269	2.1054	2.1693
x_3^*	2.3510	2.8859	3.3730
$Z_1^E(\mathbf{x}^*)$	0.5358	0.4825	0.5214
$Z_2^E(\mathbf{x}^*)$	0.5358	0.6325	0.6714
$Z_3^E(\mathbf{x}^*)$	0.5358	0.4825	0.4214
$\mu_1^{CV}(Z_1^{CV}(\mathbf{x}^*))$	0.5358	0.5255	0.6209
$\mu_2^{CV}(Z_2^{CV}(\mathbf{x}^*))$	0.6270	0.6325	0.6714
$\mu_3^{CV}(Z_3^{CV}(\mathbf{x}^*))$	0.8009	0.7630	0.6732

tions $\mu_{\tilde{G}_i}(\cdot)$, $\mu_i^{CV}(\cdot)$, $i=1,2,3$ as follows.

$$\begin{aligned}\mu_{\tilde{G}_i}(y) &= \frac{y-z_i^0}{z_i^1-z_i^0},\ z_i^1 \le y \le z_i^0, i=1,2,3\\ \mu_i^{CV}(s) &= \frac{s-q_i^0}{q_i^1-q_i^0},\ q_i^1 \le s \le q_i^0, i=1,2,3,\end{aligned}$$

where parameters z_i^0, z_i^1, q_i^0, and q_i^1 are given in Table 3.5. The interactive processes given the hypothetical decision maker are summarized in Table 3.6.

3.4 Interactive decision making for MOFRPs with simple recourse

Two-stage programming methods have been applied to various types of water resource allocation problems with random inflows [96, 104]; however, if the probability density functions of random variables are unknown or the problem is large scale with random variables, it may be extremely difficult to solve the corresponding two-stage programming problem. From such a perspective, inexact two-stage programming methods have been proposed [27, 55].

As an extension of two-stage programming methods for multiobjective programming problems, Sakawa et al. [75] proposed an interactive fuzzy decision making method for multiobjective stochastic programming problems with simple recourse; however, in real-world decision making situations, it seems natural to consider that uncertainty is expressed not only by fuzziness but also by randomness. From such a perspective, interactive decision making methods for MOFRPs have been proposed [38, 39] in which chance constraint methods and a possibility measure [18] are applied to handle fuzzy random variable coefficients [46, 65, 77].

In this section, we focus on multiobjective fuzzy random simple recourse programming problems in which the coefficients of equality constraints are defined by fuzzy random variables; we propose an interactive decision making method [135] for obtaining a satisfactory solution from a Pareto optimal solution set. In Section 3.4.1, using a possibility measure [18] and a two-stage programming method, we transform multiobjective programming problems with fuzzy random variable coefficients into multiobjective fuzzy random simple recourse programming problems; further, we introduce γ-Pareto optimality in which γ is a permissible possibility level specified by the decision maker for fuzzy random variable coefficients. To obtain a candidate for a satisfactory solution from a γ-Pareto optimal solution set, we also develop an interactive algorithm. In Section 3.4.2, we further consider a generalized multiobjective fuzzy random simple recourse programming problem in which the coefficients of equality constraints are fuzzy random variables and the coefficients of objective functions are random variables. To handle such a problem, we simultaneously apply both a two-stage programming method and a chance constraint method.

3.4.1 An interactive algorithm for MOFRPs with simple recourse

In this subsection, we focus on multiobjective programming problems involving fuzzy random variable coefficients on the right-hand-side of equality constraints, described as follows.

MOP 3.41

$$\min_{\mathbf{x}\in X} (\mathbf{c}_1\mathbf{x}, \cdots, \mathbf{c}_k\mathbf{x}) \tag{3.164}$$

subject to

$$A\mathbf{x} = \tilde{\bar{\mathbf{d}}} \tag{3.165}$$

Here, $\mathbf{c}_\ell = (c_{\ell 1}, \cdots, c_{\ell n}), \ell = 1, \cdots, k$ are n-dimensional coefficient row vectors of the objective function, $\mathbf{x} = (x_1, \cdots, x_n)^T \geq \mathbf{0}$ is an n-dimensional decision variable column vector, X is a linear constraint set with respect to $\mathbf{x}$, A is an $(m \times n)$-

dimensional coefficient matrix, and $\overline{\tilde{\tilde{\mathbf{d}}}} = (\overline{\tilde{\tilde{d}}}_1, \cdots, \overline{\tilde{\tilde{d}}}_m)^T$ is an m-dimensional coefficient column vector whose elements are fuzzy random variables [46].

To handle fuzzy random variables efficiently, Katagiri et al. [38, 39] defined an LR-type fuzzy random variable, which is a special type of a fuzzy random variable. Given the occurrence of each elementary event ω, $\overline{\tilde{\tilde{d}}}_i(\omega)$ is a realization of an LR-type fuzzy random variable $\overline{\tilde{\tilde{d}}}_i$, which is an LR fuzzy number [18] whose membership function is defined as

$$\mu_{\overline{\tilde{\tilde{d}}}_i(\omega)}(s) = \begin{cases} L\left(\frac{b_i(\omega)-s}{\alpha_i}\right), & s \leq b_i(\omega) \\ R\left(\frac{s-b_i(\omega)}{\beta_i}\right), & s > b_i(\omega), \end{cases} \tag{3.166}$$

where function $L(t) \stackrel{\text{def}}{=} \max\{0, l(t)\}$ is a real-valued continuous function from $[0, \infty)$ to $[0, 1]$ and $l(t)$ is a strictly decreasing continuous function satisfying $l(0) = 1$. Further, $R(t) \stackrel{\text{def}}{=} \max\{0, r(t)\}$ satisfies the same conditions. Here, $\alpha_i (> 0)$ and $\beta_i (> 0)$ are called left and right spreads [18], respectively. Mean value $\overline{b}_i$ is a random variable, whose probability density function and corresponding distribution function are defined as $f_i(\cdot)$ and $F_i(\cdot)$, respectively. We assume that random variables $\overline{b}_i, i = 1, \cdots, m$ are independent of one another.

Since it is difficult to work with MOP3.41 directly, we introduce permissible possibility level $\gamma (0 < \gamma \leq 1)$ based on the concept of a possibility measure [18] for equality constraints (3.165), i.e.,

$$\text{Pos}(\mathbf{a}_i\mathbf{x} = \overline{\tilde{\tilde{\mathbf{d}}}}_i(\omega)) \geq \gamma, i = 1, \cdots, m, \tag{3.167}$$

where $\mathbf{a}_i = (a_{i1}, \cdots, a_{in}), i = 1, \cdots, m$ are n-dimensional row vectors of A. From the property of LR fuzzy numbers, the i-th inequality condition of (3.167) can be transformed into

$$b_i(\omega) - L^{-1}(\gamma)\alpha_i \leq \mathbf{a}_i\mathbf{x} \leq b_i(\omega) + R^{-1}(\gamma)\beta_i. \tag{3.168}$$

For the above two inequalities (3.168), we introduce two vectors

$$\mathbf{y}^+ = (y_1^+, \cdots, y_m^+)^T \geq \mathbf{0}, \mathbf{y}^- = (y_1^-, \cdots, y_m^-)^T \geq \mathbf{0},$$

where (y_i^+, y_i^-) represent the shortage and the excess for interval (3.168), respectively, and the following relations hold [135].

(1) For $b_i(\omega) - L^{-1}(\gamma)\alpha_i > \mathbf{a}_i\mathbf{x}$, it holds that $y_i^+ = b_i(\omega) - L^{-1}(\gamma)\alpha_i - \mathbf{a}_i\mathbf{x} > 0, y_i^- = 0$.

(2) For $b_i(\omega) + R^{-1}(\gamma)\beta_i < \mathbf{a}_i\mathbf{x}$, it holds that $y_i^+ = 0, y_i^- = \mathbf{a}_i\mathbf{x} - (b_i(\omega) + R^{-1}(\gamma)\beta_i) > 0$.

(3) For $b_i(\omega) - L^{-1}(\gamma)\alpha_i \leq \mathbf{a}_i\mathbf{x} \leq b_i(\omega) + R^{-1}(\gamma)\beta_i$, it holds that $y_i^+ = 0, y_i^- = 0$.

From such relationships with respect to $(y_i^+, y_i^-), i = 1, \cdots, m$, we formulate a multiobjective fuzzy random simple recourse programming problem for MOP3.41 as follows, depending on permissible possibility level $\gamma (0 < \gamma \leq 1)$ specified by the decision maker.

MOP 3.42 (γ)

$$\begin{array}{l}\min\limits_{\mathbf{x}\in X}\ \mathbf{c}_1\mathbf{x}+E\left[\min\limits_{\mathbf{y}^+,\mathbf{y}^-}\left(\mathbf{q}_1^+\mathbf{y}^+ +\mathbf{q}_1^-\mathbf{y}^-\right)\right]\\ \cdots\cdots\cdots\cdots\cdots\cdots\\ \min\limits_{\mathbf{x}\in X}\ \mathbf{c}_k\mathbf{x}+E\left[\min\limits_{\mathbf{y}^+,\mathbf{y}^-}\left(\mathbf{q}_k^+\mathbf{y}^+ +\mathbf{q}_k^-\mathbf{y}^-\right)\right]\end{array}$$

subject to

$$\begin{array}{l}\mathbf{a}_i\mathbf{x}+y_i^+\geq b_i(\omega)-L^{-1}(\gamma)\alpha_i, i=1,\cdots,m\\ \mathbf{a}_i\mathbf{x}-y_i^-\leq b_i(\omega)+R^{-1}(\gamma)\beta_i, i=1,\cdots,m\\ \mathbf{x}\in X,\mathbf{y}^+\geq\mathbf{0},\mathbf{y}^-\geq\mathbf{0}\end{array}$$

where

$$\mathbf{q}_\ell^+ = (q_{\ell 1}^+,\cdots,q_{\ell m}^+)\geq\mathbf{0},\ell=1,\cdots,k \tag{3.169}$$

$$\mathbf{q}_\ell^- = (q_{\ell 1}^-,\cdots,q_{\ell m}^-)\geq\mathbf{0},\ell=1,\cdots,k, \tag{3.170}$$

are m-dimensional weighting row vectors for $\mathbf{y}^+$ and $\mathbf{y}^-$ respectively. For the ℓ-th objective function in MOP3.42(γ), the second term can be transformed into

$$\begin{aligned}&E\left[\min_{\mathbf{y}^+,\mathbf{y}^-}\left(\mathbf{q}_\ell^+\mathbf{y}^+ +\mathbf{q}_\ell^-\mathbf{y}^-\right)\right]\\ =\ &\sum_{i=1}^m q_{\ell i}^+\left(E[\bar{b}_i]-\mathbf{a}_i\mathbf{x}-L^{-1}(\gamma)\alpha_i\right)\\ &+\sum_{i=1}^m q_{\ell i}^+\left\{(\mathbf{a}_i\mathbf{x}+L^{-1}(\gamma)\alpha_i)F_i(\mathbf{a}_i\mathbf{x}+L^{-1}(\gamma)\alpha_i)\right.\\ &\left.-\int_{-\infty}^{\mathbf{a}_i\mathbf{x}+L^{-1}(\gamma)\alpha_i} b_i f_i(b_i)db_i\right\}\\ &+\sum_{i=1}^m q_{\ell i}^-\left\{(\mathbf{a}_i\mathbf{x}-R^{-1}(\gamma)\beta_i)F_i(\mathbf{a}_i\mathbf{x}-R^{-1}(\gamma)\beta_i)\right.\\ &\left.-\int_{-\infty}^{\mathbf{a}_i\mathbf{x}-R^{-1}(\gamma)\beta_i} b_i f_i(b_i)db_i\right\}\\ \stackrel{\text{def}}{=}\ &d_\ell(\mathbf{x},\gamma).\end{aligned} \tag{3.171}$$

In the following, we define the objective functions in MOP3.42(γ) as

$$z_\ell(\mathbf{x},\gamma)\stackrel{\text{def}}{=}\mathbf{c}_\ell\mathbf{x}+d_\ell(\mathbf{x},\gamma),\ell=1,\cdots,k. \tag{3.172}$$

Then, MOP3.42(γ) can be reduced to a multiobjective programming problem, in which permissible possibility level γ is a parameter specified by the decision maker.

MOP 3.43 (γ)

$$\min_{\mathbf{x}\in X} \; (z_1(\mathbf{x},\gamma),\cdots,z_k(\mathbf{x},\gamma)) \tag{3.173}$$

Then, we define the concept of a Pareto optimal solution for MOP3.43(γ).

Definition 3.8 $\mathbf{x}^* \in X$ is a γ-Pareto optimal solution to MOP3.43(γ) if and only if there does not exist another $\mathbf{x} \in X$ such that $z_\ell(\mathbf{x},\gamma) \leq z_\ell(\mathbf{x}^*,\gamma), \ell = 1,\cdots,k$, with strict inequality holding for at least one ℓ.

To generate a candidate for a satisfactory solution that is also a γ-Pareto optimal solution, the decision maker specifies permissible possibility level γ and reference objective values $\hat{z}_\ell, \ell = 1,\cdots,k$ [107, 108]. Once permissible possibility level γ and reference objective values $\hat{z}_\ell, \ell = 1,\cdots,k$ are specified, the corresponding γ-Pareto optimal solution, which is in a sense close to the decision maker's requirements (or better if reference objective values are attainable), is obtained by solving the following minmax problem.

MINMAX 3.10 ($\hat{\mathbf{z}},\gamma$)

$$\min_{\mathbf{x}\in X,\lambda\in \mathrm{R}^1} \; \lambda \tag{3.174}$$

subject to

$$z_\ell(\mathbf{x},\gamma) - \hat{z}_\ell \leq \lambda, \ell = 1,\cdots,k \tag{3.175}$$

The relationships between optimal solution $\mathbf{x}^* \in X, \lambda^* \in \mathrm{R}^1$ of MINMAX3.10 ($\hat{\mathbf{z}},\gamma$) and γ-Pareto optimal solutions to MOP3.43(γ) are characterized by the following theorem.

Theorem 3.10

(1) If $\mathbf{x}^* \in X, \lambda^* \in \mathrm{R}^1$ is a unique optimal solution of MINMAX3.10($\hat{\mathbf{z}},\gamma$), then $\mathbf{x}^* \in X$ is a γ-Pareto optimal solution to MOP3.43(γ).

(2) If $\mathbf{x}^* \in X$ is a γ-Pareto optimal solution to MOP3.43(γ), then $\mathbf{x}^* \in X$ $\lambda^* \stackrel{\text{def}}{=} z_\ell(\mathbf{x}^*,\gamma) - \hat{z}_\ell, \ell = 1,\cdots,k$ is an optimal solution of MINMAX3.10($\hat{\mathbf{z}},\gamma$) for some reference objective values $\hat{\mathbf{z}} = (\hat{z}_1,\cdots,\hat{z}_k)$.

(Proof)
(1) Assume that $\mathbf{x}^* \in X$ is not a γ-Pareto optimal solution to MOP3.43(γ). Then, there exists $\mathbf{x} \in X$ such that $z_\ell(\mathbf{x},\gamma) \leq z_\ell(\mathbf{x}^*,\gamma), \ell = 1,\cdots,k$ with strict inequality holding for at least one ℓ, meaning that $z_\ell(\mathbf{x},\gamma) - \hat{z}_\ell \leq z_\ell(\mathbf{x}^*,\gamma) - \hat{z}_\ell \leq \lambda^*, \ell = 1,\cdots,k$, which contradicts the fact that $\mathbf{x}^* \in X$ is a unique optimal solution to MINMAX3.10($\hat{\mathbf{z}},\gamma$).

(2) Assume that $\mathbf{x}^* \in X, \lambda^* \in \mathrm{R}^1$ is not an optimal solution of MINMAX3.10($\hat{\mathbf{z}},\gamma$) for any reference objective values $\hat{\mathbf{z}} = (\hat{z}_1, \cdots, \hat{z}_k)$, that satisfy equalities $\lambda^* = z_\ell(\mathbf{x}^*, \gamma) - \hat{z}_\ell, \ell = 1, \cdots, k$. Then, there exists some $\mathbf{x} \in X, \lambda < \lambda^*$ such that $z_\ell(\mathbf{x}, \gamma) - \hat{z}_\ell \leq \lambda, \ell = 1, \cdots, k$, meaning that $z_\ell(\mathbf{x}, \gamma) < z_\ell(\mathbf{x}^*, \gamma), \ell = 1, \cdots, k$, which contradicts the fact that $\mathbf{x}^* \in X$ is a γ-Pareto optimal solution to MOP3.43(γ). □

Unfortunately, there is no guarantee that optimal solution $\mathbf{x}^* \in X, \lambda^* \in \mathrm{R}^1$ of MINMAX3.10($\hat{\mathbf{z}},\gamma$) is γ-Pareto optimal to MOP3.43(γ) if $\mathbf{x}^* \in X, \lambda^* \in \mathrm{R}^1$ is not unique. To guarantee γ-Pareto optimality, we must solve a γ-Pareto optimality test problem for optimal solution $\mathbf{x}^* \in X, \lambda^* \in \mathrm{R}^1$ of MINMAX3.10($\hat{\mathbf{z}},\gamma$).

Test 3.4

$$w \stackrel{\text{def}}{=} \max_{\mathbf{x} \in X, \varepsilon = (\varepsilon_1, \cdots, \varepsilon_k) \geq \mathbf{0}} \sum_{\ell=1}^{k} \varepsilon_\ell \tag{3.176}$$

subject to

$$z_\ell(\mathbf{x}, \gamma) - \hat{z}_\ell + \varepsilon_\ell \leq \lambda^*, \ell = 1, \cdots, k$$

For the optimal solution of TEST3.4, the following theorem holds.

Theorem 3.11

Let $\mathbf{x}^* \in X$, $\lambda^* \in \mathrm{R}^1$ be an optimal solution of MINMAX3.10($\hat{\mathbf{z}},\gamma$) in which $\lambda^* = z_\ell(\mathbf{x}^*, \gamma) - \hat{z}_\ell, \ell = 1, \cdots, k$. Corresponding to optimal solution $\mathbf{x}^* \in X$, let $\check{\mathbf{x}} \in X, \check{\varepsilon}_\ell \geq 0, \ell = 1, \cdots, k$ be an optimal solution of TEST3.4. If $w = 0$, then $\mathbf{x}^* \in X$ is a γ-Pareto optimal solution to MOP3.43(γ).

Conversely, the partial differentiation of $z_\ell(\mathbf{x}, \gamma), \ell = 1, \cdots, k$ for $x_s, s = 1, \cdots, n$ and $x_t, t = 1, \cdots, n$ can be calculated as

$$\begin{aligned} \frac{\partial z_\ell(\mathbf{x}, \gamma)}{\partial x_s \partial x_t} &= \sum_{i=1}^{m} q_{\ell i}^{+} a_{is} a_{it} f_i(\mathbf{a}_i \mathbf{x} + L^{-1}(\gamma) \alpha_i) \\ &+ \sum_{i=1}^{m} q_{\ell i}^{-} a_{is} a_{it} f_i(\mathbf{a}_i \mathbf{x} - R^{-1}(\gamma) \beta_i). \end{aligned} \tag{3.177}$$

The Hessian matrix for $z_\ell(\mathbf{x}, \gamma), \ell = 1, \cdots, k$ can be written as

$$\begin{aligned} \nabla^2 z_\ell(\mathbf{x}, \gamma) &= \sum_{i=1}^{m} q_{\ell i}^{+} f_i(\mathbf{a}_i \mathbf{x} + L^{-1}(\gamma) \alpha_i) \cdot A_i \\ &+ \sum_{i=1}^{m} q_{\ell i}^{-} f_i(\mathbf{a}_i \mathbf{x} - R^{-1}(\gamma) \beta_i) \cdot A_i, \ell = 1, \cdots, k, \end{aligned} \tag{3.178}$$

where $A_i, i = 1, \cdots, m$ are $(n \times n)$-dimensional matrices defined as follows.

$$A_i \stackrel{\text{def}}{=} \begin{pmatrix} a_{i1}^2 & \cdots & a_{i1}a_{in} \\ \vdots & \ddots & \vdots \\ a_{in}a_{i1} & \cdots & a_{in}^2 \end{pmatrix}, i = 1, \cdots, m \tag{3.179}$$

Because of the property of the Hessian matrix for $z_\ell(\mathbf{x}, \gamma), \ell = 1, \cdots, k$, the following theorem holds.

Theorem 3.12
MINMAX3.10($\hat{\mathbf{z}}$,γ) is a convex programming problem.

(Proof)
From definition (3.179), it holds that $A_i = \mathbf{a}_i^T \cdot \mathbf{a}_i, i = 1, \cdots, m$. Therefore, the following relation holds for any n-dimensional column vector $\mathbf{y} \in \mathrm{R}^1$.

$$\begin{aligned} \mathbf{y}^T A_i \mathbf{y} &= \mathbf{y}^T \cdot (\mathbf{a}_i^T \cdot \mathbf{a}_i) \cdot \mathbf{y} \\ &= (\mathbf{y}^T \cdot \mathbf{a}_i^T) \cdot (\mathbf{a}_i \cdot \mathbf{y}) \\ &= (\mathbf{a}_i \cdot \mathbf{y})^T \cdot (\mathbf{a}_i \cdot \mathbf{y}) \geq 0, i = 1, \cdots, m \end{aligned}$$

This means that matrices $A_i, i = 1, \cdots, m$ are positive semidefinite. Because of the assumptions that probability density functions $f_i(\cdot) \geq 0, i = 1, \cdots, m$, and $q_{\ell i}^+ \geq 0, q_{\ell i}^- \geq 0, \ell = 1, \cdots, k, i = 1, \cdots, m$, the following relation holds for each Hessian matrix $\nabla^2 z_\ell(\mathbf{x}, \gamma), \ell = 1, \cdots, k$.

$$\begin{aligned} \mathbf{y}^T \nabla^2 z_\ell(\mathbf{x}, \gamma) \mathbf{y} &= \sum_{i=1}^m q_{\ell i}^+ f_i(\mathbf{a}_i \mathbf{x} + L^{-1}(\gamma)\alpha_i) \cdot \mathbf{y}^T A_i \mathbf{y} \\ &\quad + \sum_{i=1}^m q_{\ell i}^- f_i(\mathbf{a}_i \mathbf{x} - R^{-1}(\gamma)\beta_i) \cdot \mathbf{y}^T A_i \mathbf{y} \geq 0 \end{aligned}$$

This then means that MINMAX3.10($\hat{\mathbf{z}}$,γ) is a convex programming problem. □

The relationships between permissible possibility level γ and optimal objective function value $z_\ell(\mathbf{x}^*, \gamma)$ of MINMAX3.10($\hat{\mathbf{z}}$,γ) are characterized by the following theorem.

Theorem 3.13
For the optimal objective function value $z_\ell(\mathbf{x}^*, \gamma)$ of MINMAX3.10($\hat{\mathbf{z}}$,γ), the follow-

ing relation holds.

$$\begin{aligned}\frac{\partial z_\ell(\mathbf{x}^*,\gamma)}{\partial \gamma} &= -\sum_{i=1}^{m} q_{\ell i}^{+}\frac{\partial L^{-1}(\gamma)}{\partial \gamma}\alpha_i \\ &\quad +\sum_{i=1}^{m} q_{\ell i}^{+}\frac{\partial L^{-1}(\gamma)}{\partial \gamma}\alpha_i F_i(\mathbf{a}_i\mathbf{x}^* + L^{-1}(\gamma)\alpha_i) \\ &\quad -\sum_{i=1}^{m} q_{\ell i}^{-}\frac{\partial R^{-1}(\gamma)}{\partial \gamma}\beta_i F_i(\mathbf{a}_i\mathbf{x}^* - R^{-1}(\gamma)\beta_i) \qquad (3.180)\end{aligned}$$

Given the above, we present an interactive algorithm for deriving a satisfactory solution from a γ-Pareto optimal solution set corresponding to MOP3.43(γ).

Algorithm 3.7

Step 1: *Set permissible possibility level* $\gamma = 1$.
Step 2: *The decision maker sets initial reference objective values* $\hat{z}_\ell$ *for* $z_\ell(\mathbf{x},\gamma), \ell = 1,\cdots,k$.
Step 3: *Solve MINMAX3.10(*$\hat{\mathbf{z}},\gamma$*) obtaining corresponding optimal solution* $\mathbf{x}^* \in X, \lambda^* \in \mathrm{R}^1$. *For optimal solution* $\mathbf{x}^* \in X$*, TEST3.4 is solved.*
Step 4: *If the decision maker is satisfied with the current value of* γ*-Pareto optimal solution* $z_\ell(\mathbf{x}^*,\gamma), \ell = 1,\cdots,k$*, then stop. Otherwise, the decision maker updates his or her reference objective values* $\hat{z}_\ell, \ell = 1,\cdots,k$*, and/or permissible possibility level* γ*, and returns to Step 3.*

3.4.2 An interactive algorithm for MOFRPs with simple recourse using a fractile model

In this subsection, we further consider multiobjective fuzzy random simple recourse programming problems using a fractile optimization model in which both fuzzy random variable coefficients and random variables are incorporated into the constraints and objective functions, as shown below.

MOP 3.44

$$\min_{\mathbf{x}\in X}\ (\bar{\mathbf{c}}_1\mathbf{x},\cdots,\bar{\mathbf{c}}_k\mathbf{x}) \qquad (3.181)$$

subject to

$$A\mathbf{x} = \tilde{\bar{\mathbf{d}}} \qquad (3.182)$$

Here, decision variable vector $\mathbf{x}$, a constraint set X, a coefficient matrix A, and an LR-type fuzzy random variable vector $\tilde{\bar{\mathbf{d}}}$ are defined as in Section 3.4.1. Further,

$\bar{\mathbf{c}}_\ell = (\bar{c}_{\ell 1}, \cdots, \bar{c}_{\ell n}), \ell = 1, \cdots, k$ consists of n-dimensional random variable coefficient row vectors of objective function $\bar{\mathbf{c}}_\ell \mathbf{x}$. Let us assume that each element $\bar{c}_{\ell j}$ is a Gaussian random variable, i.e.,

$$\bar{c}_{\ell j} \sim \mathrm{N}(E[\bar{c}_{\ell j}], \sigma_{\ell jj}), \ell = 1, \cdots, k, j = 1, \cdots, n,$$

and positive definite variance covariance matrices $V_\ell, \ell = 1, \cdots, k$ between Gaussian random variables $\bar{c}_{\ell j}, j = 1, \cdots, n$ are given as

$$V_\ell = \begin{pmatrix} \sigma_{\ell 11} & \sigma_{\ell 12} & \cdots & \sigma_{\ell 1n} \\ \sigma_{\ell 21} & \sigma_{\ell 22} & \cdots & \sigma_{\ell 2n} \\ \vdots & \vdots & \ddots & \vdots \\ \sigma_{\ell n1} & \sigma_{\ell n2} & \cdots & \sigma_{\ell nn} \end{pmatrix}, i = 1, \cdots, k. \tag{3.183}$$

We denote the vectors of the expectation for random variable row vector $\bar{\mathbf{c}}_\ell$ as

$$\mathbf{E}[\bar{\mathbf{c}}_\ell] = (E[\bar{c}_{\ell 1}], \cdots, E[\bar{c}_{\ell n}]), \ell = 1, \cdots, k.$$

Then, using variance covariance matrix V_ℓ, objective function $\bar{\mathbf{c}}_\ell \mathbf{x}$ becomes a Gaussian random variable, i.e.,

$$\bar{\mathbf{c}}_\ell \mathbf{x} \sim \mathrm{N}(\mathbf{E}[\bar{\mathbf{c}}_\ell]\mathbf{x}, \mathbf{x}^T V_\ell \mathbf{x}), \ell = 1, \cdots, k \tag{3.184}$$

According to the discussion from the previous subsection, for a permissible possibility level, MOP3.44 can be reduced to the following multiobjective simple recourse stochastic programming problem.

MOP 3.45 (γ)

$$\min_{\mathbf{x} \in X} \; (\bar{\mathbf{c}}_1 \mathbf{x} + d_1(\mathbf{x}, \gamma), \cdots, \bar{\mathbf{c}}_k \mathbf{x} + d_k(\mathbf{x}, \gamma)) \tag{3.185}$$

Here, $d_\ell(\mathbf{x}, \gamma), \ell = 1, \cdots, k$ are penalty costs defined as follows, which matches (3.171).

$$\begin{aligned} d_\ell(\mathbf{x}, \gamma) \stackrel{\text{def}}{=} & \sum_{i=1}^{m} q_{\ell i}^{+} \left(E[\bar{b}_i] - \mathbf{a}_i \mathbf{x} - L^{-1}(\gamma)\alpha_i \right) \\ & + \sum_{i=1}^{m} q_{\ell i}^{+} \{ (\mathbf{a}_i \mathbf{x} + L^{-1}(\gamma)\alpha_i) F_i(\mathbf{a}_i \mathbf{x} + L^{-1}(\gamma)\alpha_i) \\ & - \int_{-\infty}^{\mathbf{a}_i \mathbf{x} + L^{-1}(\gamma)\alpha_i} b_i f_i(b_i) db_i \Bigg\} \\ & + \sum_{i=1}^{m} q_{\ell i}^{-} \{ (\mathbf{a}_i \mathbf{x} - R^{-1}(\gamma)\beta_i) F_i(\mathbf{a}_i \mathbf{x} - R^{-1}(\gamma)\beta_i) \\ & - \int_{-\infty}^{\mathbf{a}_i \mathbf{x} - R^{-1}(\gamma)\beta_i} b_i f_i(b_i) db_i \Bigg\} \end{aligned} \tag{3.186}$$

If the decision maker specifies permissible probability levels $\hat{p}_\ell, \ell = 1, \cdots, k$ for $\bar{\mathbf{c}}_\ell \mathbf{x}$, MOP3.45$(\gamma)$ can be transformed into the following multiobjective programming problem using a fractile optimization model [77, 87].

MOP 3.46 $(\gamma, \hat{\mathbf{p}})$

$$\min_{\mathbf{x} \in X}(f_1(\mathbf{x}, \gamma, \hat{p}_1), \cdots, f_k(\mathbf{x}, \gamma, \hat{p}_k)), \tag{3.187}$$

where $f_\ell(\mathbf{x}, \gamma, \hat{p}_\ell), \ell = 1, \cdots, k$ are defined as

$$f_\ell(\mathbf{x}, \gamma, \hat{p}_\ell) \stackrel{\text{def}}{=} \mathbf{E}[\bar{\mathbf{c}}_\ell]\mathbf{x} + \Phi^{-1}(\hat{p}_\ell) \cdot \sqrt{\mathbf{x}^T V_\ell \mathbf{x}} + d_\ell(\mathbf{x}, \gamma), \ell = 1, \cdots, k, \tag{3.188}$$

where $\Phi^{-1}(\cdot)$ is an inverse function of a standard normal distribution. Note that MOP3.46$(\gamma, \hat{\mathbf{p}})$ can be viewed as a generalized version of MOP3.43(γ), since $f_\ell(\mathbf{x}, \gamma, 0.5)$ is equivalent to $z_\ell(\mathbf{x}, \gamma)$, as defined in (3.172) if $\mathbf{E}[\bar{\mathbf{c}}_\ell]$ is replaced by $\mathbf{c}_\ell$.

Similar to Definition 3.8, we define the concept of a Pareto optimal solution for MOP3.46$(\gamma, \hat{\mathbf{p}})$ as follows.

Definition 3.9 $\mathbf{x}^* \in X$ is a $(\gamma, \hat{\mathbf{p}})$-Pareto optimal solution to MOP3.46$(\gamma, \hat{\mathbf{p}})$ if and only if there does not exist another $\mathbf{x} \in X$ such that $f_\ell(\mathbf{x}, \gamma, \hat{p}_\ell) \leq f_\ell(\mathbf{x}^*, \gamma, \hat{p}_\ell), \ell = 1, \cdots, k$, with strict inequality holding for at least one ℓ.

For reference objective values $\hat{f}_\ell, \ell = 1, \cdots, k$ specified by the decision maker, the corresponding $(\gamma, \hat{\mathbf{p}})$-Pareto optimal solution is obtained by solving the following minmax problem.

MINMAX 3.11 $(\hat{\mathbf{f}}, \gamma, \hat{\mathbf{p}})$

$$\min_{\mathbf{x} \in X, \lambda \in \mathrm{R}^1} \lambda$$

subject to

$$f_\ell(\mathbf{x}, \gamma, \hat{\mathbf{p}}) - \hat{f}_\ell \leq \lambda, \ell = 1, \cdots, k \tag{3.189}$$

Note that, similar to Theorem 3.12, MINMAX3.11$(\hat{\mathbf{f}}, \gamma, \hat{\mathbf{p}})$ becomes a convex programming problem under the conditions that $\hat{p}_\ell > 0.5, \ell = 1, \cdots, k$. The relationships between optimal solution $\mathbf{x}^* \in X, \lambda^* \in \mathrm{R}^1$ of MINMAX3.11$(\hat{\mathbf{f}}, \gamma, \hat{\mathbf{p}})$ and $(\gamma, \hat{\mathbf{p}})$-Pareto optimal solutions to MOP3.46$(\gamma, \hat{\mathbf{p}})$ are characterized by the following theorem.

Theorem 3.14

(1) If $\mathbf{x}^* \in X, \lambda^* \in \mathrm{R}^1$ is a unique optimal solution of MINMAX3.11$(\hat{\mathbf{f}}, \gamma, \hat{\mathbf{p}})$, then $\mathbf{x}^* \in X$ is a $(\gamma, \hat{\mathbf{p}})$-Pareto optimal solution to MOP3.46$(\gamma, \hat{\mathbf{p}})$.

(2) If $\mathbf{x}^* \in X$ *is a* (γ, $\hat{\mathbf{p}}$)*-Pareto optimal solution to MOP3.46*($\gamma, \hat{\mathbf{p}}$)*, then* $\mathbf{x}^* \in X$ $\lambda^* \stackrel{\text{def}}{=} f_\ell(\mathbf{x}^*, \gamma, \hat{\mathbf{p}}) - \hat{f}_\ell, \ell = 1, \cdots, k$ *is an optimal solution of MINMAX3.11*($\hat{\mathbf{f}}$, γ, $\hat{\mathbf{p}}$) *for some reference objective values* $\hat{\mathbf{f}} = (\hat{f}_1, \cdots, \hat{f}_k)$.

Unfortunately, there is no guarantee that optimal solution $\mathbf{x}^* \in X, \lambda^* \in \mathrm{R}^1$ of MINMAX3.11($\hat{\mathbf{f}}$, γ, $\hat{\mathbf{p}}$) is (γ, $\hat{\mathbf{p}}$)-Pareto optimal to MOP3.46($\gamma, \hat{\mathbf{p}}$), if $\mathbf{x}^* \in X, \lambda^* \in \mathrm{R}^1$ is not unique. To guarantee (γ, $\hat{\mathbf{p}}$)-Pareto optimality, we must define the following (γ, $\hat{\mathbf{p}}$)-Pareto optimality test problem for optimal solution $\mathbf{x}^* \in X, \lambda^* \in \mathrm{R}^1$ of MINMAX3.11($\hat{\mathbf{f}}$, γ, $\hat{\mathbf{p}}$).

Test 3.5

$$w \stackrel{\text{def}}{=} \max_{\mathbf{x} \in X, \varepsilon = (\varepsilon_1, \cdots, \varepsilon_k) \geq \mathbf{0}} \sum_{\ell=1}^{k} \varepsilon_\ell \tag{3.190}$$

subject to

$$f_\ell(\mathbf{x}, \gamma) - \hat{f}_\ell + \varepsilon_\ell \leq \lambda^*, \ell = 1, \cdots, k$$

For the optimal solution of TEST3.5, the following theorem holds.

Theorem 3.15

Let $\mathbf{x}^* \in X$, $\lambda^* \in \mathrm{R}^1$ *be an optimal solution of MINMAX3.11*($\hat{\mathbf{f}}$, γ, $\hat{\mathbf{p}}$) *in which* $\lambda^* = f_\ell(\mathbf{x}^*, \gamma) - \hat{f}_\ell, \ell = 1, \cdots, k$*. Corresponding to optimal solution* $\mathbf{x}^* \in X$*, let* $\check{\mathbf{x}} \in X, \check{\varepsilon}_\ell \geq 0, \ell = 1, \cdots, k$ *be an optimal solution of TEST3.5. If* $w = 0$*, then* $\mathbf{x}^* \in X$ *is a* (γ, $\hat{\mathbf{p}}$)*-Pareto optimal solution to MOP3.46*($\gamma, \hat{\mathbf{p}}$).

We next present our interactive algorithm for obtaining a satisfactory solution from a (γ, $\hat{\mathbf{p}}$)-Pareto optimal solution set.

Algorithm 3.8

Step 1: *Set permissible possibility level* $\gamma = 1$*; the decision maker subjectively sets permissible probability levels* $\hat{p}_\ell (> 0.5), \ell = 1, \cdots, k$.
Step 2: *The decision maker sets initial reference objective values* $\hat{f}_\ell, \ell = 1, \cdots, k$ *for* $f_\ell(\mathbf{x}, \gamma, \hat{\mathbf{p}}), \ell = 1, \cdots, k$.
Step 3: *Solve MINMAX3.11*($\hat{\mathbf{f}}$, γ, $\hat{\mathbf{p}}$)*, thus obtaining corresponding optimal solution* $\mathbf{x}^* \in X, \lambda^* \in \mathrm{R}^1$*. For optimal solution* $\mathbf{x}^* \in X$*, TEST3.5 is solved.*
Step 4: *If the decision maker is satisfied with current value* $f_\ell(\mathbf{x}^*, \gamma, \hat{\mathbf{p}}), \ell = 1, \cdots, k$ *then stop. Otherwise, the decision maker updates his or her reference objective values* $\hat{f}_\ell, \ell = 1, \cdots, k$*, permissible possibility level* $0 \leq \gamma \leq 1$*, and/or permissible probability levels* $\hat{p}_\ell (> 0.5), \ell = 1, \cdots, k$*, then returns to Step 3.*

Chapter 4

Hierarchical Multiobjective Programming Problems (HMOPs) Involving Uncertainty Conditions

CONTENTS

In real-world decision making, the goal of the overall system is often achieved via a hierarchical decision structure in which numerous decision makers belonging to various sections or divisions take actions in pursuit of their own goals independently from one another but also affecting one another. We can view the Stackelberg games [2, 105] as multilevel programming problems with multiple decision makers. Although many techniques have been proposed for obtaining a Stackelberg solution, almost all the techniques are unfortunately inefficient in terms of their computational complexity. As a set of examples, consider decentralized large organizations with divisional independence, such as those found in public and managerial sectors. In such hierarchical organizations, multiple decision makers have cooperative relationships with one another requiring communication. To best handle such hierarchical decision situations, Lai [47], Shih et al. [91], and Lee et al. [49] introduced concepts of decision powers that reflect the hierarchical decision structure between multiple decision makers and proposed fuzzy approaches for obtaining a satisfactory solution for multiple decision makers for multilevel linear programming problems. To overcome some of the disadvantages of their approaches, Sakawa et al. [57, 78, 79] proposed interactive algorithms for two-level multiobjective programming problems.

In this chapter, to distinguish between a Stackelberg solution for noncooperative decision makers and a satisfactory solution for cooperative decision makers in multilevel programming problems, we call multilevel programming problems involving cooperative decision makers hierarchical multiobjective programming problems (HMOPs) [109]. We focus on HMOPs given uncertainty conditions in which random variable coefficients or fuzzy random variable coefficients are incorporated into the objective functions and constraints.

In Section 4.1, we formulate hierarchical multiobjective stochastic programming problems (HMOSPs) using a probability maximization model in which multiple decision makers specify their own permissible objective levels in advance. We also propose an interactive linear programming algorithm [110] for obtaining a satisfactory solution for multiple decision makers from a P-Pareto optimal solution set to HMOSPs. Here, a satisfactory solution reflects not only the hierarchical decision structure of multiple decision makers but also the preferences governed by their own objective functions.

Similarly, in Section 4.2, we formulate HMOSPs through a fractile optimization model in which multiple decision makers specify their own permissible probability levels in advance. We then propose an interactive linear programming algorithm [112] for obtaining a satisfactory solution for multiple decision makers from an F-Pareto optimal solution set to HMOSPs.

In Section 4.3, we formulate HMOSPs based on the fuzzy decision [6, 71, 140] in which permissible objective levels of a probability maximization model or permissible probability levels of a fractile optimization model are automatically computed through the fuzzy decision. We propose an interactive linear programming algorithm for HMOSPs [113, 115, 118] for obtaining a satisfactory solution for multiple decision makers from an MF-Pareto optimal solution set to HMOSPs.

Finally, in Section 4.4, we formulate hierarchical multiobjective fuzzy random programming problems (HMOFRPs). Similar to the above, we propose an interactive linear programming algorithm [126] for obtaining a satisfactory solution for multiple decision makers from an MF-Pareto optimal solution set to HMOFRPs.

4.1 Hierarchical multiobjective stochastic programming problems (HMOSPs) using a probability model

In this section, we cover hierarchical multiobjective stochastic programming problems (HMOSPs) using a probability maximization model [110]. More specifically, in Section 4.1.1, after multiple decision makers specify their own permissible objective levels in advance, we formulate HMOSPs using a probability maximization model and introduce a P-Pareto optimal solution. In Section 4.1.2, we propose an interactive linear programming algorithm in which multiple decision makers iteratively update their reference membership values and decision powers until their satisfactory solution is obtained from a P-Pareto optimal solution set. In Section 4.1.3, we apply our interactive algorithm to a numerical example to illustrate the interactive processes given two hypothetical decision makers.

4.1.1 *A formulation using a probability model*

We consider the following HMOSP in which each of the decision makers (i.e., $DM_r, r = 1, \cdots, q$) has his or her own multiple objective functions together with common linear constraints; further, random variable coefficients are included in each objective function and on the right-hand-side of the constraints.

HMOP 4.1
first-level decision maker : $\mathbf{DM}_1$

$$\min \bar{\mathbf{z}}_1(\mathbf{x}) = (\bar{z}_{11}(\mathbf{x}), \cdots, \bar{z}_{1k_1}(\mathbf{x}))$$

$$\cdots\cdots\cdots\cdots\cdots\cdots\cdots$$

q-th-level decision maker : $\mathbf{DM}_q$

$$\min \bar{\mathbf{z}}_q(\mathbf{x}) = (\bar{z}_{q1}(\mathbf{x}), \cdots, \bar{z}_{qk_q}(\mathbf{x}))$$

subject to

$$A\mathbf{x} \leq \bar{\mathbf{b}},\ \mathbf{x} \geq \mathbf{0}$$

where $\mathbf{x} = (x_1, \cdots, x_n)^T$ is n-dimensional decision column vector, A is $(m \times n)$-dimensional coefficient matrix, and each objective function of $\mathrm{DM}_r, r = 1, \cdots, q$ is defined by

$$\bar{z}_{r\ell}(\mathbf{x}) = \bar{\mathbf{c}}_{r\ell}\mathbf{x} + \bar{\alpha}_{r\ell}, \ell = 1, \cdots, k_r \tag{4.1}$$

$$\bar{\mathbf{c}}_{r\ell} = \mathbf{c}^1_{r\ell} + \bar{t}_{r\ell}\mathbf{c}^2_{r\ell}, \ell = 1, \cdots, k_r \tag{4.2}$$

$$\bar{\alpha}_{r\ell} = \alpha^1_{r\ell} + \bar{t}_{r\ell}\alpha^2_{r\ell}, \ell = 1, \cdots, k_r, \tag{4.3}$$

where $\bar{\mathbf{c}}_{r\ell}, \ell = 1, \cdots, k_r$ are n-dimensional random variable row vectors, $\bar{\alpha}_{r\ell}, \ell = 1, \cdots, k_r$ are random variables, and $\bar{t}_{r\ell}, \ell = 1, \cdots, k_r$ are random variables, whose distribution functions $T_{r\ell}(\cdot), \ell = 1, \cdots, k_r$ are assumed to be strictly monotone increasing and continuous. On the right-hand-side of the constraints of HMOP4.1, $\bar{\mathbf{b}} = (\bar{b}_1, \cdots, \bar{b}_m)^T$ is a random variable column vector whose elements are independent of one another and distribution functions $F_i(\cdot), i = 1, \cdots, m$ are assumed to be strictly monotone increasing and continuous.

Similar to the formulations of multilevel linear programming problems first formulated by Lee and Shih [49], we assume that upper-level decision makers make their decisions considering the overall benefits in HMOP4.1, although they can also claim priority for their objective functions over lower-level decision makers.

To handle HMOP4.1, we apply a probability maximization model to the objective functions of HMOP4.1, and adopt chance constrained conditions with permissible constraint levels $\beta_i, i = 1, \cdots, m$ for constraints $\mathbf{a}_i\mathbf{x} \leq \bar{b}_i, i = 1, \cdots, k$ in HMOP4.1, i.e.,

$$\Pr(\omega \mid \mathbf{a}_i\mathbf{x} \leq b_i(\omega)) \geq \beta_i, i = 1, \cdots, m.$$

For the objective functions of HMOP4.1, we substitute the minimization of $\bar{z}_{r\ell}(\mathbf{x})$ for the maximization of the probability that $\bar{z}_{r\ell}(\mathbf{x})$ is less than or equal to a certain permissible objective level $\hat{f}_{r\ell}$ specified subjectively by the decision maker (i.e., DM_r), i.e.,

$$p_{r\ell}(\mathbf{x}, f_{r\ell}) \stackrel{\text{def}}{=} \Pr(\omega \mid z_{r\ell}(\mathbf{x}, \omega) \leq \hat{f}_{r\ell}), r = 1, \cdots, q, \ell = 1, \cdots, k_r.$$

We can then transform HMOP4.1 into the following hierarchical multiobjective programming problem HMOP4.2($\hat{\mathbf{f}}, \beta$), where permissible constraint levels $\beta = (\beta_1, \cdots, \beta_k)$ are specified by the first-level decision maker (DM_1) and permissible objective levels

$$\hat{\mathbf{f}}_r \stackrel{\text{def}}{=} (\hat{f}_{r1}, \cdots, \hat{f}_{rk_r}), r = 1, \cdots, q, \ \hat{\mathbf{f}} \stackrel{\text{def}}{=} (\hat{\mathbf{f}}_1, \cdots, \hat{\mathbf{f}}_q),$$

are specified subjectively by each of the decision makers ($DM_r, r = 1, \cdots, q$).

HMOP 4.2 ($\hat{\mathbf{f}}, \beta$)
first-level decision maker : DM_1

$$\max(p_{11}(\mathbf{x}, \hat{f}_{11}), \cdots, p_{1k_1}(\mathbf{x}, \hat{f}_{1k_1}))$$

....

q-th-level decision maker: DM_q

$$\max(p_{q1}(\mathbf{x}, \hat{f}_{q1}), \cdots, p_{qk_q}(\mathbf{x}, \hat{f}_{qk_q}))$$

subject to

$$\Pr(\omega \mid \mathbf{a}_i\mathbf{x} \leq b_i(\omega)) \geq \beta_i, i = 1, \cdots, m, \mathbf{x} \geq \mathbf{0}$$

By using distribution function $F_i(\cdot)$, constraint $\Pr(\omega \mid \mathbf{a}_i\mathbf{x} \leq b_i(\omega)) \geq \beta_i$ can be expressed as

$$\begin{aligned}
& \Pr(\omega \mid \mathbf{a}_i\mathbf{x} \leq b_i(\omega)) \geq \beta_i \\
\Leftrightarrow \quad & 1 - \Pr(\omega \mid b_i(\omega) \leq \mathbf{a}_i\mathbf{x}) \geq \beta_i \\
\Leftrightarrow \quad & 1 - F_i(\mathbf{a}_i\mathbf{x}) \geq \beta_i \\
\Leftrightarrow \quad & F_i(\mathbf{a}_i\mathbf{x}) \leq 1 - \beta_i \\
\Leftrightarrow \quad & \mathbf{a}_i\mathbf{x} \leq F_i^{-1}(1 - \beta_i), i = 1, \cdots, m,
\end{aligned} \tag{4.4}$$

where $F_i^{-1}(\cdot)$ represents an inverse function with respect to $F_i(\cdot)$. Similarly, under the assumption that $\mathbf{c}_{r\ell}^2\mathbf{x} + \alpha_{r\ell}^2 \geq 0, r = 1, \cdots, q, \ell = 1, \cdots, k_r$, using distribution function $T_{r\ell}(\cdot)$, objective function $p_{r\ell}(\mathbf{x}, \hat{f}_{r\ell})$ in HMOP4.2($\hat{\mathbf{f}}, \beta$) is expressed as

$$\begin{aligned}
p_{r\ell}(\mathbf{x}, \hat{f}_{r\ell}) &= \Pr(\omega \mid z_{r\ell}(\mathbf{x}, \omega) \leq \hat{f}_{r\ell}) \\
&= \Pr(\omega \mid \mathbf{c}_{r\ell}(\omega)\mathbf{x} + \alpha_{r\ell}(\omega) \leq \hat{f}_{r\ell}) \\
&= \Pr(\omega \mid (\mathbf{c}_{r\ell}^1 + t_{r\ell}(\omega)\mathbf{c}_{r\ell}^2)\mathbf{x} + (\alpha_{r\ell}^1 + t_{r\ell}(\omega)\alpha_{r\ell}^2) \leq \hat{f}_{r\ell}) \\
&= \Pr(\omega \mid (\mathbf{c}_{r\ell}^2)\mathbf{x} + \alpha_{r\ell}^2)t_{r\ell}(\omega) + (\mathbf{c}_{r\ell}^1 + \alpha_{r\ell}^1) \leq \hat{f}_{r\ell}) \\
&= \Pr\left(\omega \mid t_{r\ell}(\omega) \leq \frac{\hat{f}_{r\ell} - (\mathbf{c}_{r\ell}^1\mathbf{x} + \alpha_{r\ell}^1)}{\mathbf{c}_{r\ell}^2\mathbf{x} + \alpha_{r\ell}^2}\right) \\
&= T_{r\ell}\left(\frac{\hat{f}_{r\ell} - (\mathbf{c}_{r\ell}^1\mathbf{x} + \alpha_{r\ell}^1)}{\mathbf{c}_{r\ell}^2\mathbf{x} + \alpha_{r\ell}^2}\right).
\end{aligned} \tag{4.5}$$

Therefore, we can convert HMOP4.2($\hat{\mathbf{f}}, \beta$) into the following problem, where distribution functions $F_i(\cdot), i = 1, \cdots, m, T_{r\ell}(\cdot), r = 1, \cdots, q, \ell = 1, \cdots, k_r$ are explicitly involved.

HMOP 4.3 ($\hat{\mathbf{f}}, \beta$)
first-level decision maker: DM$_1$

$$\max \left(T_{11}\left(\frac{\hat{f}_{11} - (\mathbf{c}_{11}^1 \mathbf{x} + \alpha_{11}^1)}{\mathbf{c}_{11}^2 \mathbf{x} + \alpha_{11}^2}\right), \cdots, T_{1k_1}\left(\frac{\hat{f}_{1k_1} - (\mathbf{c}_{1k_1}^1 \mathbf{x} + \alpha_{1k_1}^1)}{\mathbf{c}_{1k_1}^2 \mathbf{x} + \alpha_{1k_1}^2}\right)\right)$$

....

q-th-level decision maker: DM$_q$

$$\max \left(T_{q1}\left(\frac{\hat{f}_{q1} - (\mathbf{c}_{q1}^1 \mathbf{x} + \alpha_{q1}^1)}{\mathbf{c}_{q1}^2 \mathbf{x} + \alpha_{q1}^2}\right), \cdots, T_{qk_q}\left(\frac{\hat{f}_{qk_q} - (\mathbf{c}_{qk_q}^1 \mathbf{x} + \alpha_{qk_q}^1)}{\mathbf{c}_{qk_q}^2 \mathbf{x} + \alpha_{qk_q}^2}\right)\right)$$

subject to

$$\mathbf{x} \in X(\beta) \stackrel{\text{def}}{=} \{\mathbf{x} \geq \mathbf{0} \mid \mathbf{a}_i \mathbf{x} \leq F_i^{-1}(1 - \beta_i), i = 1, \cdots, m\}$$

To handle HMOP4.3($\hat{\mathbf{f}}, \beta$), we next introduce a P-Pareto optimal solution.

Definition 4.1 $\mathbf{x}* \in X(\beta)$ is a P-Pareto optimal solution to HMOP4.3($\hat{\mathbf{f}}, \beta$) if and only if there is no $\mathbf{x} \in X(\beta)$ such that $p_{r\ell}(\mathbf{x}, \hat{f}_{r\ell}) \geq p_{r\ell}(\mathbf{x}^*, \hat{f}_{r\ell})$, $r = 1, \cdots, q, \ell = 1, \cdots, k_r$, with strict inequality holding for at least one ℓ and r.

Below, we assume that q decision makers in HMOP4.3($\hat{\mathbf{f}}, \beta$) reach an agreement and choose a satisfactory solution from the P-Pareto optimal solution set. Then, to generate a candidate for a satisfactory solution from a P-Pareto optimal solution set, each decision maker (i.e., DM$_r$) is asked to specify his or her reference probability values $\hat{\mathbf{p}}_r \stackrel{\text{def}}{=} (\hat{p}_{r1}, \cdots, \hat{p}_{1k_r})$, which are reference levels of achievement of distribution function $T_{r\ell}(\cdot)$. Once reference probability values $\hat{\mathbf{p}} \stackrel{\text{def}}{=} (\hat{\mathbf{p}}_1, \cdots, \hat{\mathbf{p}}_q)$ are specified, the corresponding P-Pareto optimal solution, which is in a sense close to each decision maker's requirement, is obtained by solving the following minmax problem.

MINMAX 4.1 ($\hat{\mathbf{f}}, \beta, \hat{\mathbf{p}}$)

$$\min_{\mathbf{x} \in X(\beta), \lambda \in R^1} \lambda$$

subject to

$$\hat{p}_{r\ell} - T_{r\ell}\left(\frac{\hat{f}_{r\ell} - (\mathbf{c}_{r\ell}^1 \mathbf{x} + \alpha_{r\ell}^1)}{\mathbf{c}_{r\ell}^2 \mathbf{x} + \alpha_{r\ell}^2}\right) \leq \lambda, r = 1, \cdots, q, \ell = 1, \cdots, k_r \quad (4.6)$$

The relationships between optimal solution $\mathbf{x}^* \in X(\beta), \lambda^* \in \mathrm{R}^1$ of MINMAX4.1$(\hat{\mathbf{f}}, \beta, \hat{\mathbf{p}})$ and P-Pareto optimal solutions to HMOP4.3$(\hat{\mathbf{f}}, \beta)$ are characterized by the following theorem.

Theorem 4.1
(1) If $\mathbf{x}^* \in X(\beta), \lambda^* \in \mathrm{R}^1$ *is a unique optimal solution of MINMAX4.1*$(\hat{\mathbf{f}}, \beta, \hat{\mathbf{p}})$*, then* $\mathbf{x}^* \in X(\beta)$ *is a P-Pareto optimal solution to HMOP4.3*$(\hat{\mathbf{f}}, \beta)$.

(2) If $\mathbf{x}^* \in X(\beta)$ *is a P-Pareto optimal solution to HMOP4.3*$(\hat{\mathbf{f}}, \beta)$*, then* $\mathbf{x}^* \in X(\beta), \lambda^* \in \mathrm{R}^1$ *is an optimal solution of MINMAX4.1*$(\hat{\mathbf{f}}, \beta, \hat{\mathbf{p}})$ *for some reference probability values* $\hat{\mathbf{p}}$*, where* $\hat{p}_{r\ell} \stackrel{\text{def}}{=} p_{r\ell}(\mathbf{x}^*, \hat{f}_{r\ell}) + \lambda^*, r = 1, \cdots, q, \ell = 1, \cdots, k_r$.

(Proof)
(1) Assume that $\mathbf{x}^* \in X(\beta)$ is not a P-Pareto optimal solution to HMOP4.3$(\hat{\mathbf{f}}, \beta)$. Then, there exists $\mathbf{x} \in X(\beta)$ such that $p_{r\ell}(\mathbf{x}, \hat{f}_{r\ell}) \geq p_{r\ell}(\mathbf{x}^*, \hat{f}_{r\ell})$, $r = 1, \cdots, q, \ell = 1, \cdots, k_r$, with strict inequality holding for at least one ℓ and r. Then it holds that

$$\lambda^* \geq \hat{p}_{r\ell} - p_{r\ell}(\mathbf{x}^*, \hat{f}_{r\ell}) \geq \hat{p}_{r\ell} - p_{r\ell}(\mathbf{x}, \hat{f}_{r\ell}), r = 1, \cdots, q, \ell = 1, \cdots, k_r,$$

with strict inequality holding for at least one ℓ and r. This contradicts the fact that $\mathbf{x}^* \in X(\beta), \lambda^* \in \mathrm{R}^1$ is a unique optimal solution to MINMAX4.1$(\hat{\mathbf{f}}, \beta, \hat{\mathbf{p}})$.

(2) Assume that $\mathbf{x}^* \in X(\beta), \lambda^* \in \mathrm{R}^1$ is not an optimal solution to MINMAX4.1$(\mathbf{f}, \beta, \hat{\mathbf{p}})$ for any reference probability values $\hat{p}_{r\ell} \stackrel{\text{def}}{=} p_{r\ell}(\mathbf{x}^*, f_{r\ell}) + \lambda^*, r = 1, \cdots, q, \ell = 1, \cdots, k_r$. Then, there exists some $\mathbf{x} \in X(\beta)$, $\lambda < \lambda^*$such that

$$\begin{aligned} \lambda &\geq \hat{p}_{r\ell} - p_{r\ell}(\mathbf{x}, f_{r\ell}) \\ &= (p_{r\ell}(\mathbf{x}^*, f_{r\ell}) + \lambda^*) - p_{r\ell}(\mathbf{x}, f_{r\ell}), r = 1, \cdots, q, \ell = 1, \cdots, k_r. \end{aligned}$$

Therefore, it holds that

$$0 > \lambda - \lambda^* \geq p_{r\ell}(\mathbf{x}^*, f_{r\ell}) - p_{r\ell}(\mathbf{x}, f_{r\ell}), r = 1, \cdots, q, \ell = 1, \cdots, k_r.$$

This contradicts the fact that $\mathbf{x}^* \in X(\beta)$ is a P-Pareto optimal solution to HMOP4.3$(\hat{\mathbf{f}}, \beta)$. □

Note that, in general, P-Pareto optimal solutions obtained by solving MINMAX4.1$(\hat{\mathbf{f}}, \beta, \hat{\mathbf{p}})$ do not reflect the hierarchical structure between q decision makers in which an upper-level decision maker can claim priority for his or her distribution functions over lower-level decision makers. To cope with such a hierarchical preference structure between q decision makers in MINMAX4.1$(\hat{\mathbf{f}}, \beta, \hat{\mathbf{p}})$, we introduce decision power $w_r, r = 1, \cdots, q$ [47, 49] for MINMAX4.1$(\hat{\mathbf{f}}, \beta, \hat{\mathbf{p}})$, where the r-th level decision maker (i.e., DM_r) can subjectively specify decision power w_{r+1} and the last decision maker (DM_q) has no decision power whatsoever. Therefore, decision powers $\mathbf{w} = (w_1, w_2, \cdots, w_q)^T$ must satisfy inequality condition, i.e.,

$$1 = w_1 \geq \cdots \geq w_q > 0. \tag{4.7}$$

The corresponding modified minmax problem is then reformulated as shown below, where

$$\mathbf{w} \stackrel{\text{def}}{=} (w_1, \cdots, w_q), \tag{4.8}$$

are decision powers specified by decision makers ($\text{DM}_r, r = 1, \cdots, q-1$).

MINMAX 4.2 $(\hat{\mathbf{f}}, \beta, \hat{\mathbf{p}}, \mathbf{w})$

$$\min_{\mathbf{x} \in X(\beta), \lambda \in R^1} \lambda$$

subject to

$$\hat{p}_{r\ell} - T_{r\ell}\left(\frac{\hat{f}_{r\ell} - (\mathbf{c}^1_{r\ell}\mathbf{x} + \alpha^1_{r\ell})}{\mathbf{c}^2_{r\ell}\mathbf{x} + \alpha^2_{r\ell}}\right) \leq \lambda / w_r, r = 1, \cdots, q, \ell = 1, \cdots, k_r \tag{4.9}$$

Since each distribution function $T_{r\ell}(\cdot)$ is strictly monotone increasing and continuous, MINMAX4.2$(\hat{\mathbf{f}}, \beta, \hat{\mathbf{p}})$ can be equivalently transformed into the following form.

MINMAX 4.3 $(\hat{\mathbf{f}}, \beta, \hat{\mathbf{p}}, \mathbf{w})$

$$\min_{\mathbf{x} \in X(\beta), \lambda \in R^1} \lambda$$

subject to

$$\begin{aligned} &\hat{f}_{r\ell} - (\mathbf{c}^1_{r\ell}\mathbf{x} + \alpha^1_{r\ell}) \geq T^{-1}_{r\ell}(\hat{p}_{r\ell} - \lambda / w_r) \cdot (\mathbf{c}^2_{r\ell}\mathbf{x} + \alpha^2_{r\ell}), \\ &r = 1, \cdots, q, \ell = 1, \cdots, k_r \end{aligned} \tag{4.10}$$

Note that constraints (4.10) for a fixed value of λ can be reduced to a set of linear inequalities. Therefore, optimal value λ^* of MINMAX4.3$(\hat{\mathbf{f}}, \beta, \hat{\mathbf{p}}, \mathbf{w})$ can be obtained as the maximum value of λ so that there exists an admissible set satisfying the linear constraints of MINMAX4.3$(\hat{\mathbf{f}}, \beta, \hat{\mathbf{p}}, \mathbf{w})$ for some fixed value λ. From property of $T^{-1}_{r\ell}(\cdot)$, inequality conditions $0 < \hat{p}_{r\ell} - \lambda / w_r < 1$, $r = 1, \cdots, q, \ell = 1, \cdots, k_r$ must be satisfied. As a result, λ satisfies inequalities

$$\max_{r=1,\cdots,q,\ell=1,\cdots,k_r} w_r(\hat{p}_{r\ell} - 1) < \lambda < \max_{r=1,\cdots,q,\ell=1,\cdots,k_r} w_r \hat{p}_{r\ell}.$$

Therefore, we can obtain the minimum value λ^* of MINMAX4.3$(\hat{\mathbf{f}}, \beta, \hat{\mathbf{p}}, \mathbf{w})$ by a combination of the bisection method and phase one of a linear programming technique.

After calculating the minimum value λ^* of MINMAX4.3$(\hat{\mathbf{f}}, \beta, \hat{\mathbf{p}}, \mathbf{w})$ via linear programming, one of the corresponding optimal solutions $\mathbf{x}^* \in X(\beta)$ can be obtained by solving the following linear fractional programming problem.

LFP 4.1 $(\hat{\mathbf{f}}, \beta, \hat{\mathbf{p}}, \mathbf{w}, \lambda^*)$

$$\min_{\mathbf{x} \in X(\beta)} -\left(\frac{f_{11} - (\mathbf{c}_{11}^1 \mathbf{x} + \alpha_{11}^1)}{\mathbf{c}_{11}^2 \mathbf{x} + \alpha_{11}^2} \right)$$

subject to

$$\begin{aligned} & f_{r\ell} - (\mathbf{c}_{r\ell}^1 \mathbf{x} + \alpha_{r\ell}^1) \geq T_{r\ell}^{-1}(\hat{p}_{r\ell} - \lambda^*/w_r) \cdot (\mathbf{c}_{r\ell}^2 \mathbf{x} + \alpha_{r\ell}^2) \\ & r = 1, \cdots, q, \ell = 1, \cdots, k_r, (r, \ell) \neq (1,1) \end{aligned} \tag{4.11}$$

By applying the Charnes-Cooper transformation [12], i.e.,

$$s \stackrel{\text{def}}{=} \frac{1}{\mathbf{c}_{11}^2 \mathbf{x} + \alpha_{11}^2}, \quad \mathbf{y} \stackrel{\text{def}}{=} s \cdot \mathbf{x}, \ s > 0, \tag{4.12}$$

to LFP4.1$(\hat{\mathbf{f}}, \beta, \hat{\mathbf{p}}, \mathbf{w}, \lambda^*)$, we can transform LFP1$(\hat{\mathbf{f}}, \beta, \hat{\mathbf{p}}, \mathbf{w}, \lambda^*)$ into the following linear programming problem.

LP 4.1 $(\hat{\mathbf{f}}, \beta, \hat{\mathbf{p}}, \mathbf{w}, \lambda^*)$

$$\min_{\mathbf{y} \geq \mathbf{0}, s > \delta} \mathbf{c}_{11}^1 \mathbf{y} + (\alpha_{11}^1 - f_{11}) \cdot s \tag{4.13}$$

subject to

$$\begin{aligned} & T_{r\ell}^{-1}(\hat{p}_{r\ell} - \lambda^*/w_r) \cdot (\mathbf{c}_{r\ell}^2 \mathbf{y} + \alpha_{r\ell}^2 \cdot s) + \mathbf{c}_{r\ell}^1 \mathbf{y} + (\alpha_{r\ell}^1 - f_{r\ell}) \cdot s \leq 0, \\ & r = 1, \cdots, q, \ell = 1, \cdots, k_r, (r, \ell) \neq (1,1) \end{aligned} \tag{4.14}$$

$$A\mathbf{y} - F_i^{-1}(1 - \beta_i) \cdot s \leq \mathbf{0} \tag{4.15}$$

where $\delta > 0$ is a sufficiently small and positive constant.

4.1.2 An interactive linear programming algorithm

After obtaining P-Pareto optimal solution $\mathbf{x}^*$ by solving MINMAX4.3$(\hat{\mathbf{f}}, \beta, \hat{\mathbf{p}}, \mathbf{w})$ via linear programming, each decision maker (i.e., DM$_r$) must either be satisfied with the current values of his or her distribution functions $p_{r\ell}(\mathbf{x}, f_{r\ell}), \ell = 1, \cdots, k_r$, or update his or her decision power [47, 49] w_{r+1} and/or his or her reference probability values $\hat{\mathbf{p}}_\mathbf{r} = (\hat{p}_{r1}, \cdots, \hat{p}_{rk_r})$.

To help each decision maker update his or her reference probability values, tradeoff information is rather useful. Such tradeoff information is obtainable since it relates to the simplex multipliers of constraints (4.14) in LP1$(\hat{\mathbf{f}}, \beta, \hat{\mathbf{p}}, \mathbf{w}, \lambda^*)$ [73, 86].

Theorem 4.2
Let $(\mathbf{y}^*, s^*) \stackrel{\text{def}}{=} \left(\frac{\mathbf{x}^*}{\mathbf{c}_{11}^2\mathbf{x}^* + \alpha_{11}^2}, \frac{1}{\mathbf{c}_{11}^2\mathbf{x}^* + \alpha_{11}^2}\right)$ *be a unique and non degenerate optimal solution of LP4.1*$(\hat{\mathbf{f}}, \beta, \hat{\mathbf{p}}, \mathbf{w}, \lambda^*)$*, and let constraints (4.14) be active. Then, the tradeoff rates between* $p_{11}(\mathbf{x}, \hat{f}_{11})$ *and* $p_{r\ell}(\mathbf{x}, \hat{f}_{rl})$ *at* $\mathbf{x} = \mathbf{x}^*$ *are obtained as*

$$-\left.\frac{\partial p_{11}(\mathbf{x}, f_{11})}{\partial p_{r\ell}(\mathbf{x}, f_{r\ell})}\right|_{\mathbf{x}=\mathbf{x}^*} = \pi_{r\ell}^* \cdot \left(\frac{\mathbf{c}_{r\ell}^2\mathbf{x}^* + \alpha_{r\ell}^2}{\mathbf{c}_{11}^2\mathbf{x}^* + \alpha_{11}^2}\right) \cdot \left(\frac{T'_{11}(\mathbf{x}^*)}{T'_{r\ell}(\mathbf{x}^*)}\right), \tag{4.16}$$

where $\pi_{r\ell}^* \geq 0, r = 1, \cdots, q, \ell = 1, \cdots, k_r$ are corresponding simplex multipliers for constraints (4.14) and $T'_{r\ell}(\mathbf{x}^*), r = 1, \cdots, q, \ell = 1, \cdots, k_r$ are differential coefficients for $T_{r\ell}(\cdot)$ at $\mathbf{x} = \mathbf{x}^*$.

Now, we can construct the interactive algorithm to derive a satisfactory solution of multiple decision makers ($\text{DM}_r, r = 1, \cdots, q$) in a hierarchical organization from a P-Pareto optimal solution set to HMOP4.3$(\hat{\mathbf{f}}, \beta)$.

Algorithm 4.1

Step 1: *The first level decision maker* (DM_1) *subjectively sets satisfactory constraint levels* $0 < \beta_i < 1, i = 1, \cdots, m$ *for constraints* $\mathbf{a}_i\mathbf{x} \leq \bar{b}_i, i = 1, \cdots, m$ *of HMOP4.1.*
Step 2: *Under feasible set* $X(\beta)$*, calculate the individual minimum and maximum of expected values of the objective functions in HMOP4.1. Considering such values, each decision maker (i.e.,* $DM_r, r = 1, \cdots, q$*) sets permissible objective levels* $\hat{f}_{r\ell}, \ell = 1, \cdots, k_r$ *for objective functions* $\bar{z}_{r\ell}(\mathbf{x}), \ell = 1, \cdots, k_r$ *in HMOP4.1.*
Step 3: *In HMOP4.3*$(\hat{\mathbf{f}}, \beta)$*, initial decision powers are set to* $w_r = 1, r = 1, \cdots, q$*, and initial reference probability values are set to* $\hat{p}_{r\ell} = 1, r = 1, \cdots, q, \ell = 1, \cdots, k_r$.
Step 4: *For the given parameters* $(\hat{\mathbf{f}}, \beta, \hat{\mathbf{p}}, \mathbf{w})$*, using the bisection method and the phase one of the two-phase simplex method, solve MINMAX4.3*$(\hat{\mathbf{f}}, \beta, \hat{\mathbf{p}}, \mathbf{w})$ *to obtain the optimal value* λ^**. If the optimal value* $\lambda^* \geq 0$*, then go to Step 5. Otherwise, after updating the reference probability values as* $\hat{p}_{r\ell} \leftarrow \hat{p}_{r\ell} - \lambda^*/w_r$*,* $r = 1, \cdots, q, \ell = 1, \cdots, k_r$*, solve MINMAX4.3*$(\hat{\mathbf{f}}, \beta, \hat{\mathbf{p}}, \mathbf{w})$ *to guarantee optimal value* $\lambda^* \geq 0$*, and go to Step 5.*
Step 5: *For optimal value* $\lambda^* \geq 0$ *of MINMAX4.3*$(\hat{\mathbf{f}}, \beta, \hat{\mathbf{p}}, \mathbf{w})$*, solve LP4.1* $(\mathbf{f}, \beta, \hat{\mathbf{p}}, \mathbf{w}, \lambda^*)$ *and obtain the corresponding optimal solution* $(\mathbf{y}^*, s^*)$*. Here,* $(\mathbf{y}^*, s^*)$ *is equivalently transformed into optimal solution* $\mathbf{x}^*$ *of MINMAX4.3*$(\hat{\mathbf{f}}, \beta, \hat{\mathbf{p}}, \mathbf{w})$*, and we compute the tradeoff rates between* $p_{r\ell}(\mathbf{x}^*, f_{r\ell}), r = 1, \cdots, q, \ell = 1, \cdots, k_r$.
Step 6: *If each decision maker (i.e.,* $DM_r, r = 1, \cdots, q$*) is satisfied with the current values of his or her distribution functions* $p_{r\ell}(\mathbf{x}^*, f_{r\ell}), \ell = 1, \cdots, k_r$*, then stop. Otherwise, let the s-th-level decision maker (i.e.,* DM_s*) be the uppermost decision maker who is not satisfied with the current values of his or her distribution functions* $p_{s\ell}(\mathbf{x}, f_{s\ell}), \ell = 1, \cdots, k_s$*. Considering current values* $p_{s\ell}(\mathbf{x}^*, f_{s\ell}), \ell = 1, \cdots, k_s$ *and tradeoff rates,* DM_s *updates his or her decision power* w_{s+1} *and/or his or her reference probability values* $\hat{p}_{s\ell}, \ell = 1, \cdots, k_s$ *according to the following two rules and then we return to Step 4.*

Rule 1: *When the decision maker (i.e., DM_s) updates his or her decision power w_{s+1}, $w_{s+1} \leq w_s$ must be satisfied to guarantee inequality conditions (4.7). After updating w_{s+1}, if there exists some index $t > s+1$ such that $w_{s+1} < w_t$, then corresponding decision power w_t must be replaced with $w_t \leftarrow w_{s+1}$.*
Rule 2: *When the decision maker (i.e., DM_s) updates his or her reference probability values $\hat{\mathbf{p}}_s, i = 1, \cdots, k_s$, the reference probability values of the other decision makers (i.e., $DM_r, r = 1, \cdots, q, r \neq s$) must be set to the current values of distribution functions $p_{r\ell}(\mathbf{x}^*, f_{r\ell}), \ell = 1, \cdots, k_r$, i.e., $\hat{p}_{r\ell} \leftarrow p_{r\ell}(\mathbf{x}^*, f_{r\ell}), r = 1, \cdots, q, r \neq s, \ell = 1, \cdots, k_r$.*

Note that when a decision maker (i.e., DM_s) updates his or her reference probability values $\hat{p}_{s\ell}, \ell = 1, \cdots, k_s$ according to Rule 2 at Step 6, any improvement of one distribution function can be achieved only at the expense of at least one of the other functions for fixed decision powers $w_r, r = 1, \cdots, q$. Similarly, when a decision maker (i.e., DM_s) updates his or her decision power w_{s+1} according to Rule 1 at Step 6, distribution functions $p_{r\ell}(\mathbf{x}^*, f_{r\ell}), r = 1, \cdots, s, \ell = 1, \cdots, k_r$ will be improved by the lesser value of w_{s+1} at the expense of other distribution functions $p_{r\ell}(\mathbf{x}^*, f_{r\ell}), r = s+1, \cdots, q, \ell = 1, \cdots, k_r$ for fixed reference probability values $\hat{p}_{r\ell}, r = 1, \cdots, q, \ell = 1, \cdots, k_s$.

4.1.3 A numerical example

To demonstrate our proposed method and the interactive processes, we consider the following hierarchical two objective stochastic programming problem with two hypothetical decision makers (i.e., DM_1 and DM_2).

HMOP 4.4
first-level decision maker: DM_1

$$\begin{aligned} \min \bar{z}_{11}(\mathbf{x}) &= (\mathbf{c}_{11}^1 + \bar{t}_{11}\mathbf{c}_{11}^2)\mathbf{x} + (\alpha_{11}^1 + \bar{t}_{11}\alpha_{11}^2) \\ \min \bar{z}_{12}(\mathbf{x}) &= (\mathbf{c}_{12}^1 + \bar{t}_{12}\mathbf{c}_{12}^2)\mathbf{x} + (\alpha_{12}^1 + \bar{t}_{12}\alpha_{12}^2) \end{aligned}$$

second-level decision maker: DM_2

$$\begin{aligned} \min \bar{z}_{21}(\mathbf{x}) &= (\mathbf{c}_{21}^1 + \bar{t}_{21}\mathbf{c}_{21}^2)\mathbf{x} + (\alpha_{21}^1 + \bar{t}_{21}\alpha_{21}^2) \\ \min \bar{z}_{22}(\mathbf{x}) &= (\mathbf{c}_{22}^1 + \bar{t}_{22}\mathbf{c}_{22}^2)\mathbf{x} + (\alpha_{22}^1 + \bar{t}_{22}\alpha_{22}^2) \end{aligned}$$

subject to

$$\mathbf{x} \in X \stackrel{\text{def}}{=} \{\mathbf{x} \in \mathrm{R}^{10} \mid \mathbf{a}_i\mathbf{x} \leq \bar{b}_i, i = 1, \cdots, 7, \mathbf{x} \geq \mathbf{0}\}$$

where $\mathbf{a}_i, i = 1, \cdots, 7$, $c_{r\ell}^1$, $c_{r\ell}^2, r = 1, 2$, $\ell = 1, 2$ are the constant coefficient row vectors shown in Table 4.1 and $\alpha_{11}^1 = -18, \alpha_{11}^2 = 5, \alpha_{12}^1 = -27, \alpha_{12}^2 = 6, \alpha_{21}^1 =$

Table 4.1: Parameters of the objective functions and constraints in HMOP4.4

$\mathbf{x}$	x_1	x_2	x_3	x_4	x_5	x_6	x_7	x_8	x_9	x_{10}
$\mathbf{c}_{11}^1$	19	48	21	10	18	35	46	11	24	33
$\mathbf{c}_{11}^2$	3	2	2	1	4	3	1	2	4	2
$\mathbf{c}_{12}^1$	12	–46	–23	–38	–33	–48	12	8	19	20
$\mathbf{c}_{12}^2$	1	2	4	2	2	1	2	1	2	1
$\mathbf{c}_{21}^1$	–18	–26	–22	–28	–15	–29	–10	–19	–17	–28
$\mathbf{c}_{21}^2$	2	1	3	2	1	2	3	3	2	1
$\mathbf{c}_{22}^1$	–8	31	28	29	25	36	–8	–7	–13	–15
$\mathbf{c}_{22}^2$	1	2	3	2	2	1	2	1	2	1
$\mathbf{a}_1$	12	–2	4	–7	13	–1	–6	6	11	–8
$\mathbf{a}_2$	–2	5	3	16	6	–12	12	4	–7	–10
$\mathbf{a}_3$	3	–16	–4	–8	–8	2	–12	–12	4	–3
$\mathbf{a}_4$	–11	6	–5	9	–1	8	–4	6	–9	6
$\mathbf{a}_5$	–4	7	–6	–5	13	6	–2	–5	14	–6
$\mathbf{a}_6$	5	–3	14	–3	–9	–7	4	–4	–5	9
$\mathbf{a}_7$	–3	–4	–6	9	6	18	11	–9	–4	7

$-10, \alpha_{21}^2 = 4, \alpha_{22}^1 = -27, \alpha_{22}^2 = 6$, $\bar{t}_{r\ell}, r = 1,2, \ell = 1,2$ and $\bar{b}_i, i = 1, \cdots, 7$ are Gaussian random variables defined as

$$\bar{t}_{11} \sim \mathrm{N}(4,2^2), \bar{t}_{12} \sim \mathrm{N}(3,3^2), \bar{t}_{21} \sim \mathrm{N}(3,2^2), \bar{t}_{22} \sim \mathrm{N}(3,3^2),$$
$$\bar{b}_1 \sim \mathrm{N}(164,30^2), \bar{b}_2 \sim \mathrm{N}(-190,20^2), \bar{b}_3 \sim \mathrm{N}(-184,15^2), \bar{b}_4 \sim \mathrm{N}(99,22^2),$$
$$\bar{b}_5 \sim \mathrm{N}(-150,17^2), \bar{b}_6 \sim \mathrm{N}(154,35^2), \bar{b}_7 \sim \mathrm{N}(142,42^2).$$

According to Step 1 of Algorithm 4.1, the first-level decision maker (i.e., DM_1) specifies the satisfactory constraint levels for the constraints of HMOP4.4, i.e.,

$$\beta = (\beta_1, \cdots, \beta_7) = (0.85, 0.95, 0.8, 0.9, 0.85, 0.8, 0.9).$$

At Step 2, the individual minimum and maximum expected values of the objective functions of HMOP4.4 are calculated under chance constrained conditions $X(\beta)$. Considering such values, each decision maker (i.e., $\mathrm{DM}_r, r = 1,2$) sets permissible objective levels as

$$\hat{f}_{11} = 2150, \hat{f}_{12} = 450, \hat{f}_{21} = -950, \hat{f}_{22} = 90.$$

At Step 3, we formulate HMOP4.3($\hat{\mathbf{f}}, \beta$) where distribution functions $p_{r\ell}(\mathbf{x}, \hat{f}_{r\ell}), r = 1,2, \ell = 1,2$ are defined. For distribution functions $p_{r\ell}(\mathbf{x}, \hat{f}_{r\ell}), r = 1,2, \ell = 1,2$, the initial decision powers and initial reference probability values are set as $w_r = 1, \hat{p}_{r\ell} = 1, r = 1,2, \ell = 1,2$ respectively. At Step 4, we solve

MINMAX4.3($\hat{\mathbf{f}},\beta,\hat{\mathbf{p}},\mathbf{w}$) via a combination of the bisection method and phase one of the two-phase simplex method, obtaining optimal value $\lambda^* = 0.372968(> 0)$. At Step 5, for optimal value $\lambda^* = 0.372968(> 0)$ of MINMAX4.3($\hat{\mathbf{f}},\beta,\hat{\mathbf{p}},\mathbf{w}$), we solve LP4.1($\hat{\mathbf{f}},\beta,\hat{\mathbf{p}},\mathbf{w},\lambda^*$) and obtain the optimal solution and corresponding tradeoff information, i.e.,

$$(p_{11}(\mathbf{x}^*,\hat{f}_{11}), p_{12}(\mathbf{x}^*,\hat{f}_{11})) = (0.619435, 0.619435)$$
$$(p_{21}(\mathbf{x}^*,\hat{f}_{21}), p_{22}(\mathbf{x}^*,\hat{f}_{22})) = (0.619435, 0.619435)$$
$$-\frac{\partial p_{11}}{\partial p_{12}} = 0.057848,\ -\frac{\partial p_{11}}{\partial p_{21}} = 1.945122,\ -\frac{\partial p_{11}}{\partial p_{22}} = 0.0.$$

At Step 6, for the above optimal solution, the first-level decision maker (i.e., DM$_1$) updates his or her decision power as $w_2 = 0.8$ to improve $p_{11}(\mathbf{x}^*,f_{11}), p_{12}(\mathbf{x}^*,f_{11})$ at the expense of $p_{21}(\mathbf{x}^*,f_{21})$, $p_{22}(\mathbf{x}^*,f_{21})$, then we go back to Step 4. At Step 4, we solve MINMAX4.3($\hat{\mathbf{f}},\beta,\hat{\mathbf{p}},\mathbf{w}$) and obtain optimal value $\lambda^* = 0.323623(> 0)$. At Step 5, for optimal value $\lambda^* = 0.323623(> 0)$ of MINMAX4.3($\hat{\mathbf{f}},\beta,\hat{\mathbf{p}},\mathbf{w}$), we solve LP4.1($\hat{\mathbf{f}},\beta,\hat{\mathbf{p}},\mathbf{w},\lambda^*$) and obtain the corresponding optimal solution as

$$(p_{11}(\mathbf{x}^*,\hat{f}_{11}), p_{12}(\mathbf{x}^*,\hat{f}_{11})) = (0.676377, 0.676377)$$
$$(p_{21}(\mathbf{x}^*,\hat{f}_{21}), p_{22}(\mathbf{x}^*,\hat{f}_{22})) = (0.595471, 0.595471)$$
$$-\frac{\partial p_{11}}{\partial p_{12}} = 0.052162,\ -\frac{\partial p_{11}}{\partial p_{21}} = 2.936,\ -\frac{\partial p_{11}}{\partial p_{22}} = 0.0.$$

At Step 6, the first-level decision maker (i.e., DM$_1$) is satisfied with the current value of distribution functions $p_{1\ell}(\mathbf{x}^*,f_{1\ell}), \ell = 1,2$, but the second-level decision maker (i.e., DM$_2$) is not satisfied with the current value of distribution functions $p_{2\ell}(\mathbf{x}^*,f_{2\ell}), \ell = 1,2$. Therefore, the second-level decision maker updates his or her reference probability values as

$$\hat{p}_{21} = 0.61, \hat{p}_{22} = 0.59$$

to improve $p_{21}(\mathbf{x},f_{21})$ at the expense of $p_{22}(\mathbf{x},f_{22})$. According to Rule 2 at Step 4, DM$_1$'s reference probability values are fixed at the current values of the distribution functions, i.e.,

$$\hat{p}_{11} = \hat{p}_{12} = 0.676377.$$

Similar interactive processes continue, with the corresponding optimal solution obtained as

$$(p_{11}(\mathbf{x}^*,\hat{f}_{11}), p_{12}(\mathbf{x}^*,\hat{f}_{11})) = (0.668782, 0.668782)$$
$$(p_{21}(\mathbf{x}^*,\hat{f}_{21}), p_{22}(\mathbf{x}^*,\hat{f}_{22})) = (0.600507, 0.580507)$$
$$-\frac{\partial p_{11}}{\partial p_{12}} = 0.051688,\ -\frac{\partial p_{11}}{\partial p_{21}} = 2.694078,\ -\frac{\partial p_{11}}{\partial p_{22}} = 0.0.$$

4.2 Hierarchical multiobjective stochastic programming problems (HMOSPs) using a fractile model

In this section, we focus on hierarchical multiobjective stochastic programming problems (HMOSPs) using a fractal optimization model [112]. In Section 4.2.1, after multiple decision makers specify their own permissible probability levels in advance, we formulate HMOSPs using a fractile optimization model and introduce an F-Pareto optimal solution for HMOSPs. In Section 4.2.2, we propose an interactive linear programming algorithm in which multiple decision makers iteratively update their reference membership values and decision powers until satisfactory solutions are obtained. In Section 4.2.3, we apply our interactive algorithm to a numerical example to illustrate the interactive processes with three hypothetical decision makers.

4.2.1 A formulation using a fractile model

In this subsection, we consider the following HMOSP in which each decision maker (i.e., $\mathrm{DM}_r, r = 1, \cdots, q$) has his or her own multiple objective functions together with common linear constraints; further, random variable coefficients are included in each objective function and on the right-hand-side of the constraints.

HMOP 4.5
first-level decision maker : $\mathbf{DM}_1$

$$\min \bar{\mathbf{z}}_1(\mathbf{x}) = (\bar{z}_{11}(\mathbf{x}), \cdots, \bar{z}_{1k_1}(\mathbf{x}))$$

$$\cdots\cdots\cdots\cdots\cdots\cdots\cdots$$

q-th-level decision maker : $\mathbf{DM}_q$

$$\min \bar{\mathbf{z}}_q(\mathbf{x}) = (\bar{z}_{q1}(\mathbf{x}), \cdots, \bar{z}_{qk_q}(\mathbf{x}))$$

subject to

$$A\mathbf{x} \leq \bar{\mathbf{b}},\ \mathbf{x} \geq \mathbf{0}$$

where $\mathbf{x} = (x_1, \cdots, x_n)^T$ is n-dimensional decision column vector, A is $(m \times n)$-dimensional coefficient matrix, and each objective function of DM_r is defined by

$$\bar{z}_{r\ell}(\mathbf{x}) = \bar{\mathbf{c}}_{r\ell}\mathbf{x} + \bar{\alpha}_{r\ell}, \ell = 1, \cdots, k_r \quad (4.17)$$

$$\bar{\mathbf{c}}_{r\ell} = \mathbf{c}^1_{r\ell} + \bar{t}_{r\ell}\mathbf{c}^2_{r\ell}, \ell = 1, \cdots, k_r \quad (4.18)$$

$$\bar{\alpha}_{r\ell} = \alpha^1_{r\ell} + \bar{t}_{r\ell}\alpha^2_{r\ell}, \ell = 1, \cdots, k_r, \quad (4.19)$$

where $\bar{\mathbf{c}}_{r\ell}, \ell = 1, \cdots, k_r$ are n-dimensional random variable row vectors, $\bar{\alpha}_{r\ell}, \ell = 1, \cdots, k_r$ are random variables, and $\bar{t}_{r\ell}, \ell = 1, \cdots, k_r$ are random variables, whose distribution functions $T_{r\ell}(\cdot), \ell = 1, \cdots, k_r$ are assumed to be strictly monotone increasing and continuous. On the right-hand-side of the constraints of HMOP4.5, $\bar{\mathbf{b}} = (\bar{b}_1, \cdots, \bar{b}_m)^T$ is a random variable column vector whose elements are independent of one another and distribution functions $F_i(\cdot), i = 1, \cdots, m$ are assumed to be strictly monotone increasing and continuous.

Similar to the formulations of multilevel linear programming problems proposed by Lee and Shih [49], we assume that the upper-level decision makers make their decisions considering the overall benefits in HMOP4.5, although they can claim priority for their objective functions over lower-level decision makers.

To handle HMOP4.5, we apply a fractile optimization model to the objective functions of HMOP4.5 and adopt chance constrained conditions with permissible constraint levels β_i for the constraint in HMOP4.5, i.e., $\mathbf{a}_i\mathbf{x} \leq \bar{b}_i, i = 1, \cdots, k$. Here, we obtain

$$\Pr(\omega \mid \mathbf{a}_i\mathbf{x} \leq b_i(\omega)) \geq \beta_i, i = 1, \cdots, m.$$

HMOP4.5 can then be transformed into the following hierarchical multiobjective programming problem, where permissible constraint levels $\beta = (\beta_1, \cdots, \beta_k)$ are specified by the first-level decision maker (i.e., DM_1) and permissible probability levels

$$\hat{\mathbf{p}}_{\mathbf{r}} \stackrel{\text{def}}{=} (\hat{p}_{r1}, \cdots, \hat{p}_{rk_r}), r = 1, \cdots, q, \ \hat{\mathbf{p}} \stackrel{\text{def}}{=} (\hat{\mathbf{p}}_{\mathbf{1}}, \cdots, \hat{\mathbf{p}}_{\mathbf{q}})$$

are subjectively specified by each decision makers (i.e., $\mathrm{DM}_r, r = 1, \cdots, q$).

HMOP 4.6 $(\hat{\mathbf{p}}, \beta)$
first-level decision maker: $\mathbf{DM}_1$

$$\min_{\mathbf{x} \in X(\beta), \mathbf{f}_1 \in \mathrm{R}^{k_1}} \quad \mathbf{f}_1 = (f_{11}, \cdots, f_{1k_1})$$

....

q-th-level decision maker: $\mathbf{DM}_q$

$$\min_{\mathbf{x} \in X(\beta), \mathbf{f}_q \in \mathrm{R}^{k_q}} \quad \mathbf{f}_q = (f_{q1}, \cdots, f_{qk_q})$$

subject to

$$p_{r\ell}(\mathbf{x}, f_{r\ell}) \geq \hat{p}_{r\ell}, \ell = 1, \cdots, k_r, r = 1, \cdots, q \tag{4.20}$$

where

$$\begin{aligned} p_{r\ell}(\mathbf{x}, f_{r\ell}) &\stackrel{\text{def}}{=} \Pr(\omega \mid z_{r\ell}(\mathbf{x}, \omega) \leq \hat{f}_{r\ell}), r = 1, \cdots, q, \ell = 1, \cdots, k_r, \\ X(\beta) &\stackrel{\text{def}}{=} \{\mathbf{x} \geq \mathbf{0} \mid \Pr(\omega \mid \mathbf{a}_i\mathbf{x} \leq b_i(\omega)) \geq \beta_i, i = 1, \cdots, m\}. \end{aligned} \tag{4.21}$$

By using distribution function $F_i(\cdot)$, constraint (4.21) can be expressed as

$$\begin{aligned} & \Pr(\omega \mid \mathbf{a}_i\mathbf{x} \leq b_i(\omega)) \geq \beta_i \\ \Leftrightarrow \quad & 1 - F_i(\mathbf{a}_i\mathbf{x}) \geq \beta_i \\ \Leftrightarrow \quad & \mathbf{a}_i\mathbf{x} \leq F_i^{-1}(1 - \beta_i). \end{aligned}$$

Similarly, under the assumption that $\mathbf{c}_{r\ell}^2\mathbf{x}+\alpha_{r\ell}^2>0, r=1,\cdots,q, \ell=1,\cdots,k_r$, the left-hand-side of constraint (4.20) in HMOP4.6$(\hat{\mathbf{p}},\beta)$ is expressed as

$$\begin{aligned}
p_{r\ell}(\mathbf{x},f_{r\ell}) &= \Pr(\omega \mid z_{r\ell}(\mathbf{x},\omega)\leq f_{r\ell}) \\
&= \Pr(\omega \mid \mathbf{c}_{r\ell}(\omega)\mathbf{x}+\alpha_{r\ell}(\omega)\leq f_{r\ell}) \\
&= \Pr\left(\omega \mid t_{r\ell}(\omega)\leq \frac{f_{r\ell}-(\mathbf{c}_{r\ell}^1\mathbf{x}+\alpha_{r\ell}^1)}{\mathbf{c}_{r\ell}^2\mathbf{x}+\alpha_{r\ell}^2}\right) \\
&= T_{r\ell}\left(\frac{f_{r\ell}-(\mathbf{c}_{r\ell}^1\mathbf{x}+\alpha_{r\ell}^1)}{\mathbf{c}_{r\ell}^2\mathbf{x}+\alpha_{r\ell}^2}\right).
\end{aligned}$$

Therefore, HMOP4.6$(\hat{\mathbf{p}},\beta)$ can be converted into the following problem.

HMOP 4.7 $(\hat{\mathbf{p}},\beta)$
first-level decision maker: DM$_1$

$$\min_{\mathbf{x}\in X(\beta),\mathbf{f}_1\in R^{k_1}} \mathbf{f}_1=(f_{11},\cdots,f_{1k_1}) \tag{4.22}$$

$$\cdots\cdots\cdots\cdots\cdots\cdots\cdots$$

q-th-level decision maker: DM$_q$

$$\min_{\mathbf{x}\in X(\beta),\mathbf{f}_q\in R^{k_q}} \mathbf{f}_q=(f_{q1},\cdots,f_{qk_q}) \tag{4.23}$$

subject to

$$T_{r\ell}\left(\frac{f_{r\ell}-(\mathbf{c}_{r\ell}^1\mathbf{x}+\alpha_{r\ell}^1)}{\mathbf{c}_{r\ell}^2\mathbf{x}+\alpha_{r\ell}^2}\right)\geq \hat{p}_{r\ell}, \ell=1,\cdots,k_r, r=1,\cdots,q \tag{4.24}$$

Constraints (4.24) can be transformed into the follows.

$$\begin{aligned}
&\hat{p}_{r\ell}\leq T_{r\ell}\left(\frac{f_{r\ell}-(\mathbf{c}_{r\ell}^1\mathbf{x}+\alpha_{r\ell}^1)}{\mathbf{c}_{r\ell}^2\mathbf{x}+\alpha_{r\ell}^2}\right) \\
\Leftrightarrow\ & f_{r\ell}\geq T_{r\ell}^{-1}(\hat{p}_{r\ell})\cdot(\mathbf{c}_{r\ell}^2\mathbf{x}+\alpha_{r\ell}^2)+(\mathbf{c}_{r\ell}^1\mathbf{x}+\alpha_{r\ell}^1)
\end{aligned} \tag{4.25}$$

If we define the right-hand-side of constraints (4.25) as

$$f_{r\ell}(\mathbf{x},\hat{p}_{r\ell}) \stackrel{\text{def}}{=} T_{r\ell}^{-1}(\hat{p}_{r\ell})\cdot(\mathbf{c}_{r\ell}^2\mathbf{x}+\alpha_{r\ell}^2)+(\mathbf{c}_{r\ell}^1\mathbf{x}+\alpha_{r\ell}^1), \tag{4.26}$$

then, HMOP4.7$(\hat{\mathbf{p}},\beta)$ can be reduced to the following problem, in which target values $f_{r\ell}, r=1,\cdots,q, \ell=1,\cdots,k_r$ have disappeared.

HMOP 4.8 $(\hat{\mathbf{p}},\beta)$
first-level decision maker: DM$_1$

$$\min_{\mathbf{x}\in X(\beta)} (f_{11}(\mathbf{x},\hat{p}_{11}),\cdots,f_{1k_1}(\mathbf{x},\hat{p}_{1k_1}))$$

$$\cdots\cdots\cdots\cdots\cdots\cdots\cdots$$

q-th-level decision maker: DM$_q$

$$\min_{\mathbf{x}\in X(\beta)} (f_{q1}(\mathbf{x},\hat{p}_{q1}),\cdots,f_{qk_q}(\mathbf{x},\hat{p}_{qk_q}))$$

To handle HMOP4.8($\hat{\mathbf{p}},\beta$), we next introduce an F-Pareto optimal solution.

Definition 4.2 $\mathbf{x}* \in X(\beta)$ is an F-Pareto optimal solution to HMOP4.8($\hat{\mathbf{p}},\beta$) if and only if there is no $\mathbf{x} \in X(\beta)$ such that $f_{r\ell}(\mathbf{x},\hat{p}_{r\ell}) \leq f_{r\ell}(\mathbf{x}^*,\hat{p}_{r\ell})$, $r=1,\cdots,q,\ell=1,\cdots,k_r$ with strict inequality holding for at least one ℓ and r.

Below, we assume that q decision makers (DM$_r, r=1,\cdots,q$) in the hierarchical decision making situation reach an agreement such that they choose satisfactory solutions to HMOP4.8($\hat{\mathbf{p}},\beta$) from an F-Pareto optimal solution set. Then, to generate a candidate for a satisfactory solution from an F-Pareto optimal solution set, each decision maker (i.e., DM$_r, r=1,\cdots,q$) is asked to specify his or her reference objective values [107, 108] $\hat{f}_{r\ell},\ell=1,\cdots,k_r$, which are reference levels of achievement of objective function $f_{r\ell}(\mathbf{x},p_{r\ell})$. Once reference objective values

$$\hat{\mathbf{f}}_r \stackrel{\text{def}}{=} (\hat{f}_{r1},\cdots,\hat{f}_{rk_r}), r=1,\cdots,q,\ \hat{\mathbf{f}} \stackrel{\text{def}}{=} (\hat{\mathbf{f}}_1,\cdots,\hat{\mathbf{f}}_q)$$

are specified, the corresponding F-Pareto optimal solution, which is in a sense close to each decision maker's requirements, is obtained by solving the following minmax problem.

MINMAX 4.4 ($\hat{\mathbf{p}},\beta,\hat{\mathbf{f}}$)

$$\min_{\mathbf{x}\in X(\beta),\lambda\in \mathrm{R}^1} \lambda \tag{4.27}$$

subject to

$$f_{r\ell}(\mathbf{x},\hat{p}_{r\ell}) - \hat{f}_{r\ell} \leq \lambda, r=1,\cdots,q,\ell=1,\cdots,k_r \tag{4.28}$$

Note that constraints (4.28) in MINMAX4.4($\hat{\mathbf{p}},\beta,\hat{\mathbf{f}}$) become linear constraints, i.e.,

$$\begin{aligned} & f_{r\ell}(\mathbf{x},\hat{p}_{r\ell}) - \hat{f}_{r\ell} \leq \lambda \\ \Leftrightarrow\quad & T_{r\ell}^{-1}(\hat{p}_{r\ell})\cdot(\mathbf{c}_{r\ell}^2\mathbf{x}+\alpha_{r\ell}^2)+(\mathbf{c}_{r\ell}^1\mathbf{x}+\alpha_{r\ell}^1) - \hat{f}_{r\ell} \leq \lambda. \end{aligned} \tag{4.29}$$

The relationships between optimal solution $\mathbf{x}^* \in X(\beta),\lambda^* \in \mathrm{R}^1$ of MINMAX4.4 ($\hat{\mathbf{p}},\beta,\hat{\mathbf{f}}$) and F-Pareto optimal solutions to HMOP4.8($\hat{\mathbf{p}},\beta$) are characterized by the following theorem.

Theorem 4.3

(1) If $\mathbf{x}^* \in X(\beta),\lambda^* \in \mathrm{R}^1$ is a unique optimal solution of MINMAX4.4($\mathbf{p},\beta,\hat{\mathbf{f}}$), then $\mathbf{x}^* \in X(\beta),\lambda^* \in \mathrm{R}^1$ is an F-Pareto optimal solution to HMOP4.8($\hat{\mathbf{p}},\beta$).

(2) If $\mathbf{x}^* \in X(\beta)$ is an F-Pareto optimal solution to HMOP4.8($\hat{\mathbf{p}},\beta$), then $\mathbf{x}^* \in$

$X(\beta), \lambda^* \in \mathrm{R}^1$ is an optimal solution of MINMAX4.4($\hat{\mathbf{p}}, \beta, \hat{\mathbf{f}}$) for some reference objective values $\hat{\mathbf{f}}$, where $\hat{f}_{r\ell} \stackrel{\text{def}}{=} f_{r\ell}(\mathbf{x}^*, p_{r\ell}) + \lambda^*, r = 1, \cdots, q, \ell = 1, \cdots, k_r$.

Note that, in general, an F-Pareto optimal solution obtained by solving MINMAX4.4($\hat{\mathbf{p}}, \beta, \hat{\mathbf{f}}$) does not reflect the hierarchical structure between q decision makers, in which the upper-level decision maker can claim priority for his or her distribution functions over lower-level decision makers. To cope with such a hierarchical preference structure between q decision makers in MINMAX4.4($\hat{\mathbf{p}}, \beta, \hat{\mathbf{f}}$), we introduce decision power [47, 49] $w_r, r = 1, \cdots, q$ for constraints (4.28), where the r-th level decision maker (i.e., DM_r) can subjectively specify decision power w_{r+1} and the last decision maker (i.e., DM_q) has no decision power. Therefore, decision powers $\mathbf{w} = (w_1, w_2, \cdots, w_q)^T$ must satisfy inequality

$$1 = w_1 \geq \cdots \geq w_q > 0. \tag{4.30}$$

Then, we reformulate the corresponding modified minmax problem as follows.

MINMAX 4.5 ($\hat{\mathbf{p}}, \beta, \hat{\mathbf{f}}, \mathbf{w}$)

$$\min_{\mathbf{x} \in X(\beta), \lambda \in \mathrm{R}^1} \lambda \tag{4.31}$$

subject to

$$f_{r\ell}(\mathbf{x}, p_{r\ell}) - \hat{f}_{r\ell} \leq \lambda / w_r, r = 1, \cdots, q, \ell = 1, \cdots, k_r \tag{4.32}$$

The relationships between optimal solution $\mathbf{x}^* \in X(\beta), \lambda^* \in \mathrm{R}^1$ of MINMAX4.5 ($\mathbf{p}, \beta, \hat{\mathbf{f}}, \mathbf{w}$) and F-Pareto optimal solutions to HMOP4.8($\mathbf{p}, \beta$) are characterized by the following theorem.

Theorem 4.4

If $\mathbf{x}^* \in X(\beta), \lambda^* \in \mathrm{R}^1$ is a unique optimal solution of MINMAX4.5($\hat{\mathbf{p}}, \beta, \hat{\mathbf{f}}, \mathbf{w}$), then $\mathbf{x}^* \in X(\beta)$ is an F-Pareto optimal solution to HMOP4.8($\mathbf{p}, \beta$).

In Theorem 4.4, if optimal solution $\mathbf{x}^* \in X(\beta), \lambda^* \in \mathrm{R}^1$ of MINMAX4.5($\hat{\mathbf{p}}$, $\beta, \hat{\mathbf{f}}, \mathbf{w}$) is not unique, then F-Pareto optimality cannot be guaranteed. To guarantee F-Pareto optimality for $\mathbf{x}^* \in X(\beta), \lambda^* \in \mathrm{R}^1$, we formulate the F-Pareto optimality test problem as follows.

Test 4.1

$$w \stackrel{\text{def}}{=} \max_{\mathbf{x} \in X(\beta), \varepsilon \geq \mathbf{0}} \sum_{r=1}^{q} \sum_{\ell=1}^{k_r} \varepsilon_{r\ell} \tag{4.33}$$

subject to

$$f_{r\ell}(\mathbf{x}^*, \hat{p}_{r\ell}) - f_{r\ell}(\mathbf{x}, \hat{p}_{r\ell}) = \varepsilon_{r\ell}, r = 1, \cdots, q, \ell = 1, \cdots, k_r \tag{4.34}$$

where

$$\varepsilon \stackrel{\text{def}}{=} (\varepsilon_{11}, \cdots, \varepsilon_{1k_1}, \cdots, \varepsilon_{q1}, \cdots, \varepsilon_{qk_q}). \tag{4.35}$$

The following theorem describes the relationships between the optimal solution of TEST4.1 and F-Pareto optimal solutions to HMOP4.8$(\hat{\mathbf{p}}, \beta)$.

Theorem 4.5

Let $\mathbf{x}^* \in X(\beta), \check{\varepsilon} \geq \mathbf{0}$, be an optimal solution of TEST4.1 for optimal solution $\mathbf{x}^* \in X(\beta), \lambda^* \in \mathrm{R}^1$ of MINMAX4.5$(\hat{\mathbf{p}}, \beta, \hat{\mathbf{f}}, \mathbf{w})$. If $\check{\varepsilon}_{r\ell} = 0, r = 1, \cdots, q, \ell = 1, \cdots, k_r$, then $\mathbf{x}^* \in X(\beta)$ is an F-Pareto optimal solution to HMOP4.8$(\hat{\mathbf{p}}, \beta)$.

4.2.2 An interactive linear programming algorithm

After obtaining F-Pareto optimal solution $\mathbf{x}^* \in X(\beta)$ by solving MINMAX4.5 $(\hat{\mathbf{p}}, \beta, \hat{\mathbf{f}}, \mathbf{w})$ via linear programming, each decision maker (i.e., DM$_r, r = 1, \cdots, q$) must either be satisfied with the current values of his or her objective functions $f_{r\ell}(\mathbf{x}^*, p_{r\ell}), \ell = 1, \cdots, k_r$, or update his or her decision power w_{r+1} and/or his or her reference objective values $\hat{\mathbf{f}}_{\mathbf{r}}$.

To help each decision maker update his or her reference objective values, tradeoff information is quite useful. Such tradeoff information is obtainable since it relates to simplex multipliers $\pi_{r\ell}(\geq 0)$ for constraints (4.32) in MINMAX4.5$(\hat{\mathbf{p}}, \beta, \hat{\mathbf{f}}, \mathbf{w})$ [86].

Theorem 4.6

Let $\mathbf{x}^* \in X(\beta), \lambda^* \in \mathrm{R}^1$ be a unique and non degenerate optimal solution of MINMAX4.5$(\hat{\mathbf{p}}, \beta, \hat{\mathbf{f}}, \mathbf{w})$, and let constraints (4.32) be active. Then, the following relation holds.

$$-\left.\frac{\partial(f_{r_1\ell_1}(\mathbf{x}, \hat{p}_{r_1\ell_1}))}{\partial(f_{r_2\ell_2}(\mathbf{x}, \hat{p}_{r_2\ell_2}))}\right|_{\mathbf{x}=\mathbf{x}^*} = \frac{\pi^*_{r_2\ell_2}}{\pi^*_{r_1\ell_1}} \tag{4.36}$$

$$\left.\frac{\partial(f_{r\ell}(\mathbf{x}, \hat{p}_{r\ell}))}{\partial w_r}\right|_{\mathbf{w}=\mathbf{w}^*} = -\frac{\lambda^*}{w_r^{*2}} + \frac{\lambda^*}{w_r^{*3}}\left\{\sum_{i=1}^{k_r} \pi^*_{ri}\right\}, \tag{4.37}$$

where $\pi^*_{r\ell} \geq 0, r = 1, \cdots, q, \ell = 1, \cdots, k_r$ are corresponding simplex multipliers for constraints (4.32).

Next, we can construct an interactive algorithm to derive a satisfactory solution from multiple decision makers (i.e., DM$_r, r = 1, \cdots, q$) in a hierarchical organization from an F-Pareto optimal solution set corresponding to HMOP4.8$(\hat{\mathbf{p}}, \beta)$.

Algorithm 4.2

Step 1: *The first-level decision maker (i.e., DM$_1$) subjectively sets satisfactory constraint levels $0 < \beta_i < 1, i = 1, \cdots, m$ for constraint $\mathbf{a}_i\mathbf{x} \leq \bar{b}_i$ of HMOP4.5.*

Step 2: *Each decision maker (i.e., $DM_r, r = 1, \cdots, q$) sets permissible probability levels $\hat{p}_{r\ell}, \ell = 1, \cdots, k_r$ for objective functions $\bar{z}_{r\ell}(\mathbf{x}), \ell = 1, \cdots, k_r$ of HMOP4.5.*
Step 3: *In HMOP4.8($\hat{\mathbf{p}}, \beta$), set initial decision powers as $w_r = 1, r = 1, \cdots, q$, and each decision maker (i.e., $DM_r, r = 1, \cdots, q$) subjectively sets initial reference objective values $\hat{f}_{r\ell} = 1, \ell = 1, \cdots, k_r$ for objective function $f_{r\ell}(\mathbf{x}, \hat{p}_{r\ell})$.*
Step 4: *Using a combination of the bisection method and phase one of the two-phase simplex method, solve MINMAX4.5($\hat{\mathbf{p}}, \beta, \hat{\mathbf{f}}, \mathbf{w}$) to obtain optimal value λ^*. If optimal value $\lambda^* \geq 0$, then go to Step 5. Otherwise, after updating reference objective values as $\hat{f}_{r\ell} \leftarrow \hat{f}_{r\ell} + \lambda^*/w_r$, $r = 1, \cdots, q, \ell = 1, \cdots, k_r$, solve MINMAX4.5($\hat{\mathbf{p}}, \beta, \hat{\mathbf{f}}, \mathbf{w}$), and go to Step 5.*
Step 5: *For optimal solution $\mathbf{x}^* \in X(\beta), \lambda^* \in \mathrm{R}^1$ of MINMAX4.5($\hat{\mathbf{p}}, \beta, \hat{\mathbf{f}}, \mathbf{w}$), solve TEST4.1, and compute the tradeoff rates between $f_{r\ell}(\mathbf{x}^*, \hat{p}_{r\ell}), r = 1, \cdots, q, \ell = 1, \cdots, k_r$.*
Step 6: *If each decision makers (i.e., $DM_r, r = 1, \cdots, q$) is satisfied with the current values of his or her objective functions $f_{r\ell}(\mathbf{x}^*, \hat{p}_{r\ell})$, $\ell = 1, \cdots, k_r$, then stop. Otherwise, let the s-th-level decision maker (i.e., DM_s) be the uppermost decision maker who is not satisfied with the current values of his or her objective functions $f_{s\ell}(\mathbf{x}^*, \hat{p}_{s\ell})$, $\ell = 1, \cdots, k_s$. Considering current values $f_{s\ell}(\mathbf{x}^*, \hat{p}_{s\ell})$, $\ell = 1, \cdots, k_s$ and tradeoff rates, DM_s updates his or her decision power w_{s+1} and/or his or her reference objective values $\hat{f}_{s\ell}, \ell = 1, \cdots, k_s$ according to the following two rules, then we return to Step 4.*
Rule 1: *When the decision maker (i.e., DM_s) updates his or her decision power w_{s+1}, $w_{s+1} \leq w_s$ must be satisfied to guarantee inequality conditions (4.30). After updating w_{s+1}, if there exists some index $t > s+1$ such that $w_{s+1} < w_t$, then, corresponding decision power w_t must be replaced with $w_t \leftarrow w_{s+1}$.*
Rule 2: *When the decision maker (i.e., DM_s) updates his or her reference objective values $\hat{\mathbf{f}}_s, i = 1, \cdots, k_s$, the reference objective values of the other decision makers (i.e., $DM_r, r = 1, \cdots, q, r \neq s$) must be set to the current values of objective functions $f_{r\ell}(\mathbf{x}^*, \hat{p}_{r\ell}), \ell = 1, \cdots, k_r$, i.e., $\hat{f}_{r\ell} \leftarrow f_{r\ell}(\mathbf{x}^*, \hat{p}_{r\ell}), r = 1, \cdots, q, r \neq s, \ell = 1, \cdots, k_r$.*

Note that when a decision maker (i.e., DM_s) updates his or her decision power w_{s+1} according to Rule 1 at Step 6, objective functions $f_{s\ell}(\mathbf{x}^*, \hat{p}_{s\ell}), \ell = 1, \cdots, k_r$ will be improved by the lesser value of w_{s+1} at the expense of other objective functions $f_{r\ell}(\mathbf{x}^*, \hat{p}_{r\ell}), r = s+1, \cdots, q, \ell = 1, \cdots, k_r$ for fixed reference objective values $\hat{f}_{r\ell}, r = 1, \cdots, q, \ell = 1, \cdots, k_s$. Similarly, when a decision maker (i.e., DM_s) updates his or her reference objective values $\hat{f}_{s\ell}, \ell = 1, \cdots, k_s$ according to Rule 2 at Step 6, any improvements to one objective function are achieved only at the expense of at least one of the other functions for fixed decision powers $w_r, r = 1, \cdots, q$.

4.2.3 A numerical example

To demonstrate our proposed method and the interactive processes, in this subsection, we consider the following hierarchical two objective stochastic program-

Table 4.2: Parameters of the objective functions and constraints in HMOP4.9

$\mathbf{x}$	x_1	x_2	x_3	x_4	x_5	x_6	x_7	x_8	x_9	x_{10}
$\mathbf{c}_{11}^1$	19	48	21	10	18	35	46	11	24	33
$\mathbf{c}_{11}^2$	3	2	2	1	4	3	1	2	4	2
$\mathbf{c}_{12}^1$	12	–46	–23	–38	–33	–48	12	8	19	20
$\mathbf{c}_{12}^2$	1	2	4	2	2	1	2	1	2	1
$\mathbf{c}_{21}^1$	12	38	–23	33	–33	45	12	–9	19	20
$\mathbf{c}_{21}^2$	1	2	4	2	2	1	2	1	2	1
$\mathbf{c}_{22}^1$	12	–36	27	–30	–33	45	–11	12	19	–8
$\mathbf{c}_{22}^2$	1	2	4	2	2	1	2	1	2	1
$\mathbf{c}_{31}^1$	–18	–26	–22	–28	–15	–29	–10	–19	–17	–28
$\mathbf{c}_{31}^2$	2	1	3	2	1	2	3	3	2	1
$\mathbf{c}_{32}^1$	–8	31	28	29	25	36	–8	–7	–13	–15
$\mathbf{c}_{32}^2$	1	2	3	2	2	1	2	1	2	1
$\mathbf{a}_1$	12	–2	4	–7	13	–1	–6	6	11	–8
$\mathbf{a}_2$	–2	5	3	16	6	–12	12	4	–7	–10

ming problem with three hypothetical decision makers (i.e., DM_1, DM_2 and DM_3).

HMOP 4.9
first-level decision maker: DM_1

$$\begin{aligned}\min \bar{z}_{11}(\mathbf{x}) &= (\mathbf{c}_{11}^1 + \bar{t}_{11}\mathbf{c}_{11}^2)\mathbf{x} + (\alpha_{11}^1 + \bar{t}_{11}\alpha_{11}^2)\\ \min \bar{z}_{12}(\mathbf{x}) &= (\mathbf{c}_{12}^1 + \bar{t}_{12}\mathbf{c}_{12}^2)\mathbf{x} + (\alpha_{12}^1 + \bar{t}_{12}\alpha_{12}^2)\end{aligned}$$

second-level decision maker: DM_2

$$\begin{aligned}\min \bar{z}_{21}(\mathbf{x}) &= (\mathbf{c}_{21}^1 + \bar{t}_{21}\mathbf{c}_{21}^2)\mathbf{x} + (\alpha_{21}^1 + \bar{t}_{21}\alpha_{21}^2)\\ \min \bar{z}_{22}(\mathbf{x}) &= (\mathbf{c}_{22}^1 + \bar{t}_{22}\mathbf{c}_{22}^2)\mathbf{x} + (\alpha_{22}^1 + \bar{t}_{22}\alpha_{22}^2)\end{aligned}$$

third-level decision maker: DM_3

$$\begin{aligned}\min \bar{z}_{31}(\mathbf{x}) &= (\mathbf{c}_{31}^1 + \bar{t}_{31}\mathbf{c}_{31}^2)\mathbf{x} + (\alpha_{31}^1 + \bar{t}_{31}\alpha_{31}^2)\\ \min \bar{z}_{32}(\mathbf{x}) &= (\mathbf{c}_{32}^1 + \bar{t}_{32}\mathbf{c}_{32}^2)\mathbf{x} + (\alpha_{32}^1 + \bar{t}_{32}\alpha_{32}^2)\end{aligned}$$

subject to

$$\mathbf{x} \in X \stackrel{\text{def}}{=} \{\mathbf{x} \in \mathrm{R}^{10} \mid \mathbf{a}_i\mathbf{x} \leq \bar{b}_i, i = 1,2, \mathbf{x} \geq \mathbf{0}\}$$

where $\mathbf{a}_i, i = 1,2, \mathbf{c}_{r\ell}^1, \mathbf{c}_{r\ell}^2, r = 1,2,3,\ \ell = 1,2$ are the ten-dimensional coefficient row vectors which are shown in Table 4.2, $\alpha_{11}^1 = -18, \alpha_{11}^2 = 5,\ \alpha_{12}^1 =$

Table 4.3: Interactive processes in HMOP4.9 (No.1)

First iteration $(w_1, w_2, w_3) = (1,1,1)$			
$\hat{f}_{11}$	680	$f_{11}(\mathbf{x}^*, \hat{p}_{11})$	2179.10
$\hat{f}_{12}$	–1000	$f_{12}(\mathbf{x}^*, \hat{p}_{12})$	369.098
$\hat{f}_{21}$	–500	$f_{21}(\mathbf{x}^*, \hat{p}_{21})$	999.101
$\hat{f}_{22}$	–1400	$f_{22}(\mathbf{x}^*, \hat{p}_{22})$	99.1006
$\hat{f}_{31}$	–2500	$f_{31}(\mathbf{x}^*, \hat{p}_{31})$	–1079.56
$\hat{f}_{32}$	–2300	$f_{32}(\mathbf{x}^*, \hat{p}_{32})$	–800.899
Second iteration $(w_1, w_2, w_3) = (1,0.7,0.7)$ specified by DM_1			
$\hat{f}_{11}$	680	$f_{11}(\mathbf{x}^*, \hat{p}_{11})$	1779.50
$\hat{f}_{12}$	–1000	$f_{12}(\mathbf{x}^*, \hat{p}_{12})$	99.5002
$\hat{f}_{21}$	–500	$f_{21}(\mathbf{x}^*, \hat{p}_{21})$	1070.71
$\hat{f}_{22}$	–1400	$f_{22}(\mathbf{x}^*, \hat{p}_{22})$	161.291
$\hat{f}_{31}$	–2500	$f_{31}(\mathbf{x}^*, \hat{p}_{31})$	–929.285
$\hat{f}_{32}$	–2300	$f_{32}(\mathbf{x}^*, \hat{p}_{32})$	–729.285

$-27, \alpha_{12}^2 = 6,\ \alpha_{21}^1 = -12, \alpha_{21}^2 = 3,\ \alpha_{22}^1 = -15, \alpha_{22}^2 = 4,\ \alpha_{31}^1 = -10, \alpha_{31}^2 = 4,$ $\alpha_{32}^1 = -27, \alpha_{32}^2 = 6$, and

$$\bar{t}_{11} \sim \mathrm{N}(4,2^2), \bar{t}_{12} \sim \mathrm{N}(3,3^2), \bar{t}_{21} \sim \mathrm{N}(2,1^2), \bar{t}_{22} \sim \mathrm{N}(3,2^2),$$
$$\bar{t}_{31} \sim \mathrm{N}(3,2^2), \bar{t}_{32} \sim \mathrm{N}(3,3^2), \bar{b}_1 \sim \mathrm{N}(164,30^2), \bar{b}_2 \sim \mathrm{N}(-190,20^2)$$

are Gaussian random variables.

Tables 4.3 and 4.4 show the interactive processes with three hypothetical decision makers (i.e., DM_1, DM_2 and DM_3). At Step 1, the first-level decision maker (i.e., DM_1) specifies satisfactory constraint levels for the constraints in HMOP4.1 as

$$\beta = (\beta_1, \cdots, \beta_7) = (0.85, 0.95, 0.8, 0.9, 0.85, 0.8, 0.9).$$

At Step 2, each decision maker (i.e., DM_r) sets initial permissible probability levels to

$$(p_{11}, p_{12}) = (0.7, 0.7), (p_{21}, p_{22}) = (0.7, 0.7), (p_{31}, p_{32}) = (0.7, 0.7),$$

respectively. At Step 3, the initial decision powers are set to $(w_1, w_2, w_3) = (1,1,1)$ and the reference probability values are set to

$$\hat{f}_{11} = 680,\ \hat{f}_{12} = -1000, \hat{f}_{21} = -2500,\ \hat{f}_{22} = -2300, \hat{f}_{31} = -500,\ \hat{f}_{32} = -1400.$$

At Step 4, MINMAX4.5($\hat{\mathbf{p}}, \beta, \hat{\mathbf{f}}, \mathbf{w}$) is solved, with the corresponding optimal solution shown in Table 4.3. At Step 5, for the optimal solution of MINMAX4.5($\hat{\mathbf{p}}, \beta, \hat{\mathbf{f}}, \mathbf{w}$), TEST4.1 is solved. At Step 6, the first-level decision

Table 4.4: Interactive processes in HMOP4.9 (No.2)

Third iteration		$(w_1,w_2,w_3) = (1,0.7,0.6)$ specified by DM_2	
$\hat{f}_{11}$	680	$f_{11}(\mathbf{x}^*,\hat{p}_{11})$	1715.03
$\hat{f}_{12}$	–1000	$f_{12}(\mathbf{x}^*,\hat{p}_{12})$	35.0344
$\hat{f}_{21}$	–500	$f_{21}(\mathbf{x}^*,\hat{p}_{21})$	649.863
$\hat{f}_{22}$	–1400	$f_{22}(\mathbf{x}^*,\hat{p}_{22})$	78.6205
$\hat{f}_{31}$	–2500	$f_{31}(\mathbf{x}^*,\hat{p}_{31})$	–802.790
$\hat{f}_{32}$	–2300	$f_{32}(\mathbf{x}^*,\hat{p}_{32})$	–574.943
Fourth iteration		$(w_1,w_2,w_3) = (1,0.7,0.6)$	
$\hat{f}_{11}$	1715.03	$f_{11}(\mathbf{x}^*,\hat{p}_{11})$	1730.16
$\hat{f}_{12}$	35.0344	$f_{12}(\mathbf{x}^*,\hat{p}_{12})$	50.1596
$\hat{f}_{21}$	649.863	$f_{21}(\mathbf{x}^*,\hat{p}_{21})$	671.470
$\hat{f}_{22}$	78.6205	$f_{22}(\mathbf{x}^*,\hat{p}_{22})$	30.0072
$\hat{f}_{31}$	–850	$f_{31}(\mathbf{x}^*,\hat{p}_{31})$	–824.791
$\hat{f}_{32}$	–530	$f_{32}(\mathbf{x}^*,\hat{p}_{32})$	–504.79

maker (i.e., DM_1) is not satisfied with the current value objective functions $f_{1\ell}(\mathbf{x}^*,\hat{p}_{1\ell}), \ell = 1,2$. Therefore, the first-level decision maker (DM_1) updates his or her decision power $w_2 = 0.7$ to improve $f_{1\ell}(\mathbf{x}^*,\hat{p}_{1\ell}), \ell = 1,2$ at the expense of $f_{r\ell}(\mathbf{x}^*,\hat{p}_{r\ell}), r = 2,3, \ell = 1,2$. At Step 4, MINMAX4.5$(\hat{\mathbf{p}},\beta,\hat{\mathbf{f}},\mathbf{w})$ is solved, with the corresponding optimal solution shown in Table 4.3. At Step 6, the first-level decision maker (i.e., DM_1) is satisfied with the current value of the objective functions, but the second-level decision maker (i.e., DM_2) is not satisfied with the current value of the objective functions. Therefore, the second-level decision maker updates his or her decision power $w_3 = 0.6$ to improve $f_{2\ell}(\mathbf{x}^*,\hat{p}_{2\ell}), \ell = 1,2$ at the expense of $f_{3\ell}(\mathbf{x}^*,\hat{p}_{3\ell}), \ell = 1,2$. At Step 4, MINMAX4.5$(\hat{\mathbf{p}},\beta,\hat{\mathbf{f}},\mathbf{w})$ is solved, with the corresponding optimal solution shown in Table 4.4. At Step 6, the top-level decision makers (i.e., DM_1 and DM_2) are satisfied with the current values of the objective functions, but the third-level decision maker (i.e., DM_3) is not satisfied with the current value of the objective functions. Therefore, the third-level decision maker (i.e., DM_3) updates his or her reference objective values to

$$\hat{f}_{31} = -850,\ \hat{f}_{32} = -530$$

to improve $f_{31}(\mathbf{x}^*,\hat{p}_{31})$ at the expense of $f_{32}(\mathbf{x}^*,\hat{p}_{32})$. At Step 4, MINMAX4.5 $(\hat{\mathbf{p}},\beta,\hat{\mathbf{f}},\mathbf{w})$ is solved, with the corresponding optimal solution shown in Table 4.4. Similar interactive processes continue.

4.3 Hierarchical multiobjective stochastic programming problems (HMOSPs) based on the fuzzy decision

In Sections 4.1 and 4.2 above, we proposed two interactive algorithms for obtaining a satisfactory solution using a probability maximization model and a fractile optimization model, respectively. In these methods, multiple decision makers specify their own permissible objective levels or permissible probability levels in advance. To automatically set such parameters appropriately, we propose an interactive algorithm based on the fuzzy decision for HMOSPs [113, 115, 118]. In Section 4.3.1, under the assumption that multiple decision makers have fuzzy goals not only for permissible objective levels but also for corresponding distribution functions in a probability maximization model, we introduce an MP-Pareto optimal solution for HMOSPs. Similarly, in Section 4.3.2, under the assumption that multiple decision makers have fuzzy goals not only for permissible probability levels but also for corresponding objective functions in a fractile optimization model, we introduce an MF-Pareto optimal solution for HMOSPs. In Section 4.3.3, we propose an interactive linear programming algorithm to obtain a satisfactory solution for multiple decision makers from an MF-Pareto optimal solution set. In Section 4.3.4, we apply our interactive algorithm to a numerical example to illustrate the interactive processes with three hypothetical decision makers.

4.3.1 A formulation using a probability model

In this subsection, we consider the following HMOPs, where each decision maker (i.e., $\mathrm{DM}_r, r = 1, \cdots, q$) has his or her own multiple objective functions along with common linear constraints; further, random variable coefficients are included in each objective function.

HMOP 4.10

first-level decision maker: $\mathbf{DM_1}$

$$\min_{\mathbf{x} \in X} \bar{\mathbf{z}}_1(\mathbf{x}) = (\bar{z}_{11}(\mathbf{x}), \cdots, \bar{z}_{1k_1}(\mathbf{x}))$$

$$\cdots\cdots\cdots\cdots\cdots\cdots\cdots$$

q-th-level decision maker: $\mathbf{DM}_q$

$$\min_{\mathbf{x} \in X} \bar{\mathbf{z}}_q(\mathbf{x}) = (\bar{z}_{q1}(\mathbf{x}), \cdots, \bar{z}_{qk_q}(\mathbf{x}))$$

where $\mathbf{x} = (x_1, \cdots, x_n)^T$ is n-dimensional decision column vector whose elements $x_i, i = 1, \cdots, n$ are nonnegative and X is a linear constraint set with respect to $\mathbf{x}$. Each objective function of $\text{DM}_r, r = 1, \cdots, q$ is defined by

$$\begin{aligned} \bar{z}_{r\ell}(\mathbf{x}) &= \bar{\mathbf{c}}_{r\ell}\mathbf{x} + \bar{\alpha}_{r\ell}, \ell = 1, \cdots, k_r \\ \bar{\mathbf{c}}_{r\ell} &= \mathbf{c}^1_{r\ell} + \bar{t}_{r\ell}\mathbf{c}^2_{r\ell}, \ell = 1, \cdots, k_r \\ \bar{\alpha}_{r\ell} &= \alpha^1_{r\ell} + \bar{t}_{r\ell}\alpha^2_{r\ell}, \ell = 1, \cdots, k_r, \end{aligned}$$

where $\bar{\mathbf{c}}_{r\ell}, \ell = 1, \cdots, k_r$ are n-dimensional random variable row vectors, $\bar{\alpha}_{r\ell}, \ell = 1, \cdots, k_r$ are random variables, and $\bar{t}_{r\ell}$ is a random variable whose distribution function $T_{r\ell}(\cdot)$ is assumed to be strictly monotone increasing and continuous. Without loss of generality, we assume that $\mathbf{c}^2_{r\ell} > \mathbf{0}, \alpha^2_{r\ell} > 0, r = 1, \cdots, q, \ell = 1, \cdots, k_r$.

Similar to the formulations of multilevel linear programming problems proposed by Lee et al. [49], we assume that upper-level decision makers in HMOP4.10 make their decisions considering the overall benefits for the hierarchical organization, although they can claim priority for their objective functions over lower-level decision makers.

To handle HMOP4.10, using a probability maximization model [73], we substitute the minimization of objective function $\bar{z}_{r\ell}(\mathbf{x})$ in HMOP4.10 for the maximization of the probability that $\bar{z}_{r\ell}(\mathbf{x})$ is less than or equal to a certain permissible objective level $\hat{f}_{r\ell}$. Such a probability $p_{r\ell}(\mathbf{x}, \hat{f}_{r\ell})$ can be defined as

$$p_{r\ell}(\mathbf{x}, \hat{f}_{r\ell}) \stackrel{\text{def}}{=} \Pr(\omega \mid z_{r\ell}(\mathbf{x}, \omega) \leq \hat{f}_{r\ell}), \tag{4.38}$$

where $\Pr(\cdot)$ denotes a probability measure, ω is an event, and $z_{r\ell}(\mathbf{x}, \omega)$ is a realization of random objective function $\bar{z}_{r\ell}(\mathbf{x})$ for the occurrence of each elementary event ω. Each decision maker (i.e., $\text{DM}_r, r = 1, \cdots, q$) subjectively specifies certain permissible objective levels, i.e.,

$$\hat{\mathbf{f}}_r \stackrel{\text{def}}{=} (\hat{f}_{r1}, \cdots, \hat{f}_{rk_r}), r = 1, \cdots, q, \quad \hat{\mathbf{f}} \stackrel{\text{def}}{=} (\hat{\mathbf{f}}_1, \cdots, \hat{\mathbf{f}}_q).$$

Then, HMOP4.10 can be transformed into the following problem involving permissible objective levels.

HMOP 4.11 $(\hat{\mathbf{f}})$
first-level decision maker: DM_1

$$\max_{\mathbf{x} \in X} (p_{11}(\mathbf{x}, \hat{f}_{11}), \cdots, p_{1k_1}(\mathbf{x}, \hat{f}_{1k_1}))$$

....

q-th-level decision maker: DM_q

$$\max_{\mathbf{x} \in X} (p_{q1}(\mathbf{x}, \hat{f}_{q1}), \cdots, p_{qk_q}(\mathbf{x}, \hat{f}_{qk_q}))$$

Given the assumption that $\mathbf{c}_{r\ell}^2\mathbf{x}+\alpha_{r\ell}^2>0, r=1,\cdots,q,\ell=1,\cdots,k_r$, objective function $p_{r\ell}(\mathbf{x},\hat{f}_{r\ell})$ in HMOP4.11($\hat{\mathbf{f}}$) is expressed as

$$\begin{aligned} p_{r\ell}(\mathbf{x},\hat{f}_{r\ell}) &= \Pr(\omega \mid z_{r\ell}(\mathbf{x},\omega) \leq \hat{f}_{r\ell}) \\ &= T_{r\ell}\left(\frac{\hat{f}_{r\ell}-(\mathbf{c}_{r\ell}^1\mathbf{x}+\alpha_{r\ell}^1)}{\mathbf{c}_{r\ell}^2\mathbf{x}+\alpha_{r\ell}^2}\right). \end{aligned}$$

In HMOP4.11($\mathbf{f}$), each decision maker (i.e., $\mathrm{DM}_r, r=1,\cdots,q$) not only prefers smaller values of permissible objective levels $\hat{\mathbf{f}}_r$ but also prefers larger values of corresponding distribution functions $p_{r\ell}(\mathbf{x},\hat{f}_{r\ell}),\ell=1,\cdots,k_r$. Since these values conflict with one another, the smaller value of permissible objective level results in a smaller value of the corresponding probability function. Therefore, it is important for each decision maker ($\mathrm{DM}_r, r=1,\cdots,q$) to determine appropriate values of permissible objective levels $\hat{\mathbf{f}}_r$. Unfortunately, it is difficult for decision makers to find appropriate values of permissible objective levels. To circumvent such difficulties, Yano et al. [124] proposed a fuzzy approach for multiobjective stochastic linear programming problems. From a similar perspective, instead of HMOP4.11($\hat{\mathbf{f}}$), we consider the following hierarchical multiobjective programming problem in which permissible objective levels are not constant values, but instead are decision variables.

HMOP 4.12

first-level decision maker: DM_1

$$\max_{\mathbf{x}\in X,\hat{\mathbf{f}}_1\in R^{k_1}} \quad (p_{11}(\mathbf{x},\hat{f}_{11}),\cdots,p_{1k_1}(\mathbf{x},\hat{f}_{1k_1}),-\hat{f}_{11},\cdots,-\hat{f}_{1k_1})$$

....

q-th-level decision maker: DM_q

$$\max_{\mathbf{x}\in X,\hat{\mathbf{f}}_q\in R^{k_q}} \quad (p_{q1}(\mathbf{x},\hat{f}_{q1}),\cdots,p_{qk_q}(\mathbf{x},\hat{f}_{qk_q}),-\hat{f}_{q1},\cdots,-\hat{f}_{qk_q})$$

Considering the imprecise nature of a decision maker's judgment, it is natural to assume that the decision maker has a fuzzy goal for each objective function in HMOP4.12. In this subsection, we assume that such a fuzzy goal can be quantified by eliciting a corresponding membership function. Let us denote a membership function of $p_{r\ell}(\mathbf{x},\hat{f}_{r\ell})$ as $\mu_{p_{r\ell}}(p_{r\ell}(\mathbf{x},\hat{f}_{r\ell}))$ and a membership function of $\hat{f}_{r\ell}$ as $\mu_{\hat{f}_{r\ell}}(\hat{f}_{r\ell})$. Then, HMOP4.12 can be transformed into the following multiobjective programming problem.

HMOP 4.13

first-level decision maker: DM_1

$$\max_{\mathbf{x}\in X,\hat{\mathbf{f}}_1\in R^{k_1}} \left(\mu_{p_{11}}(p_{11}(\mathbf{x},\hat{f}_{11})),\cdots,\mu_{p_{1k_1}}(p_{1k_1}(\mathbf{x},\hat{f}_{1k_1})),\mu_{\hat{f}_{11}}(\hat{f}_{11}),\cdots,\mu_{\hat{f}_{1k_1}}(\hat{f}_{1k_1})\right)$$

....

q-th-level decision maker: DM$_q$

$$\max_{\mathbf{x}\in X,\hat{\mathbf{f}}_q\in R^{k_q}} \left(\mu_{p_{q1}}(p_{q1}(\mathbf{x},\hat{f}_{q1})),\cdots,\mu_{p_{qk_q}}(p_{qk_q}(\mathbf{x},\hat{f}_{qk_q})),\mu_{\hat{f}_{q1}}(\hat{f}_{q1}),\cdots,\mu_{\hat{f}_{qk_q}}(\hat{f}_{qk_q})\right)$$

Throughout this subsection, we make the following assumptions with respect to membership functions $\mu_{p_{r\ell}}(p_{r\ell}(\mathbf{x},\hat{f}_{r\ell}))$, $\mu_{\hat{f}_{r\ell}}(\hat{f}_{r\ell}), r=1,\cdots,q, \ell=1,\cdots,k_r$.

Assumption 4.1
$\mu_{\hat{f}_{r\ell}}(\hat{f}_{r\ell}), r=1,\cdots,q, \ell=1,\cdots,k_r$ are strictly increasing and continuous with respect to $\hat{f}_{r\ell}\in[f_{r\ell\min},f_{r\ell\max}]$, where $\mu_{\hat{f}_{r\ell}}(\hat{f}_{r\ell})=0$ if $\hat{f}_{r\ell}\geq f_{r\ell\max}$, and $\mu_{\hat{f}_{r\ell}}(\hat{f}_{r\ell})=1$ if $\hat{f}_{r\ell}\leq f_{r\ell\min}$.

Assumption 4.2
$\mu_{p_{r\ell}}(p_{r\ell}(\mathbf{x},\hat{f}_{r\ell})), r=1,\cdots,q, \ell=1,\cdots,k_r$ are strictly decreasing and continuous with respect to $p_{r\ell}(\mathbf{x},\hat{f}_{r\ell})\in[p_{r\ell\min},p_{r\ell\max}]$, where $\mu_{p_{r\ell}}(p_{r\ell}(\mathbf{x},\hat{f}_{r\ell}))=0$ if $p_{r\ell}(\mathbf{x},\hat{f}_{r\ell})\leq p_{r\ell\min}$, and $\mu_{p_{r\ell}}(p_{r\ell}(\mathbf{x},\hat{f}_{r\ell}))=1$ if $p_{r\ell}(\mathbf{x},\hat{f}_{r\ell})\geq p_{r\ell\max}$.

To determine these membership functions appropriately, let us assume that the decision maker sets $f_{r\ell\min}$, $f_{r\ell\max}$ $p_{r\ell\min}$ and $p_{r\ell\max}$ as follows. First, the decision maker subjectively specifies parameters $f_{r\ell\min}$ and $f_{r\ell\max}$, where $f_{r\ell\min}$ is a sufficiently satisfactory maximum value and $f_{r\ell\max}$ is an acceptable maximum value, then he or she sets the intervals as

$$F_{r\ell}\stackrel{\text{def}}{=}[f_{r\ell\min},f_{r\ell\max}], r=1,\cdots,q, \ell=1,\cdots,k_r.$$

For interval $F_{r\ell}$, $p_{r\ell\max}$ can be obtained by solving the following problems.

$$p_{r\ell\max}\stackrel{\text{def}}{=}\max_{\mathbf{x}\in X} p_{r\ell}(\mathbf{x},f_{r\ell\max}), r=1,\cdots,q, \ell=1,\cdots,k_r \tag{4.39}$$

To obtain $p_{r\ell\min}$, we first solve the following problems.

$$\max_{\mathbf{x}\in X} p_{r\ell}(\mathbf{x},f_{r\ell\min}), r=1,\cdots,q, \ell=1,\cdots,k_r \tag{4.40}$$

Let $\mathbf{x}_{r\ell}\in X, r=1,\cdots,q, \ell=1,\cdots,k_r$ be the optimal solution of (4.40). Using optimal solutions $\mathbf{x}_{r\ell}\in X$, $p_{r\ell\min}$ can be obtained as follows.

$$p_{r\ell\min}\stackrel{\text{def}}{=}\min_{s=1,\cdots,q,t=1,\cdots,k_s,s\neq r,t\neq\ell} p_{r\ell}(\mathbf{x}_{st},f_{r\ell\min}), r=1,\cdots,q, \ell=1,\cdots,k_r \tag{4.41}$$

Note that $\mu_{p_{r\ell}}(p_{r\ell}(\mathbf{x},\hat{f}_{r\ell}))$ and $\mu_{\hat{f}_{r\ell}}(\hat{f}_{r\ell})$ compete with one another for any $\mathbf{x}\in X$. Here, let us assume that the decision maker adopts the fuzzy decision [6, 71, 140] to integrate both membership functions $\mu_{p_{r\ell}}(p_{r\ell}(\mathbf{x},\hat{f}_{r\ell}))$ and $\mu_{\hat{f}_{r\ell}}(\hat{f}_{r\ell})$. Then, the integrated membership function can be defined as

$$\mu_{D_{p_{r\ell}}}(\mathbf{x},\hat{f}_{r\ell})\stackrel{\text{def}}{=}\min\{\mu_{\hat{f}_{r\ell}}(\hat{f}_{r\ell}),\mu_{p_{r\ell}}(p_{r\ell}(\mathbf{x},\hat{f}_{r\ell}))\}. \tag{4.42}$$

Using integrated membership functions $\mu_{D_{p_{r\ell}}}(\mathbf{x},\hat{f}_{r\ell})$, HMOP4.13 can be transformed into the following form.

HMOP 4.14

first-level decision maker: DM$_1$

$$\max_{\mathbf{x}\in X,\hat{f}_{1\ell}\in \mathrm{R}^1,\ell=1,\cdots,k_1} \left(\mu_{D_{p_{11}}}(\mathbf{x},\hat{f}_{11}),\cdots,\mu_{D_{p_{1k_1}}}(\mathbf{x},\hat{f}_{1k_1})\right)$$

....

q-th-level decision maker: DM$_q$

$$\max_{\mathbf{x}\in X,\hat{f}_{q\ell}\in \mathrm{R}^1,\ell=1,\cdots,k_q} \left(\mu_{D_{p_{q1}}}(\mathbf{x},\hat{f}_{q1}),\cdots,\mu_{D_{p_{qk_q}}}(\mathbf{x},\hat{f}_{qk_q})\right)$$

To handle HMOP4.14, we introduce an MP-Pareto optimal solution below.

Definition 4.3 $\mathbf{x}^* \in X, \hat{f}^*_{r\ell} \in \mathrm{R}^1, r=1,\cdots,q, \ell=1,\cdots,k_r$ is an MP-Pareto optimal solution to HMOP4.14 if and only if there does not exist another $\mathbf{x} \in X, \hat{f}_{r\ell} \in \mathrm{R}^1, r=1,\cdots,q, \ell=1,\cdots,k_r$, such that $\mu_{D_{p_{r\ell}}}(\mathbf{x},\hat{f}_{r\ell}) \geq \mu_{D_{p_{r\ell}}}(\mathbf{x}^*,\hat{f}^*_{r\ell})$, $r=1,\cdots,q,\ell=1,\cdots,k_r$, with strict inequality holding for at least one r and ℓ.

To generate a candidate for a satisfactory solution that is also MP-Pareto optimal, each decision maker (i.e., DM$_r, r=1,\cdots,q$) specifies reference membership values [71], i.e.,

$$\begin{aligned}\hat{\mu}_{\mathbf{r}} &\stackrel{\text{def}}{=} (\hat{\mu}_{r1},\cdots,\hat{\mu}_{rk_r}), r=1,\cdots,q\\ \hat{\mu} &\stackrel{\text{def}}{=} (\hat{\mu}_{\mathbf{1}},\cdots,\hat{\mu}_{\mathbf{q}}).\end{aligned}$$

Once reference membership values are specified, the corresponding MP-Pareto optimal solution is obtained by solving the following minmax problem.

MINMAX 4.6 ($\hat{\mu}$)

$$\min_{\mathbf{x}\in X,\hat{f}_{r\ell}\in \mathrm{R}^1,r=1,\cdots,q,\ell=1,\cdots,k_r,\lambda\in\Lambda} \lambda \tag{4.43}$$

subject to

$$\hat{\mu}_{r\ell} - \mu_{p_{r\ell}}(p_{r\ell}(\mathbf{x},\hat{f}_{r\ell})) \leq \lambda, r=1,\cdots,q,\ell=1,\cdots,k_r \tag{4.44}$$

$$\hat{\mu}_{r\ell} - \mu_{\hat{f}_{r\ell}}(\hat{f}_{r\ell}) \leq \lambda, r=1,\cdots,q,\ell=1,\cdots,k_r \tag{4.45}$$

where

$$\Lambda \stackrel{\text{def}}{=} [\max_{r=1,\cdots,q,\ell=1,\cdots,k_r}(\hat{\mu}_{r\ell}-1), \max_{r=1,\cdots,q,\ell=1,\cdots,k_r}\hat{\mu}_{r\ell}].$$

Note that, in general, the optimal solution of MINMAX4.6($\hat{\mu}$) does not reflect the hierarchical structure between q decision makers in which an upper-level decision maker can claim priority for his or her distribution functions over lower-level decision makers. To cope with such a hierarchical preference structure between q decision makers, we introduce the concept of decision power [47, 49] as

$$\mathbf{w} = (w_1,\cdots,w_q),$$

where the r-th level decision maker (i.e., DM_r) can subjectively specify decision power w_{r+1} and the last decision maker (i.e., DM_q) has no decision power whatsoever. To reflect the hierarchical preference structure between multiple decision makers, decision powers $\mathbf{w}$ must satisfy inequality

$$1 = w_1 \geq \cdots \geq w_q > 0. \tag{4.46}$$

Then, by using decision powers $\mathbf{w} = (w_1, \cdots, w_q)$, MINMAX4.6($\hat{\mu}$) is reformulated as follows.

MINMAX 4.7 $(\hat{\mu}, \mathbf{w})$

$$\min_{\mathbf{x}\in X, \hat{f}_{r\ell}\in \mathrm{R}^1, r=1,\cdots,q, \ell=1,\cdots,k_r, \lambda\in\Lambda(\mathbf{w})} \lambda \tag{4.47}$$

subject to

$$\hat{\mu}_{r\ell} - \mu_{p_{r\ell}}(p_{r\ell}(\mathbf{x}, \hat{f}_{r\ell})) \leq \lambda / w_r, r = 1, \cdots, q, \ell = 1, \cdots, k_r \tag{4.48}$$

$$\hat{\mu}_{r\ell} - \mu_{\hat{f}_{r\ell}}(\hat{f}_{r\ell}) \leq \lambda / w_r, r = 1, \cdots, q, \ell = 1, \cdots, k_r \tag{4.49}$$

where

$$\Lambda(\mathbf{w}) \stackrel{\text{def}}{=} [\max_{r=1,\cdots,q,\ell=1,\cdots,k_r} w_r(\hat{\mu}_{r\ell} - 1), \max_{r=1,\cdots,q,\ell=1,\cdots,k_r} w_r \hat{\mu}_{r\ell}].$$

Given Assumption 4.2 and $\mathbf{c}_{r\ell}^2\mathbf{x} + \alpha_{r\ell}^2 > 0$, constraint (4.48) can be transformed into

$$\begin{aligned}
& \hat{\mu}_{r\ell} - \mu_{p_{r\ell}}(p_{r\ell}(\mathbf{x}, \hat{f}_{r\ell})) \leq \lambda / w_r \\
\Leftrightarrow\ & p_{r\ell}(\mathbf{x}, \hat{f}_{r\ell}) \geq \mu_{p_{r\ell}}^{-1}(\hat{\mu}_{r\ell} - \lambda / w_r) \\
\Leftrightarrow\ & T_{r\ell}\left(\frac{\hat{f}_{r\ell} - (\mathbf{c}_{r\ell}^1\mathbf{x} + \alpha_{r\ell}^1)}{\mathbf{c}_{r\ell}^2\mathbf{x} + \alpha_{r\ell}^2}\right) \geq \mu_{p_{r\ell}}^{-1}(\hat{\mu}_{r\ell} - \lambda / w_r) \\
\Leftrightarrow\ & \hat{f}_{r\ell} - (\mathbf{c}_{r\ell}^1\mathbf{x} + \alpha_{r\ell}^1) \geq T_{r\ell}^{-1}(\mu_{p_{r\ell}}^{-1}(\hat{\mu}_{r\ell} - \lambda / w_r)) \cdot (\mathbf{c}_{r\ell}^2\mathbf{x} + \alpha_{r\ell}^2), \quad (4.50)
\end{aligned}$$

where $\mu_{p_{r\ell}}^{-1}(\cdot)$ and $T_{r\ell}^{-1}(\cdot)$ are inverse functions of $\mu_{p_{r\ell}}(\cdot)$ and $T_{r\ell}(\cdot)$, respectively. Moreover, from Assumption 4.1 and (4.49), it holds that $\hat{f}_{r\ell} \leq \mu_{\hat{f}_{r\ell}}^{-1}(\hat{\mu}_{r\ell} - \lambda / w_r)$. As a result, constraints (4.48) and (4.49) can be reduced to the following single inequality, where a permissible objective level $\hat{f}_{r\ell}$ is removed.

$$\mu_{\hat{f}_{r\ell}}^{-1}(\hat{\mu}_{r\ell} - \lambda / w_r) - (\mathbf{c}_{r\ell}^1\mathbf{x} + \alpha_{r\ell}^1) \geq T_{r\ell}^{-1}(\mu_{p_{r\ell}}^{-1}(\hat{\mu}_{r\ell} - \lambda / w_r)) \cdot (\mathbf{c}_{r\ell}^2\mathbf{x} + \alpha_{r\ell}^2)$$

Then, MINMAX4.7($\hat{\mu}, \mathbf{w}$) is equivalently transformed into the following problem.

MINMAX 4.8 $(\hat{\mu}, \mathbf{w})$

$$\min_{\mathbf{x}\in X, \lambda\in\Lambda(\mathbf{w})} \lambda \tag{4.51}$$

subject to

$$\mu_{\hat{f}_{r\ell}}^{-1}(\hat{\mu}_{r\ell} - \lambda/w_r) - (\mathbf{c}_{r\ell}^1\mathbf{x} + \alpha_{r\ell}^1) \geq T_{r\ell}^{-1}(\mu_{p_{r\ell}}^{-1}(\hat{\mu}_{r\ell} - \lambda/w_r)) \cdot (\mathbf{c}_{r\ell}^2\mathbf{x} + \alpha_{r\ell}^2),$$
$$r = 1, \cdots, q, \ell = 1, \cdots, k_r \tag{4.52}$$

Note that constraints (4.52) can be reduced to a set of linear inequalities for some fixed value $\lambda \in \Lambda(\mathbf{w})$. This means that an optimal solution $\mathbf{x}^* \in X, \lambda^* \in \Lambda(\mathbf{w})$ of MINMAX4.8$(\hat{\mu}, \mathbf{w})$ is obtained via a combination of the bisection method with respect to $\lambda \in \Lambda(\mathbf{w})$ and the first phase of the two-phase simplex method of linear programming.

The relationships between optimal solution $\mathbf{x}^* \in X, \lambda^* \in \Lambda(\mathbf{w})$ of MINMAX4.8$(\hat{\mu}, \mathbf{w})$ and MP-Pareto optimal solutions to HMOP4.14 are characterized by the following theorem.

Theorem 4.7

If $\mathbf{x}^* \in X, \lambda^* \in \Lambda(\mathbf{w})$ *is a unique optimal solution of MINMAX4.8*$(\hat{\mu}, \mathbf{w})$*, then* $\mathbf{x}^* \in X$, $\hat{f}_{r\ell}^* \stackrel{\text{def}}{=} \mu_{\hat{f}_{r\ell}}^{-1}(\hat{\mu}_{r\ell} - \lambda^*/w_r)$, $r = 1, \cdots, q, \ell = 1, \cdots, k_r$ *is an MP-Pareto optimal solution to HMOP4.14.*

(Proof)
Since optimal solution $\mathbf{x}^* \in X, \lambda^* \in \Lambda(\mathbf{w})$ satisfies constraints (4.52), it holds that

$$\begin{aligned}
& T_{r\ell}\left(\frac{\mu_{\hat{f}_{r\ell}}^{-1}(\hat{\mu}_{r\ell} - \lambda^*/w_r) - (\mathbf{c}_{r\ell}^1\mathbf{x}^* + \alpha_{r\ell}^1)}{\mathbf{c}_{r\ell}^2\mathbf{x}^* + \alpha_{r\ell}^2}\right) \\
= \; & p_{r\ell}(\mathbf{x}^*, \mu_{\hat{f}_{r\ell}}^{-1}(\hat{\mu}_{r\ell} - \lambda^*/w_r)) \\
\geq \; & \mu_{p_{r\ell}}^{-1}(\hat{\mu}_{r\ell} - \lambda^*/w_r), r = 1, \cdots, q, \ell = 1, \cdots, k_r.
\end{aligned}$$

Assume that $\mathbf{x}^* \in X$, $\hat{f}_{r\ell}^* = \mu_{\hat{f}_{r\ell}}^{-1}(\hat{\mu}_{r\ell} - \lambda^*/w_r)$, $r = 1, \cdots, q, \ell = 1, \cdots, k_r$ is not an MP-Pareto optimal solution to HMOP4.14, then there exists $\mathbf{x} \in X, \hat{f}_{r\ell}, r = 1, \cdots, q, \ell = 1, \cdots, k_r$, such that

$$\begin{aligned}
\mu_{D_{p_{r\ell}}}(\mathbf{x}, \hat{f}_{r\ell}) &= \min\{\mu_{\hat{f}_{r\ell}}(\hat{f}_{r\ell}), \mu_{p_{r\ell}}(p_{r\ell}(\mathbf{x}, \hat{f}_{r\ell}))\} \\
&\geq \mu_{D_{p_{r\ell}}}(\mathbf{x}^*, \mu_{\hat{f}_{r\ell}}^{-1}(\hat{\mu}_{r\ell} - \lambda^*/w_r)) \\
&= \hat{\mu}_{r\ell} - \lambda^*/w_r, r = 1, \cdots, q, \ell = 1, \cdots, k_r,
\end{aligned}$$

with strict inequality holding for at least one r and ℓ. Then it holds that

$$\mu_{\hat{f}_{r\ell}}(f_{r\ell}) \geq \hat{\mu}_{r\ell} - \lambda^*/w_r, r = 1, \cdots, q, \ell = 1, \cdots, k_r \tag{4.53}$$

$$\mu_{p_{r\ell}}(p_{r\ell}(\mathbf{x}, \hat{f}_{r\ell})) \geq \hat{\mu}_{r\ell} - \lambda^*/w_r, r = 1, \cdots, q, \ell = 1, \cdots, k_r. \tag{4.54}$$

From definition (4.38), inequalities (4.53) and (4.54) can be transformed into the following inequalities, respectively.

$$\begin{aligned} \hat{f}_{r\ell} &\leq \mu_{\hat{f}_{r\ell}}^{-1}(\hat{\mu}_{r\ell} - \lambda^*/w_r) \\ \hat{f}_{r\ell} &\geq T_{r\ell}^{-1}(\mu_{p_{r\ell}}^{-1}(\hat{\mu}_{r\ell} - \lambda^*/w_r)) \cdot (\mathbf{c}_{r\ell}^2\mathbf{x} + \alpha_{r\ell}^2) + (\mathbf{c}_{r\ell}^1\mathbf{x} + \alpha_{r\ell}^1) \end{aligned}$$

This means that there exists some $\mathbf{x} \in X$ such that

$$\mu_{\hat{f}_{r\ell}}^{-1}(\hat{\mu}_{r\ell} - \lambda^*/w_r) \geq T_{r\ell}^{-1}(\mu_{p_{r\ell}}^{-1}(\hat{\mu}_{r\ell} - \lambda^*/w_r)) \cdot (\mathbf{c}_{r\ell}^2\mathbf{x} + \alpha_{r\ell}^2) + (\mathbf{c}_{r\ell}^1\mathbf{x} + \alpha_{r\ell}^1),$$
$$r = 1, \cdots, q, \ell = 1, \cdots, k_r,$$

which contradicts the fact that $\mathbf{x}^* \in X, \lambda^* \in \Lambda(\mathbf{w})$ is a unique optimal solution to MINMAX4.8($\hat{\mu}, \mathbf{w}$). □

4.3.2 A formulation using a fractile model

If we adopt a fractile optimization model for HMOP4.10, we can convert HMOP4.10 into the following multiobjective programming problem, where

$$\hat{\mathbf{p}}_{\mathbf{r}} \stackrel{\text{def}}{=} (\hat{p}_{r1}, \cdots, \hat{p}_{rk_r}), r = 1, \cdots, q, \ \hat{\mathbf{p}} \stackrel{\text{def}}{=} (\hat{\mathbf{p}}_{\mathbf{1}}, \cdots, \hat{\mathbf{p}}_{\mathbf{q}})$$

are vectors of permissible probability levels subjectively specified by each decision maker (i.e., DM$_r, r = 1, \cdots, q$).

HMOP 4.15 ($\hat{\mathbf{p}}$)
first-level decision maker: DM$_1$

$$\min_{\mathbf{x} \in X, f_{1\ell} \in R^1, \ell = 1, \cdots, k_r} (f_{11}, \cdots, f_{1k_1})$$

$$\cdots\cdots\cdots\cdots\cdots\cdots\cdots$$

q-th-level decision maker: DM$_q$

$$\min_{\mathbf{x} \in X, f_{q\ell} \in R^1, \ell = 1, \cdots, k_q} (f_{q1}, \cdots, f_{qk_q})$$

subject to

$$p_{r\ell}(\mathbf{x}, f_{r\ell}) \geq \hat{p}_{r\ell}, r = 1, \cdots, q, \ell = 1, \cdots, k_r \tag{4.55}$$

In HMOP4.15($\hat{\mathbf{p}}$), constraints (4.55) can be transformed into

$$\begin{aligned} \hat{p}_{r\ell} &\leq p_{r\ell}(\mathbf{x}, f_{r\ell}) = T_{r\ell}\left(\frac{f_{r\ell} - (\mathbf{c}_{r\ell}^1\mathbf{x} + \alpha_{r\ell}^1)}{\mathbf{c}_{r\ell}^2\mathbf{x} + \alpha_{r\ell}^2}\right) \\ \Leftrightarrow f_{r\ell} &\geq T_{r\ell}^{-1}(\hat{p}_{r\ell}) \cdot (\mathbf{c}_{r\ell}^2\mathbf{x} + \alpha_{r\ell}^2) + (\mathbf{c}_{r\ell}^1\mathbf{x} + \alpha_{r\ell}^1). \end{aligned}$$

Let us define the right-hand-side of the above inequality as

$$f_{r\ell}(\mathbf{x},\hat{p}_{r\ell}) \stackrel{\text{def}}{=} T_{r\ell}^{-1}(\hat{p}_{r\ell})\cdot(\mathbf{c}_{r\ell}^2\mathbf{x}+\alpha_{r\ell}^2)+(\mathbf{c}_{r\ell}^1\mathbf{x}+\alpha_{r\ell}^1) \tag{4.56}$$

Then, using $f_{r\ell}(\mathbf{x},\hat{p}_{r\ell})$, HMOP4.15($\hat{\mathbf{p}}$) can be equivalently reduced to the following simple form.

HMOP 4.16 ($\hat{\mathbf{p}}$)
first-level decision maker: DM$_1$

$$\min_{\mathbf{x}\in X}\,(f_{11}(\mathbf{x},\hat{p}_{11}),\cdots,f_{1k_1}(\mathbf{x},\hat{p}_{1k_1}))$$

....

q-th-level decision maker: DM$_q$

$$\min_{\mathbf{x}\in X}\,(f_{q1}(\mathbf{x},\hat{p}_{q1}),\cdots,f_{qk_q}(\mathbf{x},\hat{p}_{qk_q}))$$

To handle HMOP4.16($\hat{\mathbf{p}}$), the decision maker must specify permissible probability levels $\hat{\mathbf{p}}$ in advance; however, in general, the decision maker prefers not only smaller values of objective function $f_{r\ell}(\mathbf{x},\hat{p}_{r\ell})$ but also larger values of permissible probability level $\hat{p}_{r\ell}$. From such a perspective, we consider the following multiobjective programming problem as a natural extension of HMOP4.16($\hat{\mathbf{p}}$).

HMOP 4.17
first-level decision maker: DM$_1$

$$\min_{\mathbf{x}\in X,\hat{\mathbf{p}}_1\in(0,1)^{k_1}}\,(f_{11}(\mathbf{x},\hat{p}_{11}),\cdots,f_{1k_1}(\mathbf{x},\hat{p}_{1k_1}),-\hat{p}_{11},\cdots,-\hat{p}_{1k_1})$$

....

q-th-level decision maker: DM$_q$

$$\min_{\mathbf{x}\in X,\hat{\mathbf{p}}_q\in(0,1)^{k_q}}\,(f_{q1}(\mathbf{x},\hat{p}_{q1}),\cdots,f_{qk_q}(\mathbf{x},\hat{p}_{qk_q}),-\hat{p}_{q1},\cdots,-\hat{p}_{qk_q})$$

Considering the imprecise nature of the decision maker's judgment, we assume that the decision maker has a fuzzy goal for each objective function in HMOP4.17. Such a fuzzy goal can be quantified by eliciting the corresponding membership function. Let us denote a membership function of objective function $f_{r\ell}(\mathbf{x},\hat{p}_{r\ell})$ as $\mu_{f_{r\ell}}(f_{r\ell}(\mathbf{x},\hat{p}_{r\ell}))$, and a membership function of permissible probability level $\hat{p}_{r\ell}$ as $\mu_{\hat{p}_{r\ell}}(\hat{p}_{r\ell})$. Then, HMOP4.17 can be transformed into the following problem.

HMOP 4.18

first-level decision maker: DM$_1$

$$\max_{\mathbf{x}\in X,\hat{\mathbf{p}}_1\in(0,1)^{k_1}} \left(\mu_{f_{11}}(f_{11}(\mathbf{x},\hat{p}_{11})),\cdots,\mu_{f_{1k_1}}(f_{1k_1}(\mathbf{x},\hat{p}_{1k_1})),\mu_{\hat{p}_{11}}(\hat{p}_{11}),\cdots,\mu_{\hat{p}_{1k_1}}(\hat{p}_{1k_1})\right)$$

. . . .

q-th-level decision maker: DM$_q$

$$\max_{\mathbf{x}\in X,\hat{\mathbf{p}}_q\in(0,1)^{k_q}} \left(\mu_{f_{q1}}(f_{q1}(\mathbf{x},\hat{p}_{q1})),\cdots,\mu_{f_{qk_q}}(f_{qk_q}(\mathbf{x},\hat{p}_{qk_q})),\mu_{\hat{p}_{q1}}(\hat{p}_{q1}),\cdots,\mu_{\hat{p}_{qk_q}}(\hat{p}_{qk_q})\right)$$

Throughout this subsection, we make the following assumptions with respect to the membership functions $\mu_{f_{r\ell}}(f_{r\ell}(\mathbf{x},\hat{p}_{r\ell}))$, $\mu_{\hat{p}_{r\ell}}(\hat{p}_{r\ell})$.

Assumption 4.3

$\mu_{\hat{p}_{r\ell}}(\hat{p}_{r\ell}), r=1,\cdots,q, \ell=1,\cdots,k_r$ are strictly increasing and continuous with respect to $\hat{p}_{r\ell}\in[p_{r\ell\min},p_{r\ell\max}]$, where $\mu_{\hat{p}_{r\ell}}(\hat{p}_{r\ell})=0$ if $\hat{p}_{r\ell}\leq p_{r\ell\min}$, and $\mu_{\hat{p}_{r\ell}}(\hat{p}_{r\ell})=1$ if $\hat{p}_{r\ell}\geq p_{r\ell\max}$.

Assumption 4.4

$\mu_{f_{r\ell}}(f_{r\ell}(\mathbf{x},\hat{p}_{r\ell})), r=1,\cdots,q, \ell=1,\cdots,k_r$ are strictly decreasing and continuous with respect to $f_{r\ell}(\mathbf{x},\hat{p}_{r\ell})\in[f_{r\ell\min},f_{r\ell\max}]$, where $\mu_{f_{r\ell}}(f_{r\ell}(\mathbf{x},\hat{p}_{r\ell}))=0$ if $f_{r\ell}(\mathbf{x},\hat{p}_{r\ell})\geq f_{r\ell\max}$, and $\mu_{f_{r\ell}}(f_{r\ell}(\mathbf{x},\hat{p}_{r\ell}))=1$ if $f_{r\ell}(\mathbf{x},\hat{p}_{r\ell})\leq f_{r\ell\min}$.

To determine these membership functions, the decision maker can set intervals $[p_{r\ell\min},p_{r\ell\max}]$ and $[f_{r\ell\min},f_{r\ell\max}], r=1,\cdots,q, \ell=1,\cdots,k_r$ as follows. First, each decision maker (i.e., DM$_r, r=1,\cdots,q$) subjectively specifies $p_{r\ell\min}$ and $p_{r\ell\max}, \ell=1,\cdots,k_r$, where $p_{r\ell\min}$ is an acceptable minimum value and $p_{r\ell\max}$ is a sufficiently satisfactory minimum value; next, the decision maker sets intervals

$$P_{r\ell}\stackrel{\text{def}}{=}[p_{r\ell\min},p_{r\ell\max}], r=1,\cdots,q, \ell=1,\cdots,k_r.$$

Corresponding to interval $P_{r\ell}$, $f_{r\ell\min}$ can be obtained by solving linear programming problems.

$$f_{r\ell\min}\stackrel{\text{def}}{=}\min_{\mathbf{x}\in X} f_{r\ell}(\mathbf{x},p_{r\ell\min}), r=1,\cdots,q, \ell=1,\cdots,k_r \tag{4.57}$$

To obtain $f_{r\ell\max}$, we first solve linear programming problems

$$\min_{\mathbf{x}\in X} f_{r\ell}(\mathbf{x},p_{r\ell\max}), r=1,\cdots,q, \ell=1,\cdots,k_r.$$

Let $\mathbf{x}_{r\ell}\in X, r=1,\cdots,q, \ell=1,\cdots,k_r$ be the above optimal solution. Using the optimal solutions $\mathbf{x}_{r\ell}\in X$, $f_{r\ell\max}$ can be computed as

$$f_{r\ell\max}\stackrel{\text{def}}{=}\max_{s=1,\cdots,q,t=1,\cdots,k_s,s\neq r,t\neq\ell} f_{r\ell}(\mathbf{x}_{st},\hat{p}_{r\ell\max}), r=1,\cdots,q, \ell=1,\cdots,k_r. \tag{4.58}$$

Note that, because of (4.56), $\mu_{f_{r\ell}}(f_{r\ell}(\mathbf{x},\hat{p}_{r\ell}))$ and $\mu_{\hat{p}_{r\ell}}(\hat{p}_{r\ell})$ conflict with one another for any $\mathbf{x} \in X$. Here, let us assume that the decision maker adopts the fuzzy decision [6, 71, 140] to integrate both membership functions $\mu_{f_{r\ell}}(f_{r\ell}(\mathbf{x}, \hat{p}_{r\ell}))$ and $\mu_{\hat{p}_{r\ell}}(\hat{p}_{r\ell})$. Then, the integrated membership function can be defined as

$$\mu_{D_{f_{r\ell}}}(\mathbf{x}, \hat{p}_{r\ell}) \stackrel{\text{def}}{=} \min\{\mu_{\hat{p}_{r\ell}}(\hat{p}_{r\ell}), \mu_{f_{r\ell}}(f_{r\ell}(\mathbf{x}, \hat{p}_{r\ell}))\}. \tag{4.59}$$

Using membership functions $\mu_{D_{f_{r\ell}}}(\mathbf{x}, \hat{p}_{r\ell})$, HMOP4.18 can be transformed into the following form.

HMOP 4.19
first-level decision maker: DM$_1$

$$\max_{\mathbf{x}\in X, \hat{p}_{1\ell}\in(0,1), \ell=1,\cdots,k_1} \left(\mu_{D_{f_{11}}}(\mathbf{x}, \hat{p}_{11}), \cdots, \mu_{D_{f_{1k_1}}}(\mathbf{x}, \hat{p}_{1k_1}))\right)$$

$$\cdots\cdots\cdots\cdots\cdots\cdots\cdots$$

q-th-level decision maker: DM$_q$

$$\max_{\mathbf{x}\in X, \hat{p}_{q\ell}\in(0,1), \ell=1,\cdots,k_q} \left(\mu_{D_{f_{q1}}}(\mathbf{x}, \hat{p}_{q1}), \cdots, \mu_{D_{f_{qk_q}}}(\mathbf{x}, \hat{p}_{qk_q}))\right)$$

To handle HMOP4.19, we next introduce an MF-Pareto optimal solution.

Definition 4.4 $\mathbf{x}^* \in X, \hat{p}^*_{r\ell} \in (0,1), r = 1, \cdots, q, \ell = 1, \cdots, k_r$ is an MF-Pareto optimal solution to HMOP4.19 if and only if there does not exist another $\mathbf{x} \in X, \hat{p}_{r\ell}, r = 1, \cdots, q,\ \ell = 1, \cdots, k_r$ such that $\mu_{D_{f_{r\ell}}}(\mathbf{x}, \hat{p}_{r\ell}) \geq \mu_{D_{f_{r\ell}}}(\mathbf{x}^*, \hat{p}^*_{r\ell})$ $r = 1, \cdots, q, \ell = 1, \cdots, k_r$, with strict inequality holding for at least one r and ℓ.

To generate a candidate for a satisfactory solution that is also MF-Pareto optimal, each decision maker (i.e., DM$_r, r = 1, \cdots, q$) specifies reference membership values [71], i.e.,

$$\hat{\mu}_{\mathbf{r}} \stackrel{\text{def}}{=} (\hat{\mu}_{r1}, \cdots, \hat{\mu}_{rk_r}), r = 1, \cdots, q$$
$$\hat{\mu} \stackrel{\text{def}}{=} (\hat{\mu}_{\mathbf{1}}, \cdots, \hat{\mu}_{\mathbf{q}}).$$

Once these reference membership values are specified, the corresponding MF-Pareto optimal solution is obtained by solving the following minmax problem.

MINMAX 4.9 ($\hat{\mu}$)

$$\min_{\mathbf{x}\in X, \hat{p}_{r\ell}\in(0,1), r=1,\cdots,q, \ell=1,\cdots,k_r, \lambda\in\Lambda} \lambda \tag{4.60}$$

subject to

$$\hat{\mu}_{r\ell} - \mu_{f_{r\ell}}(f_{r\ell}(\mathbf{x}, \hat{p}_{r\ell})) \leq \lambda, r = 1, \cdots, q, \ell = 1, \cdots, k_r \tag{4.61}$$
$$\hat{\mu}_{r\ell} - \mu_{\hat{p}_{r\ell}}(\hat{p}_{r\ell}) \leq \lambda, r = 1, \cdots, q, \ell = 1, \cdots, k_r \tag{4.62}$$

where

$$\Lambda \stackrel{\text{def}}{=} [\max_{r=1,\cdots,q,\ell=1,\cdots,k_r}(\hat{\mu}_{r\ell}-1), \max_{r=1,\cdots,q,\ell=1,\cdots,k_r}\hat{\mu}_{r\ell}].$$

Note that, in general, the optimal solution of MINMAX4.9($\hat{\mu}$) does not reflect the hierarchical structure between q decision makers in which an upper-level decision maker can claim priority for his or her distribution functions over lower-level decision makers. To cope with such a hierarchical preference structure between q decision makers, we introduce the concept of decision power [47, 49] as

$$\mathbf{w} = (w_1, \cdots, w_q),$$

where the r-th-level decision maker (i.e., DM$_r$) can subjectively specify decision power w_{r+1} and the last decision maker (i.e., DM$_q$) has no decision power whatsoever. To reflect the hierarchical preference structure between multiple decision makers, decision powers $\mathbf{w}$ must satisfy inequality

$$1 = w_1 \geq \cdots \geq w_q > 0. \tag{4.63}$$

Then, by using decision powers $\mathbf{w} = (w_1, \cdots, w_q)$, MINMAX4.9($\hat{\mu}$) is reformulated as follows.

MINMAX 4.10 ($\hat{\mu}, \mathbf{w}$)

$$\min_{\mathbf{x}\in X, \hat{p}_{r\ell}\in(0,1), r=1,\cdots,q,\ell=1,\cdots,k_r,\lambda\in\Lambda(\mathbf{w})} \lambda \tag{4.64}$$

subject to

$$\hat{\mu}_{r\ell} - \mu_{f_{r\ell}}(f_{r\ell}(\mathbf{x}, \hat{p}_{r\ell})) \leq \lambda/w_r, r = 1, \cdots, q, \ell = 1, \cdots, k_r \tag{4.65}$$

$$\hat{\mu}_{r\ell} - \mu_{\hat{p}_{r\ell}}(\hat{p}_{r\ell}) \leq \lambda/w_r, r = 1, \cdots, q, \ell = 1, \cdots, k_r \tag{4.66}$$

where

$$\Lambda(\mathbf{w}) \stackrel{\text{def}}{=} [\max_{r=1,\cdots,q,\ell=1,\cdots,k_r} w_r(\hat{\mu}_{r\ell}-1), \max_{r=1,\cdots,q,\ell=1,\cdots,k_r} w_r\hat{\mu}_{r\ell}].$$

Because of $\mathbf{c}_{r\ell}^2\mathbf{x} + \alpha_{r\ell}^2 > 0$, constraints (4.65) can be transformed as

$$\hat{p}_{r\ell} \leq T_{r\ell}\left(\frac{\mu_{\tilde{f}_{r\ell}}^{-1}(\hat{\mu}_{r\ell} - \lambda/w_r) - (\mathbf{c}_{r\ell}^1\mathbf{x} + \alpha_{r\ell}^1)}{\mathbf{c}_{r\ell}^2\mathbf{x} + \alpha_{r\ell}^2}\right)$$

$$r = 1, \cdots, q, \ell = 1, \cdots, k_r, \tag{4.67}$$

where $\mu_{\tilde{f}_{r\ell}}^{-1}(\cdot)$ is an inverse function of $\mu_{f_{r\ell}}(\cdot)$. From constraints (4.66), it holds that $\hat{p}_{r\ell} \geq \mu_{\tilde{p}_{r\ell}}^{-1}(\hat{\mu}_{r\ell} - \lambda/w_r)$, where $\mu_{\tilde{p}_{r\ell}}^{-1}(\cdot)$ is an inverse function of $\mu_{\hat{p}_{r\ell}}(\cdot)$. Therefore, constraint (4.67) can be reduced to the following inequalities, where a permissible probability level $\hat{p}_{r\ell}$ has disappeared.

$$\mu_{\tilde{f}_{r\ell}}^{-1}(\hat{\mu}_{r\ell} - \lambda/w_r) - (\mathbf{c}_{r\ell}^1\mathbf{x} + \alpha_{r\ell}^1) \geq T_{r\ell}^{-1}(\mu_{\tilde{p}_{r\ell}}^{-1}(\hat{\mu}_{r\ell} - \lambda/w_r)) \cdot (\mathbf{c}_{r\ell}^2\mathbf{x} + \alpha_{r\ell}^2)$$

$$r = 1, \cdots, q, \ell = 1, \cdots, k_r$$

Then, MINMAX4.13($\hat{\mu}, \mathbf{w}$) can be equivalently reduced to the following problem.

MINMAX 4.11 ($\hat{\mu}, \mathbf{w}$)

$$\min_{\mathbf{x}\in X, \lambda\in\Lambda(\mathbf{w})} \lambda \tag{4.68}$$

subject to

$$\mu_{f_{r\ell}}^{-1}(\hat{\mu}_{r\ell} - \lambda/w_r) - (\mathbf{c}_{r\ell}^1\mathbf{x} + \alpha_{r\ell}^1) \geq T_{r\ell}^{-1}(\mu_{\hat{p}_{r\ell}}^{-1}(\hat{\mu}_{r\ell} - \lambda/w_r)) \cdot (\mathbf{c}_{r\ell}^2\mathbf{x} + \alpha_{r\ell}^2),$$
$$r = 1, \cdots, q, \ell = 1, \cdots, k_r \tag{4.69}$$

Note that MINMAX4.11($\hat{\mu}, \mathbf{w}$) is the same as MINMAX4.8($\hat{\mu}, \mathbf{w}$). Therefore, optimal solution $\mathbf{x}^* \in X, \lambda^* \in \Lambda(\mathbf{w})$ of MINMAX4.11($\hat{\mu}, \mathbf{w}$) can be obtained via a combination of the bisection method with respect to $\lambda \in \Lambda(\mathbf{w})$ and the first phase of the two-phase simplex method of linear programming.

The relationships between optimal solution $\mathbf{x}^* \in X, \lambda^* \in \Lambda(\mathbf{w})$ of MINMAX4.11($\hat{\mu}, \mathbf{w}$) and MF-Pareto optimal solutions to HMOP4.19 are characterized by the following theorem.

Theorem 4.8
If $\mathbf{x}^* \in X, \lambda^* \in \Lambda(\mathbf{w})$ is a unique optimal solution of MINMAX4.11($\hat{\mu}, \mathbf{w}$), then $\mathbf{x}^* \in X, \hat{p}_{r\ell}^* \stackrel{\text{def}}{=} \mu_{\hat{p}_{r\ell}}^{-1}(\hat{\mu}_{r\ell} - \lambda^*/w_r), r = 1, \cdots, q, \ell = 1, \cdots, k_r$ is an MF-Pareto optimal solution to HMOP4.19.

(Proof)
From (4.69), it holds that

$$\hat{\mu}_{r\ell} - \lambda^*/w_r \leq \mu_{f_{r\ell}}(f_{r\ell}(\mathbf{x}^*, \mu_{\hat{p}_{r\ell}}^{-1}(\hat{\mu}_{r\ell} - \lambda^*/w_r))), r = 1, \cdots, q, \ell = 1, \cdots, k_r.$$

Assume that $\mathbf{x}^* \in X$, $\mu_{\hat{p}_{r\ell}}^{-1}(\hat{\mu}_{r\ell} - \lambda^*/w_r)$, $r = 1, \cdots, q, \ell = 1, \cdots, k_r$ is not an MF-Pareto optimal solution to HMOP4.19. Then, there exists $\mathbf{x} \in X, \hat{p}_{r\ell}$, $r = 1, \cdots, q, \ell = 1, \cdots, k_r$ such that

$$\begin{aligned}
\mu_{D_{f_{r\ell}}}(\mathbf{x}, \hat{p}_{r\ell}) &= \min\{\mu_{\hat{p}_{r\ell}}(\hat{p}_{r\ell}), \mu_{f_{r\ell}}(f_{r\ell}(\mathbf{x}, \hat{p}_{r\ell}))\} \\
&\geq \mu_{D_{f_{r\ell}}}(\mathbf{x}^*, \mu_{\hat{p}_{r\ell}}^{-1}(\hat{\mu}_{r\ell} - \lambda^*/w_r)) \\
&= \hat{\mu}_{r\ell} - \lambda^*/w_r, r = 1, \cdots, q, \ell = 1, \cdots, k_r,
\end{aligned}$$

with strict inequality holding for at least one r and ℓ. Then it holds that

$$\mu_{\hat{p}_{r\ell}}(\hat{p}_{r\ell}) \geq \hat{\mu}_{r\ell} - \lambda^*/w_r, r = 1, \cdots, q, \ell = 1, \cdots, k_r, \tag{4.70}$$
$$\mu_{f_{r\ell}}(f_{r\ell}(\mathbf{x}, \hat{p}_{r\ell})) \geq \hat{\mu}_{r\ell} - \lambda^*/w_r, r = 1, \cdots, q, \ell = 1, \cdots, k_r. \tag{4.71}$$

From definition (4.56), inequalities (4.70) and (4.71) can be transformed into inequalities

$$\begin{aligned}
\hat{p}_{r\ell} &\geq \mu_{\hat{p}_{r\ell}}^{-1}(\hat{\mu}_{r\ell} - \lambda^*/w_r) \\
\hat{p}_{r\ell} &\leq T_{r\ell}\left(\frac{\mu_{f_{r\ell}}^{-1}(\hat{\mu}_{r\ell} - \lambda^*/w_r) - (\mathbf{c}_{r\ell}^1\mathbf{x}^* + \alpha_{r\ell}^1)}{\mathbf{c}_{r\ell}^2\mathbf{x}^* + \alpha_{r\ell}^2}\right).
\end{aligned}$$

This means that there exists some $\mathbf{x} \in X$ such that

$$\mu_{\tilde{f}_{r\ell}}^{-1}(\hat{\mu}_{r\ell} - \lambda^*/w_r) - (\mathbf{c}_{r\ell}^1\mathbf{x} + \alpha_{r\ell}^1) \geq T_{r\ell}^{-1}(\mu_{\tilde{p}_{r\ell}}^{-1}(\hat{\mu}_{r\ell} - \lambda^*/w_r)) \cdot (\mathbf{c}_{r\ell}^2\mathbf{x} + \alpha_{r\ell}^2),$$
$$r = 1, \cdots, q, \ell = 1, \cdots, k_r,$$

which contradicts the fact that $\mathbf{x}^* \in X, \lambda^* \in \Lambda(\mathbf{w})$ is a unique optimal solution to MINMAX4.11$(\hat{\mu}, \mathbf{w})$. □

4.3.3 An interactive linear programming algorithm

In this subsection, we propose an interactive linear programming algorithm to obtain a satisfactory solution for decision makers (i.e., $\text{DM}_r, r = 1, \cdots, q$) from an MF-Pareto optimal solution set. Unfortunately, from Theorem 4.8, there is no guarantee that optimal solution $\mathbf{x}^* \in X, \lambda^* \in \Lambda(\mathbf{w})$ of MINMAX4.11$(\hat{\mu}, \mathbf{w})$ is MF-Pareto optimal with respect to HMOP4.19 if $\mathbf{x}^* \in X, \lambda^* \in \Lambda(\mathbf{w})$ is not unique. To guarantee MF-Pareto optimality, we first assume that constraints (4.69) of MINMAX4.11$(\hat{\mu}, \mathbf{w})$ are active at optimal solution $\mathbf{x}^* \in X, \lambda^* \in \Lambda(\mathbf{w})$. If one of the constraints of (4.69) is inactive, i.e.,

$$\begin{aligned} & \mu_{\tilde{f}_{r\ell}}^{-1}(\hat{\mu}_{r\ell} - \lambda^*/w_r) - (\mathbf{c}_{r\ell}^1\mathbf{x}^* + \alpha_{r\ell}^1) \\ > \; & T_{r\ell}^{-1}(\mu_{\tilde{p}_{r\ell}}^{-1}(\hat{\mu}_{r\ell} - \lambda^*/w_r)) \cdot (\mathbf{c}_{r\ell}^2\mathbf{x}^* + \alpha_{r\ell}^2), \end{aligned} \tag{4.72}$$

we can convert inactive constraint (4.72) into an active one by applying Algorithm 4.3, where

$$g_{r\ell}(\hat{\mu}_{r\ell}) \stackrel{\text{def}}{=} \mu_{\tilde{f}_{r\ell}}^{-1}(\hat{\mu}_{r\ell} - \lambda^*/w_r) - f_{r\ell}(\mathbf{x}^*, \mu_{\tilde{p}_{r\ell}}^{-1}(\hat{\mu}_{r\ell} - \lambda^*/w_r)).$$

Algorithm 4.3

Step 1: *Set* $\mu_{r\ell}^L \leftarrow \lambda^*/w_r$, $\mu_{r\ell}^R \leftarrow \lambda^*/w_r + 1$.
Step 2: *Set* $\mu_{r\ell} \leftarrow (\mu_{r\ell}^L + \mu_{r\ell}^R)/2$.
Step 3: *If* $g_{r\ell}(q_{r\ell}) > 0$ *then* $q_{r\ell}^L \leftarrow q_{r\ell}$ *and go to Step 2, else if* $g_{r\ell}(q_{r\ell}) < 0$ *then* $q_{r\ell}^R \leftarrow q_{r\ell}$ *and go to Step 2, else if* $g_{r\ell}(q_{r\ell}) = 0$*, then update reference membership value to* $\hat{\mu}_{r\ell} \leftarrow q_{r\ell}$ *and stop.*

For optimal solution $\mathbf{x}^* \in X, \lambda^* \in \Lambda(\mathbf{w})$ of MINMAX4.11$(\hat{\mu}, \mathbf{w})$, where the active conditions of constraints (4.69) are satisfied, we solve the MF-Pareto optimality test problem formulated as follows.

Test 4.2

$$w \stackrel{\text{def}}{=} \max_{\mathbf{x} \in X, \varepsilon_{r\ell} \geq 0, r=1,\cdots,q, \ell=1,\cdots,k_r} \sum_{r=1}^{q} \sum_{\ell=1}^{k_r} \varepsilon_{r\ell} \tag{4.73}$$

subject to

$$\begin{aligned} & T_{r\ell}^{-1}(\mu_{\tilde{p}_{r\ell}}^{-1}(\hat{\mu}_{r\ell}-\lambda^*/w_r))\cdot(\mathbf{c}_{r\ell}^2\mathbf{x}+\alpha_{r\ell}^2)+(\mathbf{c}_{r\ell}^1\mathbf{x}+\alpha_{r\ell}^1)+\varepsilon_{r\ell} \\ = \; & T_{r\ell}^{-1}(\mu_{\tilde{p}_{r\ell}}^{-1}(\hat{\mu}_{r\ell}-\lambda^*/w_r))\cdot(\mathbf{c}_{r\ell}^2\mathbf{x}^*+\alpha_{r\ell}^2)+(\mathbf{c}_{r\ell}^1\mathbf{x}^*+\alpha_{r\ell}^1), \\ & r=1,\cdots,q,\ell=1,\cdots,k_r \end{aligned} \tag{4.74}$$

For the optimal solution of TEST4.2, the following theorem holds.

Theorem 4.9

Let $\check{\mathbf{x}}\in X,\check{\varepsilon}_{r\ell}\geq 0, r=1,\cdots,q,\ell=1,\cdots,k_r$ *be an optimal solution of TEST4.2. If* $w=0$, $\mathbf{x}^*\in X, \mu_{\tilde{p}_{r\ell}}^{-1}(\hat{\mu}_{r\ell}-\lambda^*/w_r), r=1,\cdots,q,\ell=1,\cdots,k_r$ *is an MF-Pareto optimal solution to HMOP4.19.*

(Proof)

From the active conditions of constraints (4.69), it holds that

$$\hat{\mu}_{r\ell}-\lambda^*/w_r = \mu_{\tilde{f}_{r\ell}}(f_{r\ell}(\mathbf{x}^*,\mu_{\tilde{p}_{r\ell}}^{-1}(\hat{\mu}_{r\ell}-\lambda^*/w_r))), r=1,\cdots,q,\ell=1,\cdots,k_r. \tag{4.75}$$

If $\mathbf{x}^*\in X, \mu_{\tilde{p}_{r\ell}}^{-1}(\hat{\mu}_{r\ell}-\lambda^*/w_r), r=1,\cdots,q,\ell=1,\cdots,k_r$ is not an MF-Pareto optimal solution, there exists some $\mathbf{x}\in X, \hat{p}_{r\ell}, r=1,\cdots,q,\ell=1,\cdots,k_r$ such that

$$\begin{aligned} \mu_{D_{\tilde{f}_{r\ell}}}(\mathbf{x},\hat{p}_{r\ell}) &= \min\{\mu_{\tilde{p}_{r\ell}}(\hat{p}_{r\ell}),\mu_{\tilde{f}_{r\ell}}(f_{r\ell}(\mathbf{x},\hat{p}_{r\ell}))\} \\ &\geq \mu_{D_{\tilde{f}_{r\ell}}}(\mathbf{x}^*,\mu_{\tilde{p}_{r\ell}}^{-1}(\hat{\mu}_{r\ell}-\lambda^*/w_r)) \\ &= \hat{\mu}_{r\ell}-\lambda^*/w_r, r=1,\cdots,q,\ell=1,\cdots,k_r, \end{aligned}$$

with strict inequality holding for at least one r and ℓ. This means that the following inequalities hold.

$$\mu_{\tilde{p}_{r\ell}}(\hat{p}_{r\ell}) \geq \hat{\mu}_{r\ell}-\lambda^*/w_r, r=1,\cdots,q,\ell=1,\cdots,k_r \tag{4.76}$$

$$\mu_{\tilde{f}_{r\ell}}(f_{r\ell}(\mathbf{x},\hat{p}_{r\ell})) \geq \hat{\mu}_{r\ell}-\lambda^*/w_r, r=1,\cdots,q,\ell=1,\cdots,k_r \tag{4.77}$$

As a result, there exists some $\mathbf{x}\in X,\hat{p}_{r\ell}, r=1,\cdots,q,\ell=1,\cdots,k_r$ such that

$$\mu_{\tilde{f}_{r\ell}}^{-1}(\hat{\mu}_{r\ell}-\lambda^*/w_r)\geq(\mathbf{c}_{r\ell}^1\mathbf{x}+\alpha_{r\ell}^1)+T_{r\ell}^{-1}(\mu_{\tilde{p}_{r\ell}}^{-1}(\hat{\mu}_{r\ell}-\lambda^*/w_r))\cdot(\mathbf{c}_{r\ell}^2\mathbf{x}+\alpha_{r\ell}^2).$$

Because of active conditions (4.75), it holds that

$$\begin{aligned} & T_{r\ell}^{-1}(\mu_{\tilde{p}_{r\ell}}^{-1}(\hat{\mu}_{r\ell}-\lambda^*/w_r))\cdot(\mathbf{c}_{r\ell}^2\mathbf{x}^*+\alpha_{r\ell}^2)+(\mathbf{c}_{r\ell}^1\mathbf{x}^*+\alpha_{r\ell}^1) \\ \geq \; & T_{r\ell}^{-1}(\mu_{\tilde{p}_{r\ell}}^{-1}(\hat{\mu}_{r\ell}-\lambda^*/w_r))\cdot(\mathbf{c}_{r\ell}^2\mathbf{x}+\alpha_{r\ell}^2)+(\mathbf{c}_{r\ell}^1\mathbf{x}+\alpha_{r\ell}^1), \\ & r=1,\cdots,q,\ell=1,\cdots,k_r, \end{aligned}$$

with strict inequality holding for at least one r and ℓ. This contradicts the fact that $w=0$. □

Given the above, we next present our interactive algorithm for deriving a satisfactory solution from an MF-Pareto optimal solution set.

Algorithm 4.4

Step 1: *According to Assumptions 4.3 and 4.4, each decision maker (i.e., $DM_r, r = 1, \cdots, q$) subjectively sets his or her membership functions $\mu_{\hat{p}_{r\ell}}(\hat{p}_{r\ell})$, $\mu_{f_{r\ell}}(f_{r\ell}(\mathbf{x}, \hat{p}_{r\ell}))$, $\ell = 1, \cdots, k_r$.*
Step 2: *Set initial reference membership values to $\hat{\mu}_{r\ell} = 1, r = 1, \cdots, q, \ell = 1, \cdots, k_r$, and initial decision power to $w_r = 1, r = 1, \cdots, q$.*
Step 3: *Solve MINMAX4.11($\hat{\mu}, \mathbf{w}$) via the combination of the bisection method and the first phase of the two-phase simplex method of linear programming. For optimal solution $\mathbf{x}^* \in X, \lambda^* \in \Lambda(\mathbf{w})$ of MINMAX4.11($\hat{\mu}, \mathbf{w}$), solve TEST4.2.*
Step 4: *If each decision maker (i.e., $DM_r, r = 1, \cdots, q$) is satisfied with the current values of MF-Pareto optimal solution $\mu_{D_{f_{r\ell}}}(\mathbf{x}^*, \hat{p}^*_{r\ell}), \ell = 1, \cdots, k_r$, where $\hat{p}^*_{r\ell} \stackrel{\text{def}}{=} \mu^{-1}_{\hat{p}_{r\ell}}(\hat{\mu}_{r\ell} - \lambda^*)$, then stop. Otherwise, let the s-th-level decision maker (i.e., DM_s) be the uppermost decision maker who is not satisfied with the current values. Considering the current values of his or her membership functions, DM_s updates his or her decision power w_{s+1} and/or his or her reference membership values $\hat{\mu}_{s\ell}, \ell = 1, \cdots, k_s$ according to the following two rules, then we return to Step 3.*
Rule 1: *w_{s+1} must be set as $w_{s+1} \leq w_s$. After updating w_{s+1}, if $w_{s+1} < w_t, s+2 \leq t \leq q$, w_t is replaced by w_{s+1} ($w_t \leftarrow w_{s+1}$).*
Rule 2: *Before updating DM_s's reference membership values $\hat{\mu}_{s\ell}, \ell = 1, \cdots, k_s$, the other decision makers' reference membership values are fixed at the current values, i.e., $\hat{\mu}_{r\ell} \leftarrow \mu_{D_{f_{r\ell}}}(\mathbf{x}^*, \hat{p}^*_{r\ell}), r = 1, \cdots, q, r \neq s, \ell = 1, \cdots, k_r$.*

4.3.4 A numerical example

To demonstrate our proposed method for HMOP4.10, we consider the following hierarchical two objective stochastic programming problem with three hypothetical decision makers.

HMOP 4.20
first-level decision maker: $\mathbf{DM_1}$

$$\min \bar{z}_{11}(\mathbf{x}) = (\mathbf{c}^1_{11} + \bar{t}_{11}\mathbf{c}^2_{11})\mathbf{x} + (\alpha^1_{11} + \bar{t}_{11}\alpha^2_{11})$$

$$\min \bar{z}_{12}(\mathbf{x}) = (\mathbf{c}^1_{12} + \bar{t}_{12}\mathbf{c}^2_{12})\mathbf{x} + (\alpha^1_{12} + \bar{t}_{12}\alpha^2_{12})$$

second-level decision maker: $\mathbf{DM_2}$

$$\min \bar{z}_{21}(\mathbf{x}) = (\mathbf{c}^1_{21} + \bar{t}_{21}\mathbf{c}^2_{21})\mathbf{x} + (\alpha^1_{21} + \bar{t}_{21}\alpha^2_{21})$$

$$\min \bar{z}_{22}(\mathbf{x}) = (\mathbf{c}^1_{22} + \bar{t}_{22}\mathbf{c}^2_{22})\mathbf{x} + (\alpha^1_{22} + \bar{t}_{22}\alpha^2_{22})$$

third-level decision maker: $\mathbf{DM_3}$

$$\min \bar{z}_{31}(\mathbf{x}) = (\mathbf{c}^1_{31} + \bar{t}_{31}\mathbf{c}^2_{31})\mathbf{x} + (\alpha^1_{31} + \bar{t}_{31}\alpha^2_{31})$$

Table 4.5: Parameters of the objective functions in HMOP4.20

$\mathbf{x}$	x_1	x_2	x_3	x_4	x_5	x_6	x_7	x_8	x_9	x_{10}
$\mathbf{c}_{11}^1$	19	48	21	10	18	35	46	11	24	33
$\mathbf{c}_{11}^2$	3	2	2	1	4	3	1	2	4	2
$\mathbf{c}_{12}^1$	12	–46	–23	–38	–33	–48	12	8	19	20
$\mathbf{c}_{12}^2$	1	2	4	2	2	1	2	1	2	1
$\mathbf{c}_{21}^1$	12	38	–23	33	–33	45	12	–9	19	20
$\mathbf{c}_{21}^2$	1	2	4	2	2	1	2	1	2	1
$\mathbf{c}_{22}^1$	12	–36	27	–30	–33	45	–11	12	19	–8
$\mathbf{c}_{22}^2$	1	2	4	2	2	1	2	1	2	1
$\mathbf{c}_{31}^1$	–18	–26	–22	–28	–15	–29	–10	–19	–17	–28
$\mathbf{c}_{31}^2$	2	1	3	2	1	2	3	3	2	1
$\mathbf{c}_{32}^1$	–8	31	28	29	25	36	–8	–7	–13	–15
$\mathbf{c}_{32}^2$	1	2	3	2	2	1	2	1	2	1

Table 4.6: Parameters of the constraints in HMOP4.20

$\mathbf{x}$	x_1	x_2	x_3	x_4	x_5	x_6	x_7	x_8	x_9	x_{10}
$\mathbf{a}_1$	12	–2	4	–7	13	–1	–6	6	11	–8
$\mathbf{a}_2$	–2	5	3	16	6	–12	12	4	–7	–10
$\mathbf{a}_3$	3	–16	–4	–8	–8	2	–12	–12	4	–3
$\mathbf{a}_4$	–11	6	–5	9	–1	8	–4	6	–9	6
$\mathbf{a}_5$	–4	7	–6	–5	13	6	–2	–5	14	–6
$\mathbf{a}_6$	5	–3	14	–3	–9	–7	4	–4	–5	9
$\mathbf{a}_7$	–3	–4	–6	9	6	18	11	–9	–4	7

$$\min\ \bar{z}_{32}(\mathbf{x}) = (\mathbf{c}_{32}^1 + \bar{t}_{32}\mathbf{c}_{32}^2)\mathbf{x} + (\alpha_{32}^1 + \bar{t}_{32}\alpha_{32}^2)$$

subject to

$$\mathbf{x} \in X \stackrel{\text{def}}{=} \{\mathbf{x} \in \mathrm{R}^{10} \mid \mathbf{a}_i\mathbf{x} \leq b_i, i = 1, \cdots, 7, \mathbf{x} \geq \mathbf{0}\}$$

where $\mathbf{a}_i, i = 1, \cdots, 7, c_{r\ell}^1, c_{r\ell}^2, r = 1,2,3,\ \ell = 1,2$ are the constant coefficient row vectors which are shown in Tables 4.5 and 4.6, and $\alpha_{11}^1 = -18, \alpha_{11}^2 = 5, \alpha_{12}^1 = -27, \alpha_{12}^2 = 6, \alpha_{21}^1 = -12, \alpha_{21}^2 = 3, \alpha_{22}^1 = -15, \alpha_{22}^2 = 4, \alpha_{31}^1 = -10, \alpha_{31}^2 = 4, \alpha_{32}^1 = -27, \alpha_{32}^2 = 6$. The right-hand-side of the constraints are $b_1 = 140, b_2 = -220, b_3 = -190, b_4 = 75, b_5 = -160, b_6 = 130, b_7 = 90$. $\bar{t}_{r\ell}, r = 1,2,3, \ell = 1,2$ are Gaussian random variables defined as

$$\bar{t}_{11} \sim \mathrm{N}(4, 2^2), \bar{t}_{12} \sim \mathrm{N}(3, 3^2), \bar{t}_{21} \sim \mathrm{N}(3, 1^2),$$
$$\bar{t}_{22} \sim \mathrm{N}(3, 2^2), \bar{t}_{31} \sim \mathrm{N}(3, 2^2), \bar{t}_{32} \sim \mathrm{N}(3, 3^2).$$

According to Algorithm 4.4, at Step 1, each decision maker (i.e., $\mathrm{DM}_r, r = 1,2,3$) sets his or her membership functions $\mu_{\hat{p}_{r\ell}}(\hat{p}_{r\ell})$, $\mu_{f_{r\ell}}(f_{r\ell}(\mathbf{x},\hat{p}_{r\ell}))$, $\ell = 1,\cdots,k_r$ as

$$\begin{aligned}\mu_{\hat{p}_{r\ell}}(\hat{p}_{r\ell}) &= \frac{p_{r\ell\min} - \hat{p}_{r\ell}}{p_{r\ell\min} - p_{r\ell\max}}, r = 1,2,3, \ell = 1,2,\\ \mu_{f_{r\ell}}(f_{r\ell}(\mathbf{x},\hat{p}_{r\ell})) &= \frac{f_{r\ell}(\mathbf{x},\hat{p}_{r\ell}) - f_{r\ell\max}}{f_{r\ell\min} - f_{r\ell\max}}, r = 1,2,3, \ell = 1,2,\end{aligned}$$

where

$$\begin{aligned}&[p_{11\min}, p_{11\max}] = [0.023, 0.959], \quad [p_{12\min}, p_{12\max}] = [0.015, 0.993],\\ &[p_{21\min}, p_{21\max}] = [0.001, 0.999], \quad [p_{22\min}, p_{22\max}] = [0.259, 0.995],\\ &[p_{31\min}, p_{31\max}] = [0.136, 0.859], \quad [p_{32\min}, p_{32\max}] = [0.001, 0.987],\\ &[f_{11\min}, f_{11\max}] = [2000, 2200], \quad [f_{12\min}, f_{12\max}] = [400, 700],\\ &[f_{21\min}, f_{21\max}] = [800, 1000], \quad [f_{22\min}, f_{22\max}] = [650, 800],\\ &[f_{31\min}, f_{31\max}] = [-1050, -950], \quad [f_{32\min}, f_{32\max}] = [-200, 50].\end{aligned}$$

At Step 2, set initial reference membership values to $\hat{\mu}_{r\ell} = 1, r = 1,2,3, \ell = 1,2$, and initial decision powers to $w_r = 1, r = 1,2,3$. At Step 3, solve MINMAX4.11$(\hat{\mu},\mathbf{w})$ and TEST4.2. The corresponding MF-Pareto optimal solution is then obtained as

$$\mu_{D_{f_{r\ell}}}(\mathbf{x}^*, \mu_{\hat{p}_{r\ell}}^{-1}(\hat{\mu}_{r\ell} - \lambda^*/w_r)) = 0.5452, r = 1,2,3, \ell = 1,2.$$

For optimal solution $\mathbf{x}^* \in X, \lambda^* \in \Lambda(\mathbf{w})$, DM_1 updates his or her decision power to $w_2 = 0.8$ to improve his or her own membership functions at the expense of the membership functions of the other decision makers (i.e., Step 4), then returns to Step 3. Next, we solve MINMAX4.11$(\hat{\mu},\mathbf{w})$ and TEST4.2, and obtain the corresponding MF-Pareto optimal solution as

$$\begin{aligned}\mu_{D_{f_{1\ell}}}(\mathbf{x}^*, \mu_{\hat{p}_{1\ell}}^{-1}(\hat{\mu}_{1\ell} - \lambda^*/w_1)) &= 0.6059, \ell = 1,2,\\ \mu_{D_{f_{r\ell}}}(\mathbf{x}^*, \mu_{\hat{p}_{r\ell}}^{-1}(\hat{\mu}_{r\ell} - \lambda^*/w_r)) &= 0.5074, r = 2,3, \ell = 1,2.\end{aligned}$$

For this optimal solution, DM_1 is satisfied with the current values of the membership functions, but DM_2 is not satisfied with the current values. Therefore, DM_2 updates his or her decision power to $w_3 = 0.75$, and the MF-Pareto optimal solution is obtained as

$$\mu_{D_{f_{1\ell}}}(\mathbf{x}^*, \mu_{\hat{p}_{1\ell}}^{-1}(\hat{\mu}_{1\ell} - \lambda^*/w_1)) = 0.6220, \ell = 1,2,$$

$$\mu_{D_{f_{2\ell}}}(\mathbf{x}^*, \mu_{\hat{p}_{2\ell}}^{-1}(\hat{\mu}_{2\ell} - \lambda^*/w_2)) = 0.5275, \ell = 1,2,$$

$$\mu_{D_{f_{3\ell}}}(\mathbf{x}^*, \mu_{\hat{p}_{3\ell}}^{-1}(\hat{\mu}_{3\ell} - \lambda^*/w_3)) = 0.4960, \ell = 1,2.$$

Table 4.7: Interactive processes in HMOP4.20

	1	2	3	4
$\hat{\mu}_{11}$	1	1	1	0.6220
$\hat{\mu}_{12}$	1	1	1	0.6220
$\hat{\mu}_{21}$	1	1	1	0.5275
$\hat{\mu}_{22}$	1	1	1	0.5275
$\hat{\mu}_{31}$	1	1	1	0.53
$\hat{\mu}_{32}$	1	1	1	0.49
w_1	1	1	1	1
w_2	1	0.8	0.8	0.8
w_3	1	0.8	0.75	0.75
$\mu_{D_{f_{11}}}(\mathbf{x}^*, \hat{p}_{11}^*)$	0.5452	0.6059	0.6220	0.6206
$\mu_{D_{f_{12}}}(\mathbf{x}^*, \hat{p}_{12}^*)$	0.5452	0.6059	0.6220	0.6206
$\mu_{D_{f_{21}}}(\mathbf{x}^*, \hat{p}_{21}^*)$	0.5452	0.5074	0.5275	0.5257
$\mu_{D_{f_{22}}}(\mathbf{x}^*, \hat{p}_{22}^*)$	0.5452	0.5074	0.5275	0.5257
$\mu_{D_{f_{31}}}(\mathbf{x}^*, \hat{p}_{31}^*)$	0.5452	0.5074	0.4960	0.5281
$\mu_{D_{f_{32}}}(\mathbf{x}^*, \hat{p}_{32}^*)$	0.5452	0.5074	0.4960	0.4881
$\hat{p}_{11}^*$	0.5338	0.5907	0.6057	0.6044
$\hat{p}_{12}^*$	0.5484	0.6077	0.6235	0.6221
$\hat{p}_{21}^*$	0.5460	0.5083	0.5284	0.5266
$\hat{p}_{22}^*$	0.6603	0.6325	0.6473	0.6460
$\hat{p}_{31}^*$	0.5306	0.5032	0.4950	0.5182
$\hat{p}_{32}^*$	0.5393	0.5021	0.4908	0.4830
$f_{11}(\mathbf{x}^*, \hat{p}_{11}^*)$	2091	2079	2076	2076
$f_{12}(\mathbf{x}^*, \hat{p}_{12}^*)$	536.4	518.2	513.4	513.8
$f_{21}(\mathbf{x}^*, \hat{p}_{21}^*)$	891.0	898.5	894.5	894.9
$f_{22}(\mathbf{x}^*, \hat{p}_{22}^*)$	718.2	723.9	720.9	721.1
$f_{31}(\mathbf{x}^*, \hat{p}_{31}^*)$	−1005	−1001	−999.6	−1003
$f_{32}(\mathbf{x}^*, \hat{p}_{32}^*)$	-86.30	−76.85	−74.00	−72.02

Since decision makers (DM_1 and DM_2) are satisfied with current values of their respective membership functions, but the third-level decision maker (i.e., DM_3) is not satisfied with his or her current values, DM_3 updates his or her reference membership values to

$$(\hat{\mu}_{31}, \hat{\mu}_{32}) = (0.53, 0.49)$$

to improve $\mu_{D_{f_{31}}}(\mathbf{x}, \hat{p}_{31})$ at the expense of $\mu_{D_{f_{32}}}(\mathbf{x}, \hat{p}_{32})$. Then, since all decision makers (i.e., DM_1, DM_2, and DM_3) are satisfied with the corresponding optimal solution, the interactive processes stop. The interactive processes with the three hypothetical decision makers are summarized in Table 4.7, where $\hat{p}_{r\ell}^* \stackrel{\text{def}}{=} \mu_{\hat{p}_{r\ell}}^{-1}(\hat{\mu}_{r\ell} - \lambda^*/w_r)$.

4.4 Hierarchical multiobjective fuzzy random programming problems (HMOFRPs) based on the fuzzy decision

In this section, we formulate HMOFRPs and propose an interactive algorithm [126] for obtaining a satisfactory solution for multiple decision makers from an MF-Pareto optimal solution set. In Section 4.4.1, after multiple decision makers specify their membership functions for their objective functions, HMOFRPs are transformed into HMOSPs by using a concept called possibility measure [18]. In Section 4.4.2, transformed HMOFRPs are formulated as multiobjective programming problems through a fractile optimization model, and we introduce an MF-Pareto optimal solution for HMOFRPs. In Section 4.4.3, we propose an interactive linear programming algorithm for obtaining a satisfactory solution for multiple decision makers from an MF-Pareto optimal solution set. In Section 4.4.4, we apply our interactive algorithm to a numerical example to illustrate the interactive processes with two hypothetical decision makers.

4.4.1 HMOFRPs and a possibility measure

In this subsection, we focus on a hierarchical multiobjective programming problem involving fuzzy random variable coefficients in objective functions.

HMOP 4.21
first-level decision maker: $\mathbf{DM}_1$

$$\min_{\mathbf{x}\in X}(\tilde{\bar{\mathbf{c}}}_{11}\mathbf{x},\cdots,\tilde{\bar{\mathbf{c}}}_{1k_1}\mathbf{x})$$

$$\cdots\cdots\cdots\cdots\cdots\cdots\cdots$$

q-th-level decision maker: $\mathbf{DM}_q$

$$\min_{\mathbf{x}\in X}(\tilde{\bar{\mathbf{c}}}_{q1}\mathbf{x},\cdots,\tilde{\bar{\mathbf{c}}}_{qk_q}\mathbf{x})$$

Here, $\mathbf{x}=(x_1,\cdots,x_n)^T$ is an n-dimensional decision variable column vector, X is a linear constraint set, and $\tilde{\bar{\mathbf{c}}}_{ri}=(\tilde{\bar{c}}_{ri1},\cdots,\tilde{\bar{c}}_{rin}), r=1,\cdots,q, i=1,\cdots,k_r$ are n-dimensional coefficient vectors of objective function $\tilde{\bar{\mathbf{c}}}_{ri}\mathbf{x}$, whose elements are fuzzy random variables [46, 65, 77].

To handle objective functions $\tilde{\bar{\mathbf{c}}}_{ri}\mathbf{x}, r=1,\cdots,q, i=1,\cdots,k_r$, Katagiri et al. [38, 39] proposed an LR-type fuzzy random variable which is a special version of a fuzzy random variable. Given each elementary event ω, $\tilde{\bar{c}}_{rij}(\omega)$ is a realization of LR-type fuzzy random variable $\tilde{\bar{c}}_{rij}$, which is an LR fuzzy number [18] whose membership function is defined as

$$\mu_{\tilde{\bar{c}}_{rij}(\omega)}(s)=\begin{cases} L\left(\dfrac{\bar{d}_{rij}(\omega)-s}{\bar{\alpha}_{rij}(\omega)}\right), & s\leq\bar{d}_{rij}(\omega)\\ R\left(\dfrac{s-\bar{d}_{rij}(\omega)}{\bar{\beta}_{rij}(\omega)}\right), & s>\bar{d}_{rij}(\omega),\end{cases}$$

where function $L(t) \stackrel{\text{def}}{=} \max\{0, l(t)\}$ is a real-valued continuous function from $[0,\infty)$ to $[0,1]$ and $l(t)$ is a strictly decreasing continuous function satisfying $l(0)=1$. Further, $R(t) \stackrel{\text{def}}{=} \max\{0, r(t)\}$ satisfies the same conditions and $\bar{d}_{rij}, \bar{\alpha}_{rij}$, and $\bar{\beta}_{rij}$ are random variables expressed as

$$\begin{aligned} \bar{d}_{rij} &= d^1_{rij} + \bar{t}_{ri} d^2_{rij}, i = 1, \cdots, k_r, j = 1, \cdots, n \\ \bar{\alpha}_{rij} &= \alpha^1_{rij} + \bar{t}_{ri} \alpha^2_{rij}, i = 1, \cdots, k_r, j = 1, \cdots, n \\ \bar{\beta}_{rij} &= \beta^1_{rij} + \bar{t}_{ri} \beta^2_{rij}, i = 1, \cdots, k_r, j = 1, \cdots, n. \end{aligned}$$

Here, $\bar{t}_{ri}$ is a random variable whose distribution function is denoted by $T_{ri}(\cdot)$, which is a strictly increasing and continuous function, and d^1_{rij}, d^2_{rij}, $\alpha^1_{rij}, \alpha^2_{rij}, \beta^1_{rij}, \beta^2_{rij}$ are constants.

Similar to Katagiri et al. [38, 39], HMOP4.21 can be transformed into a hierarchical multiobjective stochastic programming problem by using a possibility measure [18]. As shown in [38], realizations $\tilde{\bar{\mathbf{c}}}_{ri}(\omega)\mathbf{x}$ become an LR fuzzy number characterized by the following membership functions based on the extension principle [18, 38].

$$\mu_{\tilde{\bar{\mathbf{c}}}_{ri}(\omega)\mathbf{x}}(y) = \begin{cases} L\left(\frac{\bar{\mathbf{d}}_{ri}(\omega)\mathbf{x} - y}{\bar{\alpha}_{ri}(\omega)\mathbf{x}}\right), & y \leq \bar{\mathbf{d}}_{ri}(\omega)\mathbf{x} \\ R\left(\frac{y - \bar{\mathbf{d}}_{ri}(\omega)\mathbf{x}}{\bar{\beta}_{ri}(\omega)\mathbf{x}}\right), & y > \bar{\mathbf{d}}_{ri}(\omega)\mathbf{x} \end{cases}$$

For realizations $\tilde{\bar{\mathbf{c}}}_{ri}(\omega)\mathbf{x}, r = 1, \cdots, q, i = 1, \cdots, k_r$, we assume that the decision maker has fuzzy goals $\tilde{G}_{ri}, r = 1, \cdots, q, i = 1, \cdots, k_r$ [71], whose membership functions $\mu_{\tilde{G}_{ri}}(y), i = 1, \cdots, k_r, r = 1, \cdots, q$ are continuous and strictly decreasing for minimization problems. By using a possibility measure [18], the degree of possibility that objective function value $\tilde{\bar{\mathbf{c}}}_{ri}\mathbf{x}$ satisfies fuzzy goal $\tilde{G}_{ri}$ is defined as follows [35].

$$\Pi_{\tilde{\bar{\mathbf{c}}}_{ri}\mathbf{x}}(\tilde{G}_{ri}) \stackrel{\text{def}}{=} \sup_y \min\{\mu_{\tilde{\bar{\mathbf{c}}}_{ri}\mathbf{x}}(y), \mu_{\tilde{G}_{ri}}(y)\} \tag{4.78}$$

Using a possibility measure (4.78), HMOP4.21 can be transformed into the following hierarchical multiobjective stochastic programming problem.

HMOP 4.22

first-level decision maker: $\mathbf{DM}_1$

$$\max_{\mathbf{x} \in X} (\Pi_{\tilde{\bar{\mathbf{c}}}_{11}\mathbf{x}}(\tilde{G}_{11}), \cdots, \Pi_{\tilde{\bar{\mathbf{c}}}_{1k_1}\mathbf{x}}(\tilde{G}_{1k_1}))$$

....

q-th-level decision maker: $\mathbf{DM}_q$

$$\max_{\mathbf{x} \in X} (\Pi_{\tilde{\bar{\mathbf{c}}}_{q1}\mathbf{x}}(\tilde{G}_{q1}), \cdots, \Pi_{\tilde{\bar{\mathbf{c}}}_{qk_q}\mathbf{x}}(\tilde{G}_{qk_q}))$$

4.4.2 A formulation using a fractile model

If we adopt a fractile optimization model for the objective functions of HMOP4.22, HMOP4.22 can be converted into the following hierarchical multiobjective programming problem.

HMOP 4.23 $(\hat{\mathbf{p}})$
first-level decision maker: DM$_1$

$$\max_{\mathbf{x}\in X, h_{1i}\in[0,1], i=1,\cdots,k_1} (h_{11},\cdots,h_{1k_1})$$

$$\cdots\cdots\cdots\cdots\cdots\cdots\cdots$$

q-th-level decision maker: DM$_q$

$$\max_{\mathbf{x}\in X, h_{qi}\in[0,1], i=1,\cdots,k_q} (h_{q1},\cdots,h_{qk_q})$$

subject to

$$\Pr(\omega \mid \Pi_{\tilde{\bar{\mathbf{c}}}_{ri}(\omega)\mathbf{x}}(\widetilde{G}_{ri}) \geq h_{ri}) \geq \hat{p}_{ri}, r=1,\cdots,q, i=1,\cdots,k_r \tag{4.79}$$

where

$$\hat{\mathbf{p}}_{\mathbf{r}} \stackrel{\text{def}}{=} (\hat{p}_{r1},\cdots,\hat{p}_{rk_r}), r=1,\cdots,q,\ \hat{\mathbf{p}} \stackrel{\text{def}}{=} (\hat{\mathbf{p}}_{\mathbf{1}},\cdots,\hat{\mathbf{p}}_{\mathbf{q}})$$

are vectors of permissible probability levels subjectively specified by decision makers (i.e., DM$_r, r=1,\cdots,q$) [39].

Since distribution function $T_{ri}(\cdot)$ is continuous and strictly increasing, constraints (4.79) can be transformed into

$$\begin{aligned}
&\hat{p}_{ri} \leq \Pr(\omega \mid \Pi_{\tilde{\bar{\mathbf{c}}}_{ri}(\omega)\mathbf{x}}(\widetilde{G}_{ri}) \geq h_{ri}) \\
&= T_{ri}\left(\frac{\mu_{\widetilde{G}_{ri}}^{-1}(h_{ri}) - (\mathbf{d}_{ri}^1\mathbf{x} - L^{-1}(h_{ri})\alpha_{ri}^1\mathbf{x})}{\mathbf{d}_{ri}^2\mathbf{x} - L^{-1}(h_{ri})\alpha_{ri}^2\mathbf{x}}\right) \\
\Leftrightarrow\ & \mu_{\widetilde{G}_{ri}}^{-1}(h_{ri}) \geq (\mathbf{d}_{ri}^1\mathbf{x} - L^{-1}(h_{ri})\alpha_{ri}^1\mathbf{x}) + T_{ri}^{-1}(\hat{p}_{ri})(\mathbf{d}_{ri}^2\mathbf{x} - L^{-1}(h_{ri})\alpha_{ri}^2\mathbf{x}).
\end{aligned} \tag{4.80}$$

Let us define the right-hand-side of inequality (4.80) as

$$\begin{aligned}
f_{ri}(\mathbf{x}, h_{ri}, \hat{p}_{ri}) \stackrel{\text{def}}{=}\ & (\mathbf{d}_{ri}^1\mathbf{x} - L^{-1}(h_{ri})\alpha_{ri}^1\mathbf{x}) \\
& + T_{ri}^{-1}(\hat{p}_{ri})(\mathbf{d}_{ri}^2\mathbf{x} - L^{-1}(h_{ri})\alpha_{ri}^2\mathbf{x})
\end{aligned} \tag{4.81}$$

Then, using $f_{ri}(\mathbf{x}, h_{ri}, \hat{p}_{ri})$, HMOP4.23$(\hat{\mathbf{p}})$ can be equivalently transformed into the following form.

HMOP 4.24 $(\hat{\mathbf{p}})$
first-level decision maker: DM_1

$$\max_{\mathbf{x}\in X, h_{1i}\in[0,1], i=1,\cdots,k_1} (h_{11},\cdots,h_{1k_1})$$

$$\cdots\cdots\cdots\cdots\cdots\cdots$$

q-th-level decision maker: DM_q

$$\max_{\mathbf{x}\in X, h_{qi}\in[0,1], i=1,\cdots,k_q} (h_{q1},\cdots,h_{qk_q})$$

subject to

$$\mu_{\tilde{G}_{ri}}(f_{ri}(\mathbf{x},h_{ri},\hat{p}_{ri})) \geq h_{ri}, r=1,\cdots,q, i=1,\cdots,k_r \tag{4.82}$$

In HMOP4.24$(\hat{\mathbf{p}})$, focusing on inequalities (4.82), $f_{ri}(\mathbf{x},h_{ri},\hat{p}_{ri})$ is continuous and strictly increasing with respect to h_{ri} for any $\mathbf{x}\in X$. This means that the left-hand-side of (4.82) is continuous and strictly decreasing with respect to h_{ri} for any $\mathbf{x}\in X$. Since the right-hand-side of (4.82) is continuous and strictly increasing with respect to h_{ri}, inequalities (4.82) must always satisfy the active condition, i.e.,

$$\mu_{\tilde{G}_{ri}}(f_{ri}(\mathbf{x},h_{ri},\hat{p}_{ri})) = h_{ri}, r=1,\cdots,q, i=1,\cdots,k_r.$$

From such a perspective, HMOP4.24$(\hat{\mathbf{p}})$ is equivalently expressed in the following form.

HMOP 4.25 $(\hat{\mathbf{p}})$
first-level decision maker: DM_1

$$\max_{\mathbf{x}\in X, h_{1i}\in[0,1], i=1,\cdots,k_1} (\mu_{\tilde{G}_{11}}(f_{11}(\mathbf{x},h_{11},\hat{p}_{11})),\cdots,\mu_{\tilde{G}_{1k_1}}(f_{1k_1}(\mathbf{x},h_{1k_1},\hat{p}_{1k_1})))$$

$$\cdots\cdots\cdots\cdots\cdots\cdots$$

q-th-level decision maker: DM_q

$$\max_{\mathbf{x}\in X, h_{qi}\in[0,1], i=1,\cdots,k_q} (\mu_{\tilde{G}_{q1}}(f_{q1}(\mathbf{x},h_{q1},\hat{p}_{q1})),\cdots,\mu_{\tilde{G}_{qk_q}}(f_{qk_q}(\mathbf{x},h_{qk_q},\hat{p}_{qk_q})))$$

subject to

$$\mu_{\tilde{G}_{ri}}(f_{ri}(\mathbf{x},h_{ri},\hat{p}_{ri})) = h_{ri}, r=1,\cdots,q, i=1,\cdots,k_r \tag{4.83}$$

To handle HMOP4.25$(\hat{\mathbf{p}})$, decision makers $(\mathrm{DM}_r, r=1,\cdots,q)$ must specify permissible probability levels $\hat{\mathbf{p}}$ in advance; however, in general, decision makers $(\mathrm{DM}_r, r=1,\cdots,q)$ not only prefer larger values of permissible probability level but also prefer larger values of corresponding membership functions $\mu_{\tilde{G}_{ri}}(\cdot)$. From such a perspective, we consider the following multiobjective programming problem as a natural extension of HMOP4.25$(\hat{\mathbf{p}})$.

HMOP 4.26
first-level decision maker: DM$_1$

$$\max_{\mathbf{x}\in X, h_{1i}\in[0,1], \hat{p}_{1i}\in(0,1), i=1,\cdots,k_1} \quad (\mu_{\widetilde{G}_{11}}(f_{11}(\mathbf{x},h_{11},\hat{p}_{11})),\cdots,$$
$$\mu_{\widetilde{G}_{1k_1}}(f_{1k_1}(\mathbf{x},h_{1k_1},\hat{p}_{1k_1})),\hat{p}_{11},\cdots,\hat{p}_{1k_1})$$

....

q-th-level decision maker: DM$_q$

$$\max_{\mathbf{x}\in X, h_{qi}\in[0,1], \hat{p}_{qi}\in(0,1), i=1,\cdots,k_q} \quad (\mu_{\widetilde{G}_{q1}}(f_{q1}(\mathbf{x},h_{q1},\hat{p}_{q1})),\cdots,$$
$$\mu_{\widetilde{G}_{qk_q}}(f_{qk_q}(\mathbf{x},h_{qk_q},\hat{p}_{qk_q})),\hat{p}_{q1},\cdots,\hat{p}_{qk_q})$$

subject to

$$\mu_{\widetilde{G}_{ri}}(f_{ri}(\mathbf{x},h_{ri},\hat{p}_{ri})) = h_{ri}, r=1,\cdots,q, i=1,\cdots,k_r \tag{4.84}$$

Note that in HMOP4.26, permissible probability levels $\hat{\mathbf{p}}$ are not fixed values but rather decision variables.

Considering the imprecise nature of the decision maker's judgment, we assume that each decision maker (DM$_r, r=1,\cdots,q$) has fuzzy goals for permissible probability levels $\hat{p}_{ri}, i=1,\cdots,k_r$. Such fuzzy goals can be quantified by eliciting the corresponding membership function. Let us denote membership functions of permissible probability levels $\hat{p}_{ri}$ as $\mu_{\hat{p}_{ri}}(\hat{p}_{ri}), i=1,\cdots,k_r$. Then, HMOP4.26 can be transformed into the following hierarchical multiobjective programming problem.

HMOP 4.27
first-level decision maker: DM$_1$

$$\max_{\mathbf{x}\in X, h_{1i}\in[0,1], \hat{p}_{1i}\in(0,1), i=1,\cdots,k_1} (\mu_{\widetilde{G}_{11}}(f_{11}(\mathbf{x},h_{11},\hat{p}_{11})),\cdots,$$
$$\mu_{\widetilde{G}_{1k_1}}(f_{1k_1}(\mathbf{x},h_{1k_1},\hat{p}_{1k_1})),\mu_{\hat{p}_{11}}(\hat{p}_{11}),\cdots,\mu_{\hat{p}_{1k_1}}(\hat{p}_{1k_1}))$$

....

q-th-level decision maker: DM$_q$

$$\max_{\mathbf{x}\in X, h_{qi}\in[0,1], \hat{p}_{qi}\in(0,1), i=1,\cdots,k_q} (\mu_{\widetilde{G}_{q1}}(f_{q1}(\mathbf{x},h_{q1},\hat{p}_{q1})),\cdots,$$
$$\mu_{\widetilde{G}_{qk_q}}(f_{qk_q}(\mathbf{x},h_{qk_q},\hat{p}_{qk_q})),\mu_{\hat{p}_{q1}}(\hat{p}_{q1}),\cdots,\mu_{\hat{p}_{qk_q}}(\hat{p}_{qk_q}))$$

subject to

$$\mu_{\widetilde{G}_{ri}}(f_{ri}(\mathbf{x},h_{ri},\hat{p}_{ri})) = h_{ri}, r=1,\cdots,q, i=1,\cdots,k_r \tag{4.85}$$

Throughout this subsection, we make the following assumption for $\mu_{\hat{p}_{ri}}(\hat{p}_{ri})$.

Assumption 4.5
$\mu_{\hat{p}_{ri}}(\hat{p}_{ri}), r=1,\cdots,q, i=1,\cdots,k_r$ are strictly increasing and continuous with respect to $\hat{p}_{ri} \in P_{ri} \stackrel{\text{def}}{=} [p_{ri\min}, p_{ri\max}]$, and $\mu_{\hat{p}_{ri}}(p_{ri\min})=0$, $\mu_{\hat{p}_{ri}}(p_{ri\max})=1$, where $p_{ri\min}$ is an unacceptable maximum value of $\hat{p}_{ri}$ and $p_{ri\max}$ is a sufficiently satisfactory minimum value of $\hat{p}_{ri}$.

Note that $\mu_{\tilde{G}_{ri}}(f_{ri}(\mathbf{x}, h_{ri}, \hat{p}_{ri}))$ is strictly decreasing with respect to $\hat{p}_{ri}$. If decision makers adopt the fuzzy decision [6, 71, 140] to integrate $\mu_{\tilde{G}_{ri}}(f_{ri}(\mathbf{x}, h_{ri}, \hat{p}_{ri}))$ and $\mu_{\hat{p}_{ri}}(\hat{p}_{ri})$, HMOP4.27 can be transformed into the following form in which

$$\mu_{D_{G_{ri}}}(\mathbf{x}, h_{ri}, \hat{p}_{ri}) \stackrel{\text{def}}{=} \min\{\mu_{\hat{p}_{ri}}(\hat{p}_{ri}), \mu_{\tilde{G}_{ri}}(f_{ri}(\mathbf{x}, h_{ri}, \hat{p}_{ri}))\} \tag{4.86}$$

is an integrated membership function of $\mu_{\hat{p}_{ri}}(\hat{p}_{ri})$ and $\mu_{\tilde{G}_{ri}}(f_{ri}(\mathbf{x}, h_{ri}, \hat{p}_{ri}))$.

HMOP 4.28
first-level decision maker: DM$_1$

$$\max_{\mathbf{x}\in X, \hat{p}_{1i}\in P_{1i}, h_{1i}\in[0,1], i=1,\cdots,k_1} \left(\mu_{D_{G_{11}}}(\mathbf{x}, h_{11}, \hat{p}_{11}), \cdots, \mu_{D_{G_{1k_1}}}(\mathbf{x}, h_{1k_1}, \hat{p}_{1k_1})\right)$$

....

q-th-level decision maker: DM$_q$

$$\max_{\mathbf{x}\in X, \hat{p}_{qi}\in P_{qi}, h_{qi}\in[0,1], i=1,\cdots,k_q} \left(\mu_{D_{G_{q1}}}(\mathbf{x}, h_{q1}, \hat{p}_{q1}), \cdots, \mu_{D_{G_{qk_q}}}(\mathbf{x}, h_{qk_q}, \hat{p}_{qk_q})\right)$$

subject to

$$\mu_{\tilde{G}_{ri}}(f_{ri}(\mathbf{x}, h_{ri}, \hat{p}_{ri})) = h_{ri}, r=1,\cdots,q, i=1,\cdots,k_r \tag{4.87}$$

To handle HMOP4.28, we next introduce an MF-Pareto optimal solution.

Definition 4.5 $\mathbf{x}^* \in X, \hat{p}^*_{ri} \in P_{ri}, h^*_{ri} \in [0,1], r=1,\cdots,q, i=1,\cdots,k_r$ is an MF-Pareto optimal solution to HMOP4.28 if and only if there does not exist another $\mathbf{x} \in X, \hat{p}_{ri} \in P_{ri}, h_{ri} \in [0,1], r=1,\cdots,q, i=1,\cdots,k_r$ such that $\mu_{D_{G_{ri}}}(\mathbf{x}, h_{ri}, \hat{p}_{ri}) \geq \mu_{D_{G_{ri}}}(\mathbf{x}^*, h^*_{ri}, \hat{p}^*_{ri}), r=1,\cdots,q, i=1,\cdots,k_r$, with strict inequality holding for at least one r and i, where $\mu_{\tilde{G}_{ri}}(f_{ri}(\mathbf{x}^*, h^*_{ri}, \hat{p}^*_{ri})) = h^*_{ri}$, $\mu_{\tilde{G}_{ri}}(f_{ri}(\mathbf{x}, h_{ri}, \hat{p}_{ri})) = h_{ri}, r=1,\cdots,q, i=1,\cdots,k_r$.

To generate a candidate for a satisfactory solution that is also MF-Pareto optimal, each decision maker (i.e., DM$_r, r=1,\cdots,q$) specifies reference membership values [71]. Once reference membership values

$$\hat{\mu}_{\mathbf{r}} = (\hat{\mu}_{r1}, \cdots, \hat{\mu}_{rk_r})$$

are specified, the corresponding MF-Pareto optimal solution is obtained by solving the following minmax problem.

MINMAX 4.12 ($\hat{\mu}$)

$$\min_{\mathbf{x}\in X, \hat{p}_{ri}\in P_{ri}, h_{ri}\in[0,1], r=1,\cdots,q, i=1,\cdots,k_r, \lambda\in\Lambda} \lambda$$

subject to

$$\hat{\mu}_{ri} - \mu_{\hat{p}_{ri}}(\hat{p}_{ri}) \leq \lambda, \quad r=1,\cdots,q, i=1,\cdots,k_r \tag{4.88}$$

$$\hat{\mu}_{ri} - h_{ri} \leq \lambda, \quad r=1,\cdots,q, i=1,\cdots,k_r \tag{4.89}$$

$$\mu_{\tilde{G}_{ri}}(f_{ri}(\mathbf{x}, h_{ri}, \hat{p}_{ri})) = h_{ri}, \quad r=1,\cdots,q, i=1,\cdots,k_r \tag{4.90}$$

where

$$\Lambda \stackrel{\text{def}}{=} [\max_{r=1,\cdots,q, i=1,\cdots,k_r} \hat{\mu}_{ri} - 1, \max_{r=1,\cdots,q, i=1,\cdots,k_r} \hat{\mu}_{ri}].$$

Note that, in general, the optimal solution of MINMAX4.12($\hat{\mu}$) does not reflect the hierarchical structure between q decision makers in which the upper-level decision maker can claim priority over lower-level decision makers. To cope with such a hierarchical preference structure between q decision makers in MINMAX4.12($\hat{\mu}$), we introduce the concept of decision powers [47, 49] as

$$\mathbf{w} = (w_1, \cdots, w_q),$$

for membership functions (4.86), where the r-th level decision maker (i.e., DM$_r$) can subjectively specify decision power w_{r+1} and the last decision maker (i.e., DM$_q$) has no decision power whatsoever. To reflect the hierarchical preference structure between multiple decision makers, decision powers $\mathbf{w}$ must satisfy inequality

$$1 = w_1 \geq \cdots \geq w_q > 0. \tag{4.91}$$

Then, MINMAX4.12($\hat{\mu}$) is reformulated as follows.

MINMAX 4.13 ($\hat{\mu}, \mathbf{w}$)

$$\min_{\mathbf{x}\in X, \hat{p}_{ri}\in P_{ri}, \lambda\in\Lambda(\mathbf{w}), h_{ri}\in[0,1], r=1,\cdots,q, i=1,\cdots,k_r} \lambda$$

subject to

$$\hat{\mu}_{ri} - \mu_{\hat{p}_{ri}}(\hat{p}_{ri}) \leq \lambda/w_r, \quad r=1,\cdots,q, i=1,\cdots,k_r \tag{4.92}$$

$$\hat{\mu}_{ri} - h_{ri} \leq \lambda/w_r, \quad r=1,\cdots,q, i=1,\cdots,k_r \tag{4.93}$$

$$\mu_{\tilde{G}_{ri}}(f_{ri}(\mathbf{x}, h_{ri}, \hat{p}_{ri})) = h_{ri}, \quad r=1,\cdots,q, i=1,\cdots,k_r \tag{4.94}$$

where

$$\Lambda(\mathbf{w}) \stackrel{\text{def}}{=} [\max_{r=1,\cdots,q, i=1,\cdots,k_r} w_i(\hat{\mu}_{ri} - 1), \max_{r=1,\cdots,q, i=1,\cdots,k_r} w_i\hat{\mu}_{ri}].$$

In constraints (4.93) and (4.94), it holds that

$$\begin{aligned} & h_{ri} = \mu_{\tilde{G}_{ri}}(f_{ri}(\mathbf{x}, h_{ri}, \hat{p}_{ri})) \geq \hat{\mu}_{ri} - \lambda/w_r \\ \Leftrightarrow \quad & \mu_{\tilde{G}_{ri}}^{-1}(h_{ri}) = (\mathbf{d}_{ri}^1\mathbf{x} - L^{-1}(h_{ri})\alpha_{ri}^1\mathbf{x}) \\ & + T_{ri}^{-1}(\hat{p}_{ri}) \cdot (\mathbf{d}_{ri}^2\mathbf{x} - L^{-1}(h_{ri})\alpha_{ri}^2\mathbf{x}) \leq \mu_{\tilde{G}_{ri}}^{-1}(\hat{\mu}_{ri} - \lambda/w_r). \end{aligned} \tag{4.95}$$

On the right-hand-side of (4.95), because of $L^{-1}(h_{ri}) \leq L^{-1}(\hat{\mu}_{ri} - \lambda/w_r)$ and $\alpha_{ri}^1\mathbf{x} + T_{ri}^{-1}(\hat{p}_{ri})\alpha_{ri}^2\mathbf{x} > 0$, it holds that

$$\begin{aligned}
& (\mathbf{d}_{ri}^1\mathbf{x} - L^{-1}(h_{ri})\alpha_{ri}^1\mathbf{x}) + T_{ri}^{-1}(\hat{p}_{ri}) \cdot (\mathbf{d}_{ri}^2\mathbf{x} - L^{-1}(h_{ri})\alpha_{ri}^2\mathbf{x}) \\
= \; & (\mathbf{d}_{ri}^1\mathbf{x} + T_{ri}^{-1}(\hat{p}_{ri})\mathbf{d}_{ri}^2\mathbf{x}) - L^{-1}(h_{ri})\left(\alpha_{ri}^1\mathbf{x} + T_{ri}^{-1}(\hat{p}_{ri})\alpha_{ri}^2\mathbf{x}\right) \\
\geq \; & (\mathbf{d}_{ri}^1\mathbf{x} + T_{ri}^{-1}(\hat{p}_{ri})\mathbf{d}_{ri}^2\mathbf{x}) \\
& -L^{-1}(\hat{\mu}_{ri} - \lambda/w_r)\left(\alpha_{ri}^1\mathbf{x} + T_{ri}^{-1}(\hat{p}_{ri})\alpha_{ri}^2\mathbf{x}\right). \qquad (4.96)
\end{aligned}$$

Using (4.95) and (4.96), it holds that

$$\begin{aligned}
\mu_{\tilde{G}_{ri}}^{-1}(\hat{\mu}_{ri} - \lambda/w_r) \; \geq \; & (\mathbf{d}_{ri}^1\mathbf{x} + T_{ri}^{-1}(\hat{p}_{ri})\mathbf{d}_{ri}^2\mathbf{x}) \\
& -L^{-1}(\hat{\mu}_{ri} - \lambda/w_r)\left(\alpha_{ri}^1\mathbf{x} + T_{ri}^{-1}(\hat{p}_{ri})\alpha_{ri}^2\mathbf{x}\right) \\
= \; & (\mathbf{d}_{ri}^1\mathbf{x} - L^{-1}(\hat{\mu}_{ri} - \lambda/w_r)\alpha_{ri}^1\mathbf{x}) \\
& +T_{ri}^{-1}(\hat{p}_{ri}) \cdot (\mathbf{d}_{ri}^2\mathbf{x} - L^{-1}(\hat{\mu}_{ri} - \lambda/w_r)\alpha_{ri}^2\mathbf{x}). \qquad (4.97)
\end{aligned}$$

Moreover, because $\hat{p}_{ri} \geq \mu_{\hat{p}_{ri}}^{-1}(\hat{\mu}_{ri} - \lambda/w_r)$, (4.97) can be transformed into

$$\begin{aligned}
& T_{ri}\left(\frac{\mu_{\tilde{G}_{ri}}^{-1}(\hat{\mu}_{ri} - \lambda/w_r) - (\mathbf{d}_{ri}^1\mathbf{x} - L^{-1}(\hat{\mu}_{ri} - \lambda/w_r)\alpha_{ri}^1\mathbf{x})}{\mathbf{d}_{ri}^2\mathbf{x} - L^{-1}(\hat{\mu}_{ri} - \lambda/w_r)\alpha_{ri}^2\mathbf{x}}\right) \\
& \geq \hat{p}_{ri} \geq \mu_{\hat{p}_{ri}}^{-1}(\hat{\mu}_{ri} - \lambda/w_r) \\
\Leftrightarrow \; & \mu_{\tilde{G}_{ri}}^{-1}(\hat{\mu}_{ri} - \lambda) \geq (\mathbf{d}_{ri}^1\mathbf{x} - L^{-1}(\hat{\mu}_{ri} - \lambda)\alpha_{ri}^1\mathbf{x}) \\
& +T_{ri}^{-1}(\mu_{\hat{p}_{ri}}^{-1}(\hat{\mu}_{ri} - \lambda)) \cdot (\mathbf{d}_{ri}^2\mathbf{x} - L^{-1}(\hat{\mu}_{ri} - \lambda)\alpha_{ri}^2\mathbf{x}). \qquad (4.98)
\end{aligned}$$

Therefore, MINMAX4.13$(\hat{\mu}, \mathbf{w})$ can be reduced to the following minmax problem, in which permissible probability levels $\hat{p}_{ri}$, $r = 1, \cdots, q, i = 1, \cdots, k_r$ have disappeared.

MINMAX 4.14 $(\hat{\mu}, \mathbf{w})$

$$\min_{\mathbf{x} \in X, \lambda \in \Lambda(\mathbf{w})} \lambda$$

subject to

$$\begin{aligned}
\mu_{\tilde{G}_{ri}}^{-1}(\hat{\mu}_{ri} - \lambda/w_r) \; \geq \; & (\mathbf{d}_{ri}^1\mathbf{x} - L^{-1}(\hat{\mu}_{ri} - \lambda/w_r)\alpha_{ri}^1\mathbf{x}) \\
& +T_{ri}^{-1}(\mu_{\hat{p}_{ri}}^{-1}(\hat{\mu}_{ri} - \lambda/w_r)) \cdot (\mathbf{d}_{ri}^2\mathbf{x} - L^{-1}(\hat{\mu}_{ri} - \lambda/w_r)\alpha_{ri}^2\mathbf{x}), \\
& r = 1, \cdots, q, i = 1, \cdots, k_r \qquad (4.99)
\end{aligned}$$

The relationships between optimal solution $\mathbf{x}^* \in X, \lambda^* \in \Lambda(\mathbf{w})$ of MINMAX4.14$(\hat{\mu}, \mathbf{w})$ and MF-Pareto optimal solutions to HMOP4.28 are characterized by the following theorem.

Theorem 4.10

If $\mathbf{x}^* \in X, \lambda^* \in \Lambda(\mathbf{w})$ *is a unique optimal solution of MINMAX4.14(*$\hat{\mu}, \mathbf{w}$*), then* $\mathbf{x}^* \in X, \hat{p}^*_{ri} \stackrel{\text{def}}{=} \mu^{-1}_{\hat{p}_{ri}}(\hat{\mu}_{ri} - \lambda^*/w_r) \in P_{ri}, h^*_{ri} \stackrel{\text{def}}{=} \hat{\mu}_{ri} - \lambda^*/w_r, r = 1, \cdots, q, i = 1, \cdots, k_r$ *is an MF-Pareto optimal solution to HMOP4.28.*

4.4.3 An interactive linear programming algorithm

In this subsection, we propose an interactive algorithm for obtaining a satisfactory solution from an MF-Pareto optimal solution set to HMOP4.28. From Theorem 4.10, there is no guarantee that optimal solution $\mathbf{x}^* \in X, \lambda^* \in \Lambda(\mathbf{w})$ of MINMAX4.14($\hat{\mu}, \mathbf{w}$) is MF-Pareto optimal if it is not unique. To guarantee MF-Pareto optimality, we first assume that all constraints (4.99) of MINMAX4.14($\hat{\mu}, \mathbf{w}$) are active at optimal solution $\mathbf{x}^* \in X, \lambda^* \in \Lambda(\mathbf{w})$, i.e.,

$$\begin{aligned} & \mu^{-1}_{\tilde{G}_{ri}}(\hat{\mu}_{ri} - \lambda^*/w_r) - (\mathbf{d}^1_{ri}\mathbf{x}^* - L^{-1}(\hat{\mu}_{ri} - \lambda^*/w_r)\alpha^1_{ri}\mathbf{x}^*) \\ = \; & T^{-1}_{ri}(\mu^{-1}_{\hat{p}_{ri}}(\hat{\mu}_{ri} - \lambda^*/w_r)) \cdot (\mathbf{d}^2_{ri}\mathbf{x}^* - L^{-1}(\hat{\mu}_{ri} - \lambda^*/w_r)\alpha^2_{ri}\mathbf{x}^*), \\ & r = 1, \cdots, q, i = 1, \cdots, k_r. \end{aligned} \tag{4.100}$$

If some constraint of (4.99) is inactive, i.e.,

$$\begin{aligned} & \mu^{-1}_{\tilde{G}_{ri}}(\hat{\mu}_{ri} - \lambda^*/w_r) - (\mathbf{d}^1_{ri}\mathbf{x}^* - L^{-1}(\hat{\mu}_{ri} - \lambda^*/w_r)\alpha^1_{ri}\mathbf{x}^*) \\ & > T^{-1}_{ri}(\mu^{-1}_{\hat{p}_{ri}}(\hat{\mu}_{ri} - \lambda^*/w_r)) \cdot (\mathbf{d}^2_{ri}\mathbf{x}^* - L^{-1}(\hat{\mu}_{ri} - \lambda^*/w_r)\alpha^2_{ri}\mathbf{x}^*) \\ \Leftrightarrow \; & \mu^{-1}_{\tilde{G}_{ri}}(\hat{\mu}_{ri} - \lambda^*/w_r) > f_{ri}(\mathbf{x}^*, \hat{\mu}_{ri} - \lambda^*/w_r, \mu^{-1}_{\hat{p}_{ri}}(\hat{\mu}_{ri} - \lambda^*/w_r)), \end{aligned} \tag{4.101}$$

then we can convert inactive constraint (4.101) into an active one by applying the bisection method for reference membership value $\hat{\mu}_{ri} \in [\lambda^*/w_r, \lambda^*/w_r + 1]$.

For optimal solution $\mathbf{x}^* \in X, \lambda^* \in \Lambda(\mathbf{w})$ of MINMAX4.14($\hat{\mu}, \mathbf{w}$), where active conditions (4.100) are satisfied, we solve the MF-Pareto optimality test problem defined as follows.

Test 4.3

$$w \stackrel{\text{def}}{=} \max_{\mathbf{x} \in X, \varepsilon_{ri} \geq 0, r=1,\cdots,q, i=1,\cdots,k_r} \sum_{r=1}^{q} \sum_{i=1}^{k_r} \varepsilon_{ri} \tag{4.102}$$

subject to

$$\begin{aligned} & T^{-1}_{ri}(\mu^{-1}_{\hat{p}_{ri}}(\hat{\mu}_{ri} - \lambda^*/w_r)) \cdot (\mathbf{d}^2_{ri}\mathbf{x} - L^{-1}(\hat{\mu}_{ri} - \lambda^*/w_r)\alpha^2_{ri}\mathbf{x}) \\ & + (\mathbf{d}^1_{ri}\mathbf{x} - L^{-1}(\hat{\mu}_{ri} - \lambda^*/w_r)\alpha^1_{ri}\mathbf{x}) + \varepsilon_{ri} \\ = \; & T^{-1}_{ri}(\mu^{-1}_{\hat{p}_{ri}}(\hat{\mu}_{ri} - \lambda^*/w_r)) \cdot (\mathbf{d}^2_{ri}\mathbf{x}^* - L^{-1}(\hat{\mu}_{ri} - \lambda^*/w_r)\alpha^2_{ri}\mathbf{x}^*) \\ & + (\mathbf{d}^1_{ri}\mathbf{x}^* - L^{-1}(\hat{\mu}_{ri} - \lambda^*/w_r)\alpha^1_{ri}\mathbf{x}^*), r = 1, \cdots, q, i = 1, \cdots, k_r \end{aligned} \tag{4.103}$$

For the optimal solution of TEST4.3, the following theorem holds.

Theorem 4.11
For optimal solution $\check{\mathbf{x}} \in X, \check{\varepsilon}_{ri} \geq 0, i = 1, \cdots, k_r, r = 1, \cdots, q$ *of TEST4.3, if* $w = 0$ *(equivalently,* $\check{\varepsilon}_{ri} = 0, i = 1, \cdots, k_r, r = 1, \cdots, q$*),* $\mathbf{x}^* \in X, \mu^{-1}_{\hat{p}_{ri}}(\hat{\mu}_{ri} - \lambda^*/w_r) \in P_{ri}, \hat{\mu}_{ri} - \lambda^*/w_r \in [0,1], r = 1, \cdots, q, i = 1, \cdots, k_r$ *is an MF-Pareto optimal solution to HMOP4.28.*

Given the above, we next present an interactive algorithm for deriving a satisfactory solution from an MF-Pareto optimal solution set to HMOP4.28.

Algorithm 4.5

Step 1: *Decision makers (i.e.,* $DM_r, r = 1, \cdots, q$*) set their membership functions* $\mu_{\tilde{G}_{ri}}(y), i = 1, \cdots, k_r$ *for the fuzzy goals of the objective functions of HMOP4.21.*
Step 2: *Each decision maker (i.e.,* $DM_r, r = 1, \cdots, q$*) sets intervals* $P_{ri} = [p_{ri\min}, p_{ri\max}], i = 1, \cdots, k_r$*. According to Assumption 4.5, each decision maker sets his or her membership function* $\mu_{\hat{p}_{ri}}(\hat{p}_{ri}), i = 1, \cdots, k_r$*.*
Step 3: *Set initial reference membership values to* $\hat{\mu}_{ri} = 1, i = 1, \cdots, k_r, r = 1, \cdots, q$*.*
Step 4: *Set initial decision powers* $w_r = 1, r = 1, \cdots, q$*.*
Step 5: *Solve MINMAX4.14(*$\hat{\mu}, \mathbf{w}$*) via a combination of the bisection method and the first phase of the two-phase simplex method of linear programming, thus obtaining optimal solution* $\mathbf{x}^* \in X, \lambda^* \in \Lambda(\mathbf{w})$*. For optimal solution* $\mathbf{x}^* \in X, \lambda^* \in \Lambda(\mathbf{w})$*, TEST4.3 is then solved.*
Step 6: *If each decision maker (i.e.,* $DM_r, r = 1, \cdots, q$*) is satisfied with the current values of MF-Pareto optimal solution* $\mu_{D_{G_{ri}}}(\mathbf{x}^*, h^*_{ri}, \hat{p}^*_{ri}), i = 1, \cdots, k_r$*, where* $\hat{p}^*_{ri} \stackrel{\text{def}}{=} \mu^{-1}_{\hat{p}_{ri}}(\hat{\mu}_{ri} - \lambda^*/w_r), h^*_{ri} \stackrel{\text{def}}{=} \hat{\mu}_{ri} - \lambda^*/w_r, r = 1, \cdots, q, i = 1, \cdots, k_r$*, then stop. Otherwise, let the s-th-level decision maker (i.e.,* DM_s*) be the uppermost decision maker who is not satisfied with the current values of his or her membership functions* $\mu_{D_{G_{si}}}(\mathbf{x}^*, h^*_{si}, \hat{p}^*_{si}), i = 1, \cdots, k_s$*. Considering current values,* DM_s *updates his or her decision power* w_{s+1} *and/or reference membership values* $\hat{\mu}_{si} = 1, i = 1, \cdots, k_s$ *according to the following two rules, then we return to Step 5.*
Rule 1: *To guarantee inequality conditions (4.91),* $w_{s+1} \leq w_s$ *must be satisfied. After updating* w_{s+1}*, if there exists some index* $t > s+1$ *such that* $w_{s+1} < w_t$*, then corresponding decision power* w_t *is set to* $w_t \leftarrow w_{s+1}$*.*
Rule 2: *Reference membership values of* $DM_r, r = 1, \cdots, q, r \neq s$ *must be set to the current values of the membership functions, i.e.,* $\hat{\mu}_{ri} \leftarrow \mu_{D_{G_{ri}}}(\mathbf{x}^*, h^*_{ri}, \hat{p}^*_{ri}), i = 1, \cdots, k_r, r = 1, \cdots, q, r \neq s$*. After that,* DM_s *updates his or her reference membership values* $\hat{\mu}_{si}, i = 1, \cdots, k_s$*.*

4.4.4 A numerical example

To demonstrate our proposed method and the interactive processes involved, we present the following hierarchical two objective programming problem with fuzzy random variable coefficients with two hypothetical decision makers.

HMOP 4.29
first-level decision maker: DM$_1$

$$\min_{\mathbf{x}\in X} \quad (\tilde{\bar{\mathbf{c}}}_{11}\mathbf{x}, \tilde{\bar{\mathbf{c}}}_{12}\mathbf{x}) = (\sum_{j=1}^{10} \tilde{\bar{c}}_{11j}x_j, \sum_{j=1}^{10} \tilde{\bar{c}}_{12j}x_j)$$

second-level decision maker: DM$_2$

$$\min_{\mathbf{x}\in X} \quad (\tilde{\bar{\mathbf{c}}}_{21}\mathbf{x}, \tilde{\bar{\mathbf{c}}}_{22}\mathbf{x}) = (\sum_{j=1}^{10} \tilde{\bar{c}}_{21j}x_j, \sum_{j=1}^{10} \tilde{\bar{c}}_{22j}x_j)$$

subject to

$$\mathbf{x} \in X \stackrel{\text{def}}{=} \{(x_1, \cdots, x_{10}) \geq \mathbf{0} \mid \sum_{j=1}^{10} a_{\ell j}x_j \leq b_\ell, 1 \leq \ell \leq 7\}$$

We assume that realization $\tilde{\bar{c}}_{rij}(\omega)$ of an LR-type fuzzy random variable $\tilde{\bar{c}}_{rij}$ is an LR fuzzy number whose membership function is defined as

$$\mu_{\tilde{\bar{c}}_{rij}(\omega)}(s) = L\left(\frac{(d^1_{rij} + t_{ri}(\omega)d^2_{rij}) - s}{\alpha^1_{rij} + t_{ri}(\omega)\alpha^2_{rij}}\right), \ s \leq d_{rij}(\omega)$$

$$\mu_{\tilde{\bar{c}}_{rij}(\omega)}(s) = R\left(\frac{s - (d^1_{rij} + t_{ri}(\omega)d^2_{rij})}{\beta^1_{rij} + t_{ri}(\omega)\beta^2_{rij}}\right), \ s > d_{rij}(\omega),$$

where $L(t) = R(t) = \max\{0, 1-t\}$ and the parameters $d^1_{rij}, d^2_{rij}, \alpha^1_{rij}, \alpha^2_{rij}, \beta^1_{rij}, \beta^2_{rij}$ are given in Table 4.8. The right-hand-side parameters of the constraints are $b_1 = 140$, $b_2 = -220$, $b_3 = -190$, $b_4 = 75$, $b_5 = -160$, $b_6 = 130$, $b_7 = 90$, while the left-hand-side parameters of the constraints are given in Table 4.9. We assume that $\bar{t}_{ri} \sim \mathrm{N}(0,1), r = 1,2, i = 1,2$ are standard Gaussian random variables. In HMOP4.29, let us assume that the hypothetical decision makers ($\mathrm{DM}_r, r = 1,2$) set their membership functions for the objective functions as follows (i.e., Step 1).

$$\mu_{\tilde{G}_{ri}}(f_{ri}(\mathbf{x}, h_{ri}, \hat{p}_{ri})) = \frac{f_{ri}(\mathbf{x}, h_{ri}, \hat{p}_{ri}) - f_{ri\max}}{f_{ri\min} - f_{ri\max}}, i = 1,2, r = 1,2$$

Table 4.8: Parameters of the objective functions in HMOP4.29

$\mathbf{x}$	x_1	x_2	x_3	x_4	x_5	x_6	x_7	x_8	x_9	x_{10}
d_{11}^1	19	48	21	10	18	35	46	11	24	33
d_{11}^2	3	2	2	1	4	3	1	2	4	2
d_{12}^1	12	46	23	38	33	48	12	8	19	20
d_{12}^2	1	2	4	2	2	1	2	1	2	1
d_{21}^1	−12	−38	−23	−33	−33	−45	−12	−9	−19	−20
d_{21}^2	1	2	4	2	2	1	2	1	2	1
d_{22}^1	−12	−36	−27	−30	−33	−45	−11	−12	−19	−8
d_{22}^2	1	2	4	2	2	1	2	1	2	1
α_{11}^1	0.312	0.759	0.225	0.990	0.248	0.951	0.643	0.984	0.340	0.465
α_{11}^2	0.098	0.042	0.074	0.052	0.045	0.024	0.06	0.036	0.082	0.035
α_{12}^1	0.756	0.704	0.295	0.508	0.216	0.859	0.692	0.741	0.641	0.569
α_{12}^2	0.042	0.058	0.06	0.095	0.081	0.068	0.096	0.091	0.011	0.09
α_{21}^1	0.605	0.107	0.564	0.885	0.652	0.957	0.464	0.733	0.230	0.416
α_{21}^2	0.083	0.074	0.058	0.06	0.032	0.096	0.042	0.076	0.081	0.083
α_{22}^1	0.272	0.313	0.442	0.583	0.341	0.641	0.892	0.432	0.611	0.147
α_{22}^2	0.085	0.056	0.065	0.096	0.066	0.043	0.026	0.091	0.052	0.035

where

$$[f_{11\min}, f_{11\max}] = [2400, 2600],\ [f_{12\min}, f_{12\max}] = [1400, 1500],$$

$$[f_{21\min}, f_{21\max}] = [-1000, -800],\ [f_{22\min}, f_{22\max}] = [-700, -500].$$

We also assume that the hypothetical decision makers (i.e., $\mathrm{DM}_r, r = 1,2$) set their membership functions for permissible probability levels as (i.e., Step 2).

$$\mu_{\hat{p}_{ri}}(\hat{p}_{ri}) = \frac{p_{ri\min} - \hat{p}_{ri}}{p_{ri\min} - p_{ri\max}}, i = 1,2, r = 1,2,$$

where

$$[p_{11\min}, p_{11\max}] = [0.3, 0.9],\ [p_{12\min}, p_{12\max}] = [0.2, 0.85],$$

$$[p_{21\min}, p_{21\max}] = [0.3, 0.87],\ [p_{22\min}, p_{22\max}] = [0.4, 0.91].$$

Next, we set initial reference membership values to $(\hat{\mu}_{11}, \hat{\mu}_{12}) = (\hat{\mu}_{21}, \hat{\mu}_{22}) = (1,1)$ (i.e., Step 3) and decision powers to $(w_1, w_2) = (1,1)$ (i.e., Step 4). At Step 5, we solve MINMAX4.14($\hat{\mu}, \mathbf{w}$) and TEST4.3, obtaining the corresponding MF-Pareto optimal solution as

$$\mu_{D_{G_{ri}}}(\mathbf{x}^*, h_{ri}^*, \hat{p}_{ri}^*) = 0.804, r = 1,2, i = 1,2.$$

Table 4.9: Parameters of the constraints in HMOP4.29

$\mathbf{x}$	x_1	x_2	x_3	x_4	x_5	x_6	x_7	x_8	x_9	x_{10}
a_1	12	−2	4	−7	13	−1	−6	6	11	−8
a_2	−2	5	3	16	6	−12	12	4	−7	−10
a_3	3	−16	−4	−8	−8	2	−12	−12	4	−3
a_4	−11	6	−5	9	−1	8	−4	6	−9	6
a_5	−4	7	−6	−5	13	6	−2	−5	14	−6
a_6	5	−3	14	−3	−9	−7	4	−4	−5	9
a_7	−3	−4	−6	9	6	18	11	−9	−4	7

Table 4.10: Interactive processes in HMOP4.29

	1	2	3
$\hat{\mu}_{11}$	1	1	0.838
$\hat{\mu}_{12}$	1	1	0.838
$\hat{\mu}_{21}$	1	1	0.730
$\hat{\mu}_{22}$	1	1	0.790
w_1	1	1	1
w_2	1	0.7	0.8
$\mu_{D_{G_{11}}}(\mathbf{x}^*, h_{11}^*, \hat{p}_{11}^*)$	0.804	0.838	0.852
$\mu_{D_{G_{12}}}(\mathbf{x}^*, h_{12}^*, \hat{p}_{12}^*)$	0.804	0.838	0.852
$\mu_{D_{G_{21}}}(\mathbf{x}^*, h_{21}^*, \hat{p}_{21}^*)$	0.804	0.769	0.750
$\mu_{D_{G_{22}}}(\mathbf{x}^*, h_{22}^*, \hat{p}_{22}^*)$	0.804	0.769	0.810

For the current value of the MF-Pareto optimal solution, DM_1 updates his or her decision power to $w_2 = 0.7$ to improve his or her own membership functions $\mu_{D_{G_{1i}}}(\cdot), i = 1,2$ at the expense of DM_2's membership functions $\mu_{D_{G_{2i}}}(\cdot), i = 1,2$ (i.e., Step 6). Next, the corresponding MF-Pareto optimal solution is obtained as (i.e., Step 5)

$$\mu_{D_{G_{1i}}}(\mathbf{x}^*, h_{1i}^*, \hat{p}_{1i}^*) = 0.838, i = 1,2, \mu_{D_{G_{2i}}}(\mathbf{x}^*, h_{2i}^*, \hat{p}_{2i}^*) = 0.769, i = 1,2.$$

Then, with DM_1 satisfied with the current value of the membership functions, but DM_2 not satisfied with the current values, to improve $\mu_{D_{G_{22}}}(\cdot)$ at the expense of $\mu_{D_{G_{21}}}(\cdot)$, DM_2 updates his or her reference membership values as shown below, in which $(\hat{\mu}_{11}, \hat{\mu}_{12})$ are automatically set according to Rule 2 (i.e., Step 6).

$$(\hat{\mu}_{11}, \hat{\mu}_{12}) = (0.838, 0.838), \ (\hat{\mu}_{21}, \hat{\mu}_{22}) = (0.730, 0.790)$$

Then, the MF-Pareto optimal solution is obtained as follows (i.e., Step 5).

$$\mu_{D_{G_{1i}}}(\mathbf{x}^*, h_{1i}^*, \hat{p}_{1i}^*) = 0.852, i = 1,2$$
$$\mu_{D_{G_{21}}}(\mathbf{x}^*, h_{21}^*, \hat{p}_{21}^*) = 0.750,\ \mu_{D_{G_{22}}}(\mathbf{x}^*, h_{22}^*, \hat{p}_{22}^*) = 0.810$$

Since both DM_1 and DM_2 are satisfied with current values of the above membership functions, the interactive processes stop (i.e., Step 6).

Chapter 5

Multiobjective Two-Person Zero-Sum Games

CONTENTS

In this chapter, we propose interactive algorithms for two-person zero-sum games with vector payoffs and vector fuzzy payoffs on the assumption that each player has fuzzy goals for his or her multiple expected payoffs.

Shapley [90] first defined the Pareto equilibrium solution concept for two-person zero-sum games with vector payoffs, proving the existence of a Pareto equilibrium solution by utilizing the weighting method for multiobjective optimization. Zeleny [138] formulated a two-person zero-sum game with vector payoffs as a single objective optimization problem to obtain the minmax solution. Cook [15] also formulated a two-person zero-sum game with vector payoffs as a goal programming problem, in which each player sets goals for multiple expected payoffs and the distances between them are minimized. It has been shown that such a goal programming problem is reduced to a linear programming problem. Moreover, Ghose and Prasad [23] proposed a solution incorporating not only Pareto optimality but also security levels. The concept of security levels is inherent in the definition of maximin solutions in two-person zero-sum games. Sakawa and Nishizaki [76] proposed a fuzzy approach for two-person zero-sum games with vector payoffs to obtain maximin solutions that are defined from the viewpoint of maximization of the degree of minimal goal attainment [6, 140]. They showed that such a problem is reduced to a linear programming problem.

Conversely, Campos [8] first formulated two-person zero-sum games with fuzzy payoffs as fuzzy linear programming problems to obtain maximin solutions. Li [51, 52] also formulated special types of two-person zero-sum games with fuzzy payoffs represented by triangular fuzzy numbers as three objective linear programming problems and proposed the corresponding computation method. Bector et al. [4, 5], Vidyottama et al. [99], and Vijay et al. [100] proposed computational methods for solving not only two-person zero-sum games with fuzzy payoffs but also two-person nonzero-sum games with fuzzy payoffs, which are based on the duality of mathematical programming techniques. Maeda [54] introduced an order relationship between fuzzy numbers with respect to two-person zero-sum games with fuzzy payoffs, and proposed a solution. By introducing fuzzy goals, Nishizaki and Sakawa [60, 61, 62] formulated two-person zero-sum games with vector fuzzy payoffs as a linear programming problem to obtain maximin solutions under the assumption that fuzzy goals are defined as linear membership functions. They also investigated equilibrium solutions in two-person nonzero-sum games with fuzzy goals and/or vector fuzzy payoffs.

In this chapter, we focus on two-person zero-sum games with vector payoffs and vector fuzzy payoffs under the assumption that a player has fuzzy goals for expected payoffs that are defined as nonlinear membership functions [129]. In Section 5.1, we introduce the pessimistic Pareto optimal solution by assuming that a player an opponent adopts the most disadvantageous strategy, then translate two-person zero-sum games with vector payoffs into corresponding multiobjective programming problems. Further, we propose an interactive algorithm based on a linear programming technique to obtain a pessimistic compromise solution from a pessimistic Pareto optimal solution set. In Section 5.2, we also consider two-person zero-sum games with vector fuzzy payoffs and propose an interactive algorithm for obtaining a pessimistic compromise solution from a pessimistic Pareto optimal solution set defined based on the possibility measure [18]. In Section 5.3, to demonstrate our proposed interactive algorithm and the interactive processes involved, we consider a numerical example with hypothetical players, formulated as a two-person zero-sum game with vector payoffs.

5.1 Two-person zero-sum games with vector payoffs

In this subsection, we consider two-person zero-sum games with multiple payoffs defined by $(m \times n)$-dimensional matrices, i.e.,

$$A^k \in \mathrm{R}^{m \times n}, k = 1, \ldots, K. \tag{5.1}$$

For each (i, j)-element a_{ij}^k of payoff matrices A^k, $k = 1, \ldots, K$, row $i \in \{1, 2, \ldots, m\}$ is interpreted as a pure strategy of Player 1 and column $j \in \{1, 2, \ldots, n\}$ is similarly a pure strategy of Player 2. When Player 1 chooses pure strategy i and Player 2 chooses pure strategy j, Players 1 and 2 receive K-dimensional payoff vectors $(a_{ij}^1, \ldots, a_{ij}^K)$ and $(-a_{ij}^1, \ldots, -a_{ij}^K)$, respectively. Let

$$\mathbf{x} \in X \stackrel{\text{def}}{=} \left\{ \mathbf{x} = (x_1, \ldots, x_m) \mid \sum_{i=1}^{m} x_i = 1, x_i \geq 0, i = 1, \ldots, m \right\} \tag{5.2}$$

be a mixed strategy for Player 1 and let

$$\mathbf{y} \in Y \stackrel{\text{def}}{=} \left\{ \mathbf{y} = (y_1, \ldots, y_n) \mid \sum_{j=1}^{n} y_j = 1, y_j \geq 0, j = 1, \ldots, n \right\} \tag{5.3}$$

be a mixed strategy for Player 2.

In this subsection, we assume that each player has fuzzy goals for his or her expected payoffs, i.e.,

$$\mathbf{x}A^k\mathbf{y}^T, k = 1, \ldots, K, \tag{5.4}$$

where $\mathbf{x}$ and $\mathbf{y}$ are mixed strategies specified by two players.

Assumption 5.1

Let

$$D_1^k \stackrel{\text{def}}{=} \{\mathbf{x}A^k\mathbf{y}^T \in \mathrm{R}^1 \mid \mathbf{x} \in X, \mathbf{y} \in Y\}, k = 1, \ldots, K \tag{5.5}$$

be the set of Player 1's payoffs. Then, Player 1's fuzzy goal $\widetilde{G}_1^k$ for the k-th payoff is a fuzzy set defined on set D_1^k characterized by strictly increasing and continuous membership functions

$$\mu_{\widetilde{G}_{1k}} : D_1^k \to [0, 1], \ k = 1, \ldots, K. \tag{5.6}$$

Similarly, membership functions $\mu_{\widetilde{G}_{2k}}(\cdot), k = 1, \ldots, K$ of Player 2's fuzzy goals are defined on

$$D_2^k \stackrel{\text{def}}{=} \{-\mathbf{x}A^k\mathbf{y}^T \in \mathrm{R}^1 \mid \mathbf{x} \in X, \mathbf{y} \in Y\}, k = 1, \ldots, K, \tag{5.7}$$

and they are strictly increasing and continuous.

Given the assumption that Player 1 supposes Player 2 adopts the most disadvantageous strategy, the multiobjective programming problem for Player 1 is formulated as follows.

MOP 5.1

$$\max_{\mathbf{x} \in X} \left(\min_{\mathbf{y} \in Y} \mu_{\widetilde{G}_{11}}(\mathbf{x}A^1\mathbf{y}^T), \cdots, \min_{\mathbf{y} \in Y} \mu_{\widetilde{G}_{1K}}(\mathbf{x}A^K\mathbf{y}^T) \right)$$

Each objective function in MOP5.1 can be transformed into the following form.

$$\begin{aligned}\max_{\mathbf{x}\in X}\min_{\mathbf{y}\in Y}\mu_{\tilde{G}_{1k}}(\mathbf{x}A^k\mathbf{y}^T) &= \max_{\mathbf{x}\in X}\left\{\min_{\mathbf{y}\in Y}\mu_{\tilde{G}_{1k}}\left(\sum_{i=1}^{m}\sum_{j=1}^{n}x_i a_{ij}^k y_j\right)\right\}\\ &= \max_{\mathbf{x}\in X}\left\{\mu_{\tilde{G}_{1k}}\left(\min_{\mathbf{y}\in Y}\sum_{j=1}^{n}\left(\sum_{i=1}^{m}x_i a_{ij}^k\right)y_j\right)\right\}\\ &= \max_{\mathbf{x}\in X}\left\{\mu_{\tilde{G}_{1k}}\left(\min_{j=1,\ldots,n}\left(\sum_{i=1}^{m}x_i a_{ij}^k\right)\right)\right\},\\ & k=1,\ldots,K \end{aligned} \tag{5.8}$$

Therefore, MOP5.1 is equivalently transformed into the following form.

MOP 5.2

$$\max_{\mathbf{x}\in X}\left(\mu_{\tilde{G}_{11}}\left(\min_{j=1,\ldots,n}\left(\sum_{i=1}^{m}x_i a_{ij}^1\right)\right),\cdots,\mu_{\tilde{G}_{1K}}\left(\min_{j=1,\ldots,n}\left(\sum_{i=1}^{m}x_i a_{ij}^K\right)\right)\right)$$

To handle MOP5.1, we define the following Pareto optimal solution.

Definition 5.1 $\mathbf{x}^* \in X$ is a Player 1's pessimistic Pareto optimal solution to MOP5.1 if and only if there does not exist another $\mathbf{x} \in X$ such that

$$\min_{\mathbf{y}\in Y}\mu_{\tilde{G}_{1k}}(\mathbf{x}^*A^k\mathbf{y}^T) \leq \min_{\mathbf{y}\in Y}\mu_{\tilde{G}_{1k}}(\mathbf{x}A^k\mathbf{y}^T),\ k=1,\cdots,K,$$

with strict inequality holding for at least one k.

We assume that Player 1 can find a compromise solution from the given pessimistic Pareto optimal solution set. Different from the satisfactory solution concept [71] for usual multiobjective programming problems, we call this solution a pessimistic compromise solution from the perspective of the maximin solution.

To generate a candidate for a pessimistic compromise solution, Player 1 specifies reference membership values [71]. Once reference membership values

$$\hat{\mu}_1 \stackrel{\text{def}}{=} (\hat{\mu}_1^1,\cdots,\hat{\mu}_1^K)$$

are specified by Player 1, the corresponding pessimistic Pareto optimal solution is obtained by solving the minmax problem.

MINMAX 5.1 ($\hat{\mu}_1$)

$$\min_{\mathbf{x}\in X,\lambda\in\Lambda} \lambda$$

subject to

$$\hat{\mu}_1^k - \mu_{\tilde{G}_{1k}}\left(\min_j\left(\sum_{i=1}^m x_i a_{ij}^k\right)\right) \le \lambda,\ k = 1,\ldots,K \tag{5.9}$$

where

$$\Lambda \stackrel{\text{def}}{=} \left[\max_{k=1,\ldots,K} \hat{\mu}_1^k - 1,\ \max_{k=1,\ldots,K} \hat{\mu}_1^k\right].$$

Since inverse functions $(\mu_{\tilde{G}_{1k}})^{-1}(\cdot), k = 1,\cdots,K$ always exist because of Assumption 5.1, constraints (5.9) are transformed into equivalent inequalities

$$\begin{aligned}
& \mu_{\tilde{G}_{1k}}\left(\min_j\left(\sum_{i=1}^m x_i a_{ij}^k\right)\right) \ge \hat{\mu}_1^k - \lambda \\
\Leftrightarrow\quad & \min_j\left(\sum_{i=1}^m x_i a_{ij}^k\right) \ge (\mu_{\tilde{G}_{1k}})^{-1}(\hat{\mu}_1^k - \lambda) \\
\Leftrightarrow\quad & \sum_{i=1}^m x_i a_{ij}^k \ge (\mu_{\tilde{G}_{1k}})^{-1}(\hat{\mu}_1^k - \lambda), j = 1,\cdots,n,\ k = 1,\cdots,K.
\end{aligned} \tag{5.10}$$

As a result, MINMAX5.1($\hat{\mu}_1$) is expressed as the following simple problem.

MINMAX 5.2 ($\hat{\mu}_1$)

$$\min_{\mathbf{x}\in X, \lambda\in\Lambda} \lambda$$

subject to

$$\sum_{i=1}^m x_i a_{ij}^k \ge (\mu_{\tilde{G}_{1k}})^{-1}(\hat{\mu}_1^k - \lambda), j = 1,\cdots,n,\ k = 1,\ldots,K \tag{5.11}$$

Note that MINMAX5.2($\hat{\mu}_1$) can be easily solved via a combination of the bisection method and the first phase of the two-phase simplex method of linear programming.

The relationships between optimal solution $\mathbf{x}^* \in X, \lambda^* \in \Lambda$ of MINMAX5.2 ($\hat{\mu}_1$) and pessimistic Pareto optimal solutions to MOP5.1 are characterized by the following theorem.

Theorem 5.1

(1) If $\mathbf{x}^* \in X, \lambda^* \in \Lambda$ is a unique optimal solution of MINMAX5.2($\hat{\mu}_1$), then $\mathbf{x}^* \in X$ is a pessimistic Pareto optimal solution to MOP5.1.

(2) If $\mathbf{x}^* \in X$ is a pessimistic Pareto optimal solution to MOP5.1, then $\mathbf{x}^* \in X$, $\lambda^* \stackrel{\text{def}}{=} \hat{\mu}_1^k - \mu_{\tilde{G}_{1k}}\left(\min_j\left(\sum_{i=1}^m x_i^* a_{ij}^k\right)\right), k = 1,\cdots,K$ is an optimal solution of MINMAX5.2($\hat{\mu}_1$) for some reference membership values $\hat{\mu}_1 = (\hat{\mu}_1^1,\cdots,\mu_1^K)$.

(Proof)
(1) Since $\mathbf{x}^* \in X, \lambda^* \in \Lambda$ is an optimal solution of MINMAX5.2($\hat{\mu}_1$), the following inequalities hold.

$$\mu_{\widetilde{G}_{1k}}\left(\sum_{i=1}^{m} x_i^* a_{ij}^k\right) \geq \hat{\mu}_1^k - \lambda^*, j = 1,\cdots,n, k = 1,\ldots,K$$

Assume that $\mathbf{x}^* \in X$ is not a pessimistic Pareto optimal solution to MOP5.1. Then, there exists $\mathbf{x} \in X$ such that

$$\mu_{\widetilde{G}_{1k}}\left(\min_j\left(\sum_{i=1}^{m} x_i^* a_{ij}^k\right)\right) \leq \mu_{\widetilde{G}_{1k}}\left(\min_j\left(\sum_{i=1}^{m} x_i a_{ij}^k\right)\right), \; k = 1,\cdots,K,$$

with strict inequality holding for at least one k. From Assumption 5.1, it holds that

$$\mu_{\widetilde{G}_{1k}}\left(\sum_{i=1}^{m} x_i a_{ij}^k\right) \geq \hat{\mu}_1^k - \lambda^*, j = 1,\cdots,n, k = 1,\ldots,K.$$

This contradicts the fact that $\mathbf{x}^* \in X, \lambda^* \in \Lambda$ is a unique optimal solution of MINMAX5.2($\hat{\mu}_1$).

(2) Assume that $\mathbf{x}^* \in X$, $\lambda^* \in \Lambda$ is not an optimal solution for any reference membership values $\hat{\mu}_1 = (\hat{\mu}_1^1, \cdots, \mu_1^K)$ that satisfy inequalities

$$\lambda^* \stackrel{\text{def}}{=} \hat{\mu}_1^k - \mu_{\widetilde{G}_{1k}}\left(\min_j\left(\sum_{i=1}^{m} x_i^* a_{ij}^k\right)\right), k = 1,\cdots,K.$$

Then, there exists some $\mathbf{x} \in X, \lambda < \lambda^*$ such that

$$\mu_{\widetilde{G}_{1k}}\left(\sum_{i=1}^{m} x_i a_{ij}^k\right) \geq \hat{\mu}_1^k - \lambda, j = 1,\cdots,n, k = 1,\ldots,K.$$

From Assumption 5.1 and the fact that $\hat{\mu}_1^k - \lambda > \hat{\mu}_1^k - \lambda^*, k = 1,\ldots,K$, the following relation holds.

$$\mu_{\widetilde{G}_{1k}}\left(\min_j\left(\sum_{i=1}^{m} x_i^* a_{ij}^k\right)\right) < \mu_{\widetilde{G}_{1k}}\left(\min_j\left(\sum_{i=1}^{m} x_i a_{ij}^k\right)\right), \; k = 1,\cdots,K$$

This contradicts that the fact that $\mathbf{x}^* \in X$, $\lambda^* \in \Lambda$ is a pessimistic Pareto optimal solution to MOP5.1. □

Unfortunately, from Theorem 5.1, there is no guarantee that optimal solution $\mathbf{x}^* \in X, \lambda^* \in \Lambda$ of MINMAX5.2($\hat{\mu}_1$) is pessimistic Pareto optimal, if $\mathbf{x}^* \in X, \lambda^* \in \Lambda$ is not unique. To guarantee pessimistic Pareto optimality, we assume the following K constraints of (5.11) are active at the optimal solution, i.e.,

$$\lambda^*=\hat{\mu}_1^k-\mu_{\widetilde{G}_{1k}}\left(\min_j\left(\sum_{i=1}^m x_i^* a_{ij}^k\right)\right),k=1,\cdots,K, \tag{5.12}$$

simultaneously hold. For optimal solution $\mathbf{x}^*\in X,\lambda^*\in\Lambda$ of MINMAX5.2($\hat{\mu}_1$) in which the active conditions (5.12) are satisfied, we solve the following pessimistic Pareto optimality test problem.

Test 5.1

$$w\stackrel{\text{def}}{=}\max\sum_{k=1}^K \varepsilon_k$$

subject to

$$\sum_{i=1}^m x_i a_{ij}^k \geq (\mu_{\widetilde{G}_{1k}})^{-1}(\hat{\mu}_1^k-\lambda^*)+\varepsilon_k, j=1,\cdots,n,k=1,\ldots,K$$
$$\mathbf{x}\in X,\lambda\in\Lambda,\ \varepsilon_k\geq 0,k=1,\cdots,K$$

For the optimal solution of TEST5.1, the following theorem holds.

Theorem 5.2
For optimal solution $\check{\mathbf{x}}\in X,\check{\varepsilon}_k\geq 0,k=1,\cdots,K$ of TEST5.1, if $w=0$, then $\mathbf{x}^*\in X$ is a pessimistic Pareto optimal solution to MOP5.1.

Given the above, we present an interactive algorithm for deriving a pessimistic compromise solution from the pessimistic Pareto optimal solution set.

Algorithm 5.1

Step 1: *Player 1 sets his or her membership functions $\mu_{\widetilde{G}_{1k}}(\cdot),k=1,\ldots,K$ for the expected payoffs that satisfy Assumption 5.1.*
Step 2: *Set initial reference membership values to $\hat{\mu}_1^k=1,k=1,\ldots,K$.*
Step 3: *Solve MINMAX5.2($\hat{\mu}_1$) via a combination of the bisection method and the first phase of the two-phase simplex method of linear programming. For an optimal solution $\mathbf{x}^*\in X,\lambda^*\in\Lambda$, solve TEST5.1.*
Step 4: *If Player 1 agrees to the current pessimistic Pareto optimal solution $\mu_{\widetilde{G}_{1k}}\left(\min_j\left(\sum_{i=1}^m x_i^* a_{ij}^k\right)\right),k=1,\cdots,K$, then stop. Otherwise, Player 1 updates his or her reference membership values $\hat{\mu}_1^k,k=1,\cdots,K$, and we return to Step 3.*

5.2 Two-person zero-sum games with vector fuzzy payoffs

In this section, we consider two-person zero-sum games with vector fuzzy payoffs defined by $(m \times n)$-dimensional fuzzy matrices $\widetilde{A}^k, k = 1, \cdots, K$, whose (i, j)-element $\widetilde{a}_{ij}^k, i = 1, \cdots, m, j = 1, \cdots, n$ is an LR fuzzy number [18] and the corresponding membership function is

$$\mu_{\widetilde{a}_{ij}^k}(s) = \begin{cases} L\left(\dfrac{a_{ij}^k - s}{\alpha_{ij}^k}\right), & s \leq a_{ij}^k \\ R\left(\dfrac{s - a_{ij}^k}{\beta_{ij}^k}\right), & s > a_{ij}^k, \end{cases} \tag{5.13}$$

where function $L(t) \stackrel{\text{def}}{=} \max\{0, l(t)\}$ is a real-valued continuous function from $[0, \infty)$ to $[0, 1]$ and $l(t)$ is a strictly decreasing continuous function satisfying $l(0) = 1$. Further, function $R(t) \stackrel{\text{def}}{=} \max\{0, r(t)\}$ satisfies the same conditions, a_{ij}^k is the mean value, and $\alpha_{ij}^k > 0$ and $\beta_{ij}^k > 0$ are the left and right spreads, respectively [18]. Similar to the previous section, let

$$\mathbf{x} \in X \stackrel{\text{def}}{=} \left\{ \mathbf{x} = (x_1, \cdots, x_m) \mid \sum_{i=1}^m x_i = 1, x_i \geq 0, i = 1, \ldots, m \right\} \tag{5.14}$$

$$\mathbf{y} \in Y \stackrel{\text{def}}{=} \left\{ \mathbf{y} = (y_1, \ldots, y_n) \mid \sum_{j=1}^n y_j = 1, y_j \geq 0, j = 1, \cdots, n \right\}, \tag{5.15}$$

be mixed strategies for Players 1 and 2, respectively. Then, according to operations of fuzzy numbers based on the extension principle [18], the k-th fuzzy expected payoff of Player 1 is an LR fuzzy number whose membership function is defined as

$$\mu_{\mathbf{x}\widetilde{A}^k\mathbf{y}^T}(t) = \begin{cases} L\left(\dfrac{\sum_{i=1}^m \sum_{j=1}^n a_{ij}^k x_i y_j - t}{\sum_{i=1}^m \sum_{j=1}^n \alpha_{ij}^k x_i y_j}\right), & t \leq \sum_{i=1}^m \sum_{j=1}^n a_{ij}^k x_i y_j \\ R\left(\dfrac{t - \sum_{i=1}^m \sum_{j=1}^n a_{ij}^k x_i y_j}{\sum_{i=1}^m \sum_{j=1}^n \beta_{ij}^k x_i y_j}\right), & t > \sum_{i=1}^m \sum_{j=1}^n a_{ij}^k x_i y_j. \end{cases} \tag{5.16}$$

In this section, we assume that Player 1 has fuzzy goals for his or her fuzzy expected payoffs $\mathbf{x}\widetilde{A}^k\mathbf{y}^T, k = 1, \ldots, K$, whose membership functions are defined as follows.

Assumption 5.2
Let

$$\widetilde{D}_1^k \stackrel{\text{def}}{=} \{\mathbf{x}\widetilde{A}^k\mathbf{y}^T \mid \mathbf{x} \in X, \mathbf{y} \in Y\}, k = 1, \cdots, K \tag{5.17}$$

be the set of Player 1's fuzzy payoffs. Then, Player 1's fuzzy goal $\widetilde{G}_1^k$ for the k-th fuzzy payoff is a fuzzy set defined on set $(\widetilde{D}_1^k)_{\alpha=0}$ characterized by strictly increasing and continuous membership functions

$$\mu_{\widetilde{G}_{1k}} : (\widetilde{D}_1^k)_{\alpha=0} \to [0,1],\ k = 1, \cdots, K, \tag{5.18}$$

where $(\cdot)_\alpha$ represents an α-cut set for fuzzy sets [18]. Similarly, Player 2's membership functions $\mu_{\widetilde{G}_{2k}}(\cdot), k = 1, \cdots, K$ are defined on set $(\widetilde{D}_2^k)_{\alpha=0}$, where

$$\widetilde{D}_2^k \stackrel{\text{def}}{=} \{-\mathbf{x}\widetilde{A}^k\mathbf{y}^T \mid \mathbf{x} \in X, \mathbf{y} \in Y\}, k = 1, \cdots, K, \tag{5.19}$$

which are strictly increasing and continuous.

Using the concept of a possibility measure [18], we define the value of membership function $\mu_{\widetilde{G}_{1k}}(\mathbf{x}\widetilde{A}^k\mathbf{y}^T)$ as

$$\begin{aligned} \mu_{\widetilde{G}_{1k}}(\mathbf{x}\widetilde{A}^k\mathbf{y}^T) &\stackrel{\text{def}}{=} \Pi_{\mathbf{x}\widetilde{A}^k\mathbf{y}^T}(\widetilde{G}_{1k}) \\ &= \max_u \min\{\mu_{\mathbf{x}\widetilde{A}^k\mathbf{y}^T}(u), \mu_{\widetilde{G}_{1k}}(u)\}, k = 1, \cdots, K, \end{aligned} \tag{5.20}$$

where $\mu_{\widetilde{G}_{1k}}(u)$ is a membership function of Player 1's fuzzy goal for the k-th payoff. Then, under the assumption that Player 1 supposes Player 2 adopts the most disadvantageous strategy, we can formulate the following multiobjective programming problem for Player 1.

MOP 5.3

$$\max_{\mathbf{x} \in X} \left(\min_{\mathbf{y} \in Y} \Pi_{\mathbf{x}\widetilde{A}^1\mathbf{y}^T}(\widetilde{G}_{11}), \cdots, \min_{\mathbf{y} \in Y} \Pi_{\mathbf{x}\widetilde{A}^K\mathbf{y}^T}(\widetilde{G}_{1K}) \right)$$

To handle MOP5.3, we introduce pessimistic Pareto optimality.

Definition 5.2 $\mathbf{x}^* \in X$ is a Player 1's pessimistic Pareto optimal solution to MOP5.3 if and only if there does not exist another $\mathbf{x} \in X$ such that

$$\min_{\mathbf{y} \in Y} \Pi_{\mathbf{x}^*\widetilde{A}^k\mathbf{y}^T}(\widetilde{G}_{1k}) \leq \min_{\mathbf{y} \in Y} \Pi_{\mathbf{x}\widetilde{A}^k\mathbf{y}^T}(\widetilde{G}_{1k}), k = 1, \cdots, K, \tag{5.21}$$

with strict inequality holding for at least one k.

Constraints (5.21) are transformed into the following forms, where $\widetilde{\mathbf{a}}_j^k, j = 1, \cdots, n$ represents the j-th column vectors of $\widetilde{A}^k$.

$$\min_{j=1,\ldots,n} \Pi_{\mathbf{x}^* \widetilde{\mathbf{a}}_j^k}(\widetilde{G}_{1k}) \leq \min_{j=1,\ldots,n} \Pi_{\mathbf{x}\widetilde{\mathbf{a}}_j^k}(\widetilde{G}_{1k}) \tag{5.22}$$

Note that decision vector $\mathbf{y} \in Y$ disappears in constraints (5.22).

Similar to the previous section, we assume that Player 1 can find a compromise solution from a pessimistic Pareto optimal solution set. We call this solution a pessimistic compromise solution.

To generate a candidate for a pessimistic compromise solution, Player 1 specifies reference membership values [71]. Once reference membership values

$$\hat{\mu}_1 \stackrel{\text{def}}{=} (\hat{\mu}_1^1, \cdots, \hat{\mu}_1^K)$$

are specified, the corresponding pessimistic Pareto optimal solution is obtained by solving the following minmax problem.

MINMAX 5.3 ($\hat{\mu}_1$)

$$\min_{\mathbf{x} \in X, \lambda \in \Lambda} \lambda$$

subject to

$$\hat{\mu}_1^k - \min_{j=1,\ldots,n} \Pi_{\mathbf{x}\widetilde{\mathbf{a}}_j^k}(\widetilde{G}_{1k}) \leq \lambda, k = 1, \cdots, K, \tag{5.23}$$

where

$$\Lambda \stackrel{\text{def}}{=} [\max_{k=1,\cdots,K} \hat{\mu}_1^k - 1, \max_{k=1,\cdots,K} \hat{\mu}_1^k].$$

Since not only the inverse functions $(\mu_{\widetilde{G}_{1k}})^{-1}(\cdot), k = 1, \cdots, K$ but also $L^{-1}(\cdot)$ and $R^{-1}(\cdot)$ always exist, the k-th constraint of (5.23) is transformed into the following.

$$\begin{aligned}
& \min_{j=1,\cdots,n} \Pi_{\mathbf{x}\tilde{\mathbf{a}}_j^k}(\tilde{G}_{1k}) \geq \hat{\mu}_{1k} - \lambda \\
\Leftrightarrow\ & \min_{j=1,\cdots,n} \max_u \min\{\mu_{\mathbf{x}\tilde{\mathbf{a}}_j^k}(u), \mu_{\tilde{G}_{1k}}(u)\} \geq \hat{\mu}_{1k} - \lambda \\
\Leftrightarrow\ & \max_u \min_{j=1,\cdots,n} \min\left\{\mu_{\mathbf{x}\tilde{\mathbf{a}}_j^k}(u), \mu_{\tilde{G}_{1k}}(u)\right\} \geq \hat{\mu}_{1k} - \lambda \\
\Leftrightarrow\ & \max_u h_k \geq \hat{\mu}_{1k} - \lambda \\
& \mu_{\mathbf{x}\tilde{\mathbf{a}}_j^k}(u) \geq h_k, j = 1, \cdots, n \\
& \mu_{\tilde{G}_{1k}}(u) \geq h_k \\
\Leftrightarrow\ & \max_u h_k \geq \hat{\mu}_{1k} - \lambda \\
& L\left(\frac{\sum_{i=1}^m a_{ij}^k x_i - u}{\sum_{i=1}^m \alpha_{ij}^k x_i}\right) \geq h_k, j = 1, \cdots, n \\
& R\left(\frac{u - \sum_{i=1}^m a_{ij}^k x_i}{\sum_{i=1}^m \beta_{ij}^k x_i}\right) \geq h_k, j = 1, \cdots, n \\
& \mu_{\tilde{G}_{1k}}(u) \geq h_k \\
\Leftrightarrow\ & h_k \geq \hat{\mu}_{1k} - \lambda \\
& \mu_{\tilde{G}_{1k}}^{-1}(h_k) \leq \sum_{i=1}^m a_{ij}^k x_i + R^{-1}(h_k) \cdot \sum_{i=1}^m \beta_{ij}^k x_i, j = 1, \cdots, n \\
\Leftrightarrow\ & \mu_{\tilde{G}_{1k}}^{-1}(\hat{\mu}_{1k} - \lambda) \leq \sum_{i=1}^m a_{ij}^k x_i + R^{-1}(\hat{\mu}_{1k} - \lambda) \cdot \sum_{i=1}^m \beta_{ij}^k x_i, j = 1, \cdots, n
\end{aligned}$$

From the above discussion, MINMAX5.3($\hat{\mu}_1$) for Player 1 can be expressed as follows.

MINMAX 5.4 ($\hat{\mu}_1$)

$$\min_{\mathbf{x} \in X, \lambda \in \Lambda} \lambda$$

subject to

$$\mu_{\tilde{G}_{1k}}^{-1}(\hat{\mu}_{1k} - \lambda) \leq \sum_{i=1}^m a_{ij}^k x_i + R^{-1}(\hat{\mu}_{1k} - \lambda) \cdot \sum_{i=1}^m \beta_{ij}^k x_i,$$
$$j = 1, \cdots, n, k = 1, \cdots, K \tag{5.24}$$

Note that MINMAX5.4($\hat{\mu}_1$) can be easily solved via a combination of the bisection method with respect to $\lambda \in \Lambda$ and the first phase of the two-phase simplex method of linear programming.

The relationships between optimal solution $\mathbf{x}^* \in X, \lambda^* \in \Lambda$ of MINMAX5.4 ($\hat{\mu}_1$) and pessimistic Pareto optimal solutions to MOP5.3 are characterized by the following theorem.

Theorem 5.3
(1) If $\mathbf{x}^* \in X, \lambda^* \in \Lambda$ is a unique optimal solution of MINMAX5.4($\hat{\mu}_1$), then $\mathbf{x}^* \in X$ is a pessimistic Pareto optimal solution to MOP5.3.

(2) If $\mathbf{x}^* \in X$ is a pessimistic Pareto optimal solution to MOP5.3, then there exists $\mathbf{x}^* \in X$, $\lambda^* \in \Lambda$ such that

$$\mu_{\widetilde{G}_{1k}}^{-1}(\hat{\mu}_1^k - \lambda^*) = \min_{j=1,\cdots,n} \left(\sum_{i=1}^{m} a_{ij}^k x_i^* + R^{-1}(\hat{\mu}_1^k - \lambda^*) \cdot \sum_{i=1}^{m} \beta_{ij}^k x_i^* \right)$$

is an optimal solution of MINMAX5.4($\hat{\mu}_1$) for some reference membership values $\hat{\mu}_1^k, k = 1, \ldots, K$.

(Proof)
(1) Since $\mathbf{x}^* \in X, \lambda^* \in \Lambda$ is an optimal solution of MINMAX5.4($\hat{\mu}_1$), the following inequalities hold.

$$\mu_{\widetilde{G}_{1k}}^{-1}(\hat{\mu}_{1k} - \lambda^*) \leq \sum_{i=1}^{m} a_{ij}^k x_i^* + R^{-1}(\hat{\mu}_{1k} - \lambda^*) \cdot \sum_{i=1}^{m} \beta_{ij}^k x_i^*,$$
$$j = 1, \cdots, n, k = 1, \cdots, K$$

Since the constraints of (5.23) are equivalent to those of (5.24), the following relations hold.

$$\hat{\mu}_{1k} - \lambda^* \leq \min_{j=1,\ldots,n} \Pi_{\mathbf{x}^* \tilde{\mathbf{a}}_j^k}(\widetilde{G}_{1k}), k = 1, \cdots, K.$$

Assume that $\mathbf{x}^* \in X$ is not a pessimistic Pareto optimal solution to MOP5.3. Then, there exists $\mathbf{x} \in X$ such that

$$\min_{j=1,\ldots,n} \Pi_{\mathbf{x}^* \tilde{\mathbf{a}}_j^k}(\widetilde{G}_{1k}) \leq \min_{j=1,\ldots,n} \Pi_{\mathbf{x} \tilde{\mathbf{a}}_j^k}(\widetilde{G}_{1k}),\ k = 1, \cdots, K,$$

with strict inequality holding for at least one k. Therefore, it holds that

$$\hat{\mu}_{1k} - \lambda^* \leq \min_{j=1,\ldots,n} \Pi_{\mathbf{x} \tilde{\mathbf{a}}_j^k}(\widetilde{G}_{1k}), k = 1, \ldots, K.$$

This contradicts the fact that $\mathbf{x}^* \in X, \lambda^* \in \Lambda$ is a unique optimal solution of MINMAX5.4($\hat{\mu}_1$).

(2) Assume that $\mathbf{x}^* \in X$, $\lambda^* \in \Lambda$ is not an optimal solution of MINMAX5.4($\hat{\mu}_1$) for any reference membership values $\hat{\mu}_1^k, k = 1, \ldots, K$ that satisfy

$$\mu_{\widetilde{G}_{1k}}^{-1}(\hat{\mu}_{1k} - \lambda^*) = \min_{j=1,\ldots,n} \left(\sum_{i=1}^{m} a_{ij}^k x_i^* + R^{-1}(\hat{\mu}_{1k} - \lambda^*) \cdot \sum_{i=1}^{m} \beta_{ij}^k x_i^* \right),\ k = 1, \ldots, K.$$

Then, there exists some $\mathbf{x} \in X, \lambda < \lambda^*$ such that

$$\mu_{\widetilde{G}_{1k}}^{-1}(\hat{\mu}_{1k} - \lambda) \leq \sum_{i=1}^{m} a_{ij}^k x_i + R^{-1}(\hat{\mu}_{1k} - \lambda) \cdot \sum_{i=1}^{m} \beta_{ij}^k x_i,$$
$$j = 1, \ldots, n, k = 1, \ldots, K.$$

This means that there exists some $\mathbf{x} \in X, \lambda < \lambda^*$ such that

$$\hat{\mu}_{1k} - \lambda \leq \min_{j=1,\ldots,n} \Pi_{\mathbf{x}\widetilde{\mathbf{a}}_j^k}(\widetilde{G}_{1k}), k = 1, \ldots, K.$$

Because of $\hat{\mu}_{1k} - \lambda > \hat{\mu}_{1k} - \lambda^*, k = 1, \ldots, K$, there exists $\mathbf{x} \in X$ such that

$$\min_{j=1,\ldots,n} \Pi_{\mathbf{x}^*\widetilde{\mathbf{a}}_j^k}(\widetilde{G}_{1k}) < \min_{j=1,\ldots,n} \Pi_{\mathbf{x}\widetilde{\mathbf{a}}_j^k}(\widetilde{G}_{1k}),\ k = 1, \ldots, K.$$

This contradicts that the fact that $\mathbf{x}^* \in X$, $\lambda^* \in \Lambda$ is a pessimistic Pareto optimal solution to MOP5.3. □

Unfortunately, from Theorem 5.3, there is no guarantee that optimal solution $\mathbf{x}^* \in X, \lambda^* \in \Lambda$ of MINMAX5.4($\hat{\mu}_1$) is pessimistic Pareto optimal, if $\mathbf{x}^* \in X, \lambda^* \in \Lambda$ is not unique. To guarantee pessimistic Pareto optimality, we assume that the following K constraints of (5.24) are active at the optimal solution, i.e.,

$$\begin{aligned} \mu_{\widetilde{G}_{1k}}^{-1}(\hat{\mu}_{1k} - \lambda^*) &= \min_{j=1,\cdots,n} \left(\sum_{i=1}^{m} a_{ij}^k x_i^* + R^{-1}(\hat{\mu}_{1k} - \lambda^*) \cdot \sum_{i=1}^{m} \beta_{ij}^k x_i^* \right), \\ & k = 1, \cdots, K \end{aligned} \tag{5.25}$$

simultaneously hold. For optimal solution $\mathbf{x}^* \in X, \lambda^* \in \Lambda$ of MINMAX5.4($\hat{\mu}_1$) that satisfies active conditions (5.25), we solve the pessimistic Pareto optimality test problem defined as follows:

Test 5.2

$$w \stackrel{\text{def}}{=} \max \sum_{k=1}^{K} \varepsilon_k$$

subject to

$$\mu_{\widetilde{G}_{1k}}^{-1}(\hat{\mu}_{1k} - \lambda^*) + \varepsilon_k \leq \sum_{i=1}^{m} a_{ij}^k x_i + R^{-1}(\hat{\mu}_{1k} - \lambda^*) \cdot \sum_{i=1}^{m} \beta_{ij}^k x_i,$$
$$j = 1, \cdots, n, k = 1, \cdots, K,$$
$$\mathbf{x} \in X, \lambda \in \Lambda, \varepsilon_k \geq 0, k = 1, \cdots, K$$

For optimal solution of TEST5.2 the following theorem holds.

Theorem 5.4
For optimal solution $\check{\mathbf{x}} \in X, \check{\varepsilon}_k \geq 0, k = 1, \cdots, K$ of TEST5.2, if $w = 0$, then $\mathbf{x}^* \in X$ is a pessimistic Pareto optimal solution to MOP5.3.

Given the above, we next present an interactive algorithm for deriving a pessimistic compromise solution from the pessimistic Pareto optimal solution set of MOP5.3.

Algorithm 5.2

Step 1: *Player 1 sets his or her membership functions $\mu_{\tilde{G}_{1k}}(\cdot), k = 1, \ldots, K$ for fuzzy expected payoffs that satisfy Assumption 5.2.*
Step 2: *Set initial reference membership values to $\hat{\mu}_1^k = 1, k = 1, \ldots, K$.*
Step 3: *For reference membership values $\hat{\mu}_1^k = 1, k = 1, \ldots, K$, solve MINMAX5.4($\hat{\mu}_1$) via a combination of the bisection method and the first phase of the two-phase simplex method of linear programming. For optimal solution $\mathbf{x}^* \in X, \lambda^* \in \Lambda$, the corresponding test problem TEST5.2 is solved.*
Step 4: *If Player 1 agrees to the current pessimistic Pareto optimal solution, then stop. Otherwise, Player 1 updates his or her reference membership values $\hat{\mu}_1^k, k = 1, \ldots, K$, and we return to Step 3.*

5.3 Numerical examples

In this section, to show the efficiency of our proposed algorithms (Algorithms 5.1 and 5.2), we first consider the following two-person zero-sum game, which is based on Cook's example [15].

$$A^1 = \begin{pmatrix} 2 & 5 & 1 \\ -1 & -2 & 6 \\ 0 & 3 & -1 \end{pmatrix}$$

$$A^2 = \begin{pmatrix} -3 & 7 & 2 \\ 0 & -2 & 0 \\ 3 & -1 & -6 \end{pmatrix}$$

$$A^3 = \begin{pmatrix} 8 & -2 & 3 \\ -5 & 6 & 0 \\ -3 & 1 & 6 \end{pmatrix}$$

Let fuzzy goals $\tilde{G}_{1k}, k = 1, 2, 3$ for the three expected payoffs $\mathbf{x}A^k\mathbf{y}^T, k = 1, 2, 3$ of Player 1 be represented by membership functions $\mu_{\tilde{G}_{1k}}(\mathbf{x}A^k\mathbf{y}^T), k = 1, 2, 3$, where $\mathbf{x} \stackrel{\text{def}}{=} (x_1, x_2, x_3)$ and $\mathbf{y} \stackrel{\text{def}}{=} (y_1, y_2, y_3)$ are mixed strategies for Player 1 and Player 2 respectively. We assume that Player 1 sets his or her membership functions

$\mu_{\widetilde{G}_{1k}}(\mathbf{x}A^k\mathbf{y}^T), k=1,2,3$ as follows (i.e., Step 1 of Algorithm 5.1).

$$\mu_{\widetilde{G}_{11}}(\mathbf{x}A^1\mathbf{y}^T) = \left(\frac{\mathbf{x}A^1\mathbf{y}^T+2}{8}\right)^{\frac{1}{2}}, \ -2 \le \mathbf{x}A^1\mathbf{y}^T \le 6 \tag{5.26}$$

$$\mu_{\widetilde{G}_{12}}(\mathbf{x}A^2\mathbf{y}^T) = \left(\frac{\mathbf{x}A^2\mathbf{y}^T+6}{13}\right)^{\frac{1}{2}}, \ -6 \le \mathbf{x}A^2\mathbf{y}^T \le 7 \tag{5.27}$$

$$\mu_{\widetilde{G}_{13}}(\mathbf{x}A^3\mathbf{y}^T) = \left(\frac{\mathbf{x}A^3\mathbf{y}^T+5}{13}\right)^{\frac{1}{2}}, \ -5 \le \mathbf{x}A^3\mathbf{y}^T \le 8 \tag{5.28}$$

Here, expected payoffs $\mathbf{x}A^k\mathbf{y}^T, k=1,2,3$ are transformed as follows, given the assumption that Player 1 supposes Player 2 adopts the most disadvantageous strategy.

$$z_k \stackrel{\text{def}}{=} \min_{j=1,2,3}\left(\sum_{i=1}^{3} x_i a_{ij}^k\right), k=1,2,3 \tag{5.29}$$

Then, the three objective two-person zero-sum game for Player 1 is formulated as the following multiobjective programming problem.

MOP 5.4

$$\min_{\mathbf{x}\in X}\left(\mu_{\widetilde{G}_{11}}(z_1), \mu_{\widetilde{G}_{12}}(z_2), \mu_{\widetilde{G}_{13}}(z_3)\right)$$

subject to

$$\mathbf{x} \in X = \left\{\mathbf{x} = (x_1, x_2, x_3) \mid \sum_{i=1}^{3} x_i = 1, x_i \ge 0, i=1,2,3\right\}$$

For MOP5.4, we set the initial reference membership values to $\hat{\mu}_1^k = 1, k = 1,2,3$ (i.e., Step 2 of Algorithm 5.1). Next, we solve MINMAX5.2($\hat{\mu}_1$) and Test5.1, obtaining the corresponding pessimistic Pareto optimal solution via

$$\begin{aligned} \mu_{\widetilde{G}_{1k}}(z_k^*) &= 0.617282, k=1,2,3 \\ (x_1^*, x_2^*, x_3^*) &= (0.599285, 0.150269, 0.250447) \end{aligned}$$

where

$$z_k^* \stackrel{\text{def}}{=} \min_{j=1,2,3}\left(\sum_{i=1}^{3} x_i^* a_{ij}^k\right), k=1,2,3.$$

The interactive processes for hypothetical Player 1 are summarized in Table 5.1. In this example, we obtain a pessimistic compromise solution for Player 1 at the third iteration.

Table 5.1: Interactive processes in MOP5.4

	1	2	3
$\hat{\mu}_1^1$	1	1	1
$\hat{\mu}_1^2$	1	0.85	0.92
$\hat{\mu}_1^3$	1	0.95	0.95
$\mu_{\widetilde{G}_{11}}(z_1^*)$	0.617282	0.649384	0.643438
$\mu_{\widetilde{G}_{12}}(z_2^*)$	0.617282	0.528162	0.563440
$\mu_{\widetilde{G}_{13}}(z_3^*)$	0.617282	0.599384	0.593440
z_1^*	1.04830	1.37360	1.31210
z_2^*	−1.04651	−2.37359	−1.87297
z_3^*	−0.04651	−0.329599	−0.421771
x_1^*	0.599285	0.791199	0.734106
x_2^*	0.150269	0.208799	0.156109
x_3^*	0.250447	0.	0.109784

As a numerical example in Section 5.2, we consider the following two-person zero-sum game, which is regarded as a fuzzy version of Cook's example [15, 61].

$$\widetilde{A}^1 = \begin{pmatrix} (2,0.2,0.2) & (5,0.5,0.5) & (1,0.8,0.8) \\ (-1,0.8,0.8) & (-2,0.4,0.4) & (6,0.1,0.1) \\ (0,0.1,0.1) & (3,0.5,0.5) & (-1,0.8,0.8) \end{pmatrix}$$

$$\widetilde{A}^2 = \begin{pmatrix} (-3,0.8,0.8) & (7,0.3,0.3) & (2,0.4,0.4) \\ (0,0.5,0.5) & (-2,0.2,0.2) & (0,0.7,0.7) \\ (3,0.4,0.4) & (-1,0.8,0.8) & (-6,0.5,0.5) \end{pmatrix}$$

$$\widetilde{A}^3 = \begin{pmatrix} (8,0.1,0.1) & (-2,0.5,0.5) & (3,0.7,0.7) \\ (-5,0.5,0.5) & (6,0.4,0.4) & (0,0.6,0.6) \\ (-3,0.8,0.8) & (1,0.6,0.6) & (6,0.1,0.1) \end{pmatrix}$$

Here, (i,j)-element of $\widetilde{A}^k$ is defined as LR fuzzy number

$$\tilde{a}_{ij}^k \stackrel{\text{def}}{=} (a_{ij}^k, \alpha_{ij}^k, \beta_{ij}^k)_{LR},$$

and $L(x) = R(x) = \max(0, 1-x), 0 \le x \le 1$. Similar to MOP5.4, we assume that Player 1 sets his or her membership functions $\mu_{\widetilde{G}_{1k}}(\mathbf{x}A^k\mathbf{y}^T), k = 1,2,3$ for fuzzy goals $\widetilde{G}_{1k}, k = 1,2,3$ of fuzzy expected payoff $\mathbf{x}\widetilde{A}^k\mathbf{y}^T$ as defined in (5.26), (5.27), and (5.28), respectively. Then, three objective programming problem for Player 1 is formulated as follows.

MOP 5.5

$$\max_{\mathbf{x}\in X} \left(\min_{j=1,2,3} \Pi_{\mathbf{x}\tilde{\mathbf{a}}_j^1}(\widetilde{G}_{11}), \min_{j=1,2,3} \Pi_{\mathbf{x}\tilde{\mathbf{a}}_j^2}(\widetilde{G}_{12}), \min_{j=1,2,3} \Pi_{\mathbf{x}\tilde{\mathbf{a}}_j^3}(\widetilde{G}_{13}) \right)$$

Table 5.2: Interactive processes in MOP5.5

	1	2	3
$\hat{\mu}_1^1$	1	1	1
$\hat{\mu}_1^2$	1	0.85	0.92
$\hat{\mu}_1^3$	1	0.95	0.95
$\mu_{\tilde{G}_{11}}(\tilde{z}_1^*)$	0.628582	0.660385	0.655455
$\mu_{\tilde{G}_{11}}(\tilde{z}_2^*)$	0.628582	0.553487	0.575455
$\mu_{\tilde{G}_{11}}(\tilde{z}_3^*)$	0.628582	0.610385	0.605455
z_1	1.16092	1.48887	1.43697
z_2	−0.86351	−2.01747	−1.69507
z_3	0.13649	−0.156589	−0.23452
x_1^*	0.607625	0.792916	0.750841
x_2^*	0.154107	0.207084	0.164781
x_3^*	0.238268	0.0	0.0843772

subject to

$$\mathbf{x} \in X = \left\{ \mathbf{x} = (x_1, x_2, x_3) \mid \sum_{i=1}^{3} x_i = 1, x_i \geq 0, i = 1,2,3 \right\}$$

By applying Algorithm 5.2 to MOP5.5 for hypothetical Player 1, we obtain the interactive processes shown in Table 5.2.

Next, let us compare the first iteration of Tables 5.1 and 5.2. We obtain the pessimistic Pareto optimal solutions of Tables 5.1 and 5.2 as $\mu_{\tilde{G}_{1k}}(z_k^*) = 0.617282, k = 1,2,3$ and $\mu_{\tilde{G}_{1k}}(\tilde{z}_k^*) = 0.628582, k = 1,2,3$. Note that these differences are due to the definition of a possibility measure and spread parameters $\beta_{ij}^k > 0, i,j,k = 1,2,3$, because MINMAX5.2($\hat{\mu}_1$) is equivalent to MINMAX5.2($\hat{\mu}_1$), if $\beta_{ij}^k = 0, i,j,k = 1,2,3$.

Chapter 6

Generalized Multiobjective Programming Problems (GMOPs)

CONTENTS

In this chapter, we focus on multiobjective programming problems involving standard coefficients, fuzzy coefficients, random variable coefficients, and/or fuzzy random variable coefficients, which collectively are called generalized multiobjective programming problems (GMOPs). Using the results from Chapters 2 and 3, we propose two interactive decision making methods for GMOPs. In Section 6.1, we focus on GMOPs with a special structure [122], transforming these GMOPs into multiobjective programming problems using a fractile optimization model and a possibility measure; we also define a generalized Pareto optimal solution. Further, we propose an interactive linear programming algorithm for obtaining a satisfactory solution from a generalized Pareto optimal solution set. In Section 6.2, we focus on GMOPs with variance covariance matrices. We transform such GMOPs into multiobjective programming problems using a fractile optimization model and a possibility measure; we also define a generalized Pareto optimal solution. Further, we propose an interactive convex programming algorithm for obtaining a satisfactory solution from a generalized Pareto optimal solution set.

6.1 GMOPs with a special structure

In this section, we focus on GMOPs with a special structure. More specifically, in Section 6.1.1, we formulate GMOPs with a special structure as multiobjective programming problems using a fractile optimization model and a possibility measure; we also introduce a generalized Pareto optimal solution [122]. In Section 6.1.2, we propose an interactive linear programming algorithm for obtaining a satisfactory solution from a generalized Pareto optimal solution set. In Section 6.1.3, we apply our interactive algorithm to a numerical example to illustrate interactive processes given a hypothetical decision maker.

6.1.1 *A formulation using a fractile model and a possibility measure*

In this subsection, we focus on a multiobjective programming problem with the special structure defined below in which fuzzy coefficients, random variable coefficients and fuzzy random variable coefficients are incorporated into the objective functions.

MOP 6.1

$$\begin{aligned}
\min_{\mathbf{x}\in X} z_i(\mathbf{x}) &\stackrel{\text{def}}{=} \mathbf{c}_i\mathbf{x}, \ i\in I_1 \\
\min_{\mathbf{x}\in X} \tilde{z}_i(\mathbf{x}) &\stackrel{\text{def}}{=} \tilde{\mathbf{c}}_i\mathbf{x}, \ i\in I_2 \\
\min_{\mathbf{x}\in X} \bar{z}_i(\mathbf{x}) &\stackrel{\text{def}}{=} \bar{\mathbf{c}}_i\mathbf{x}+\bar{\alpha}_i, \ i\in I_3 \\
\min_{\mathbf{x}\in X} \tilde{\bar{z}}_i(\mathbf{x}) &\stackrel{\text{def}}{=} \tilde{\bar{\mathbf{c}}}_i\mathbf{x}, \ i\in I_4
\end{aligned}$$

Here, $\mathbf{x}=(x_1,\cdots,x_n)^T$ is an n-dimensional decision variable column vector, X is a linear constraint set, and I_1, I_2, I_3, and I_4 are index sets of the objective functions.

All coefficients of the objective functions, i.e., $\mathbf{c_i}, i \in I_1$, $\widetilde{\mathbf{c}}_\mathbf{i}, i \in I_2$, $\bar{\mathbf{c}}_\mathbf{i}, i \in I_3$, and $\widetilde{\bar{\mathbf{c}}}_i, i \in I_4$, are n-dimensional vectors. Further, the elements of $\mathbf{c_i}, i \in I_1$, $\widetilde{\mathbf{c}}_\mathbf{i}, i \in I_2$, $\bar{\mathbf{c}}_\mathbf{i}, i \in I_3$ and $\widetilde{\bar{\mathbf{c}}}_i, i \in I_4$ are a real number, a fuzzy coefficient, a random variable coefficient, and a fuzzy random variable coefficient [46, 65, 77], respectively.

For objective function $\widetilde{z}_i(\mathbf{x}), i \in I_2$, we assume that each element of fuzzy coefficients $\widetilde{\mathbf{c}}_\mathbf{i}, i \in I_2$ is an LR fuzzy number [18] whose membership function is defined as

$$\mu_{\widetilde{c}_{ij}}(s) = \begin{cases} L\left(\frac{d_{ij}-s}{\alpha_{ij}}\right) & s \leq d_{ij}, j = 1, \cdots, n, i \in I_2 \\ R\left(\frac{s-d_{ij}}{\beta_{ij}}\right) & s > d_{ij}, j = 1, \cdots, n, i \in I_2, \end{cases}$$

where function $L(t) \stackrel{\text{def}}{=} \max\{0, l(t)\}$ is a real-valued continuous function from $[0, \infty)$ to $[0, 1]$ and $l(t)$ is a strictly decreasing continuous function satisfying $l(0) = 1$. Further, $R(t) \stackrel{\text{def}}{=} \max\{0, r(t)\}$ satisfies the same conditions. d_{ij} is the mean value, and α_{ij} and β_{ij} are called left and right spreads, respectively [18]. According to operations of fuzzy numbers based on the extension principle [18], each objective function $\widetilde{\mathbf{c}}_i\mathbf{x}, i \in I_2$ can also be expressed by an LR fuzzy number whose membership function is defined as

$$\mu_{\widetilde{\mathbf{c}}_i\mathbf{x}}(y) = \begin{cases} L\left(\frac{\mathbf{d}_i\mathbf{x}-y}{\alpha_i\mathbf{x}}\right) & y \leq \mathbf{d}_i\mathbf{x}, i \in I_2 \\ R\left(\frac{y-\mathbf{d}_i\mathbf{x}}{\beta_i\mathbf{x}}\right) & y > \mathbf{d}_i\mathbf{x}, i \in I_2, \end{cases}$$

where $\mathbf{d}_i = (d_{i1}, \cdots, d_{in})$, $\alpha_i = (\alpha_{i1}, \cdots, \alpha_{in}) > \mathbf{0}$, $\beta_i = (\beta_{i1}, \cdots, \beta_{in}) > \mathbf{0}, i \in I_2$.

For objective function $\bar{z}_i(\mathbf{x}), i \in I_3$, we assume that random variable coefficient vector $\bar{\mathbf{c}}_\mathbf{i}, i \in I_3$ and random variable coefficient $\bar{\alpha}_i, i \in I_2$ are defined as

$$\bar{\mathbf{c}}_i = \mathbf{c}_i^1 + \bar{t}_i\mathbf{c}_i^2,\ \bar{\alpha}_i = \alpha_i^1 + \bar{t}_i\alpha_i^2, i \in I_3,$$

where $\bar{t}_i, i \in I_3$ is a random variable whose distribution function $T_i(\cdot), i \in I_3$ is strictly increasing and continuous.

For objective function $\widetilde{\bar{z}}_i(\mathbf{x}), i \in I_4$, we assume that $\widetilde{\bar{c}}_{ij}$ is an LR-type fuzzy random variable [36, 38, 39, 77]. Given the occurrence of each elementary event ω, $\widetilde{\bar{c}}_{ij}(\omega)$ is a realization of LR-type fuzzy random variable $\widetilde{\bar{c}}_{ij}$, which is an LR fuzzy number [18] whose membership function is defined as

$$\mu_{\widetilde{\bar{c}}_{ij}(\omega)}(s) = \begin{cases} L\left(\dfrac{d_{ij}(\omega) - s}{\alpha_{ij}(\omega)}\right), & s \leq d_{ij}(\omega), j = 1, \cdots, n, i \in I_4 \\ R\left(\dfrac{s - d_{ij}(\omega)}{\beta_{ij}(\omega)}\right) & s > d_{ij}(\omega), j = 1, \cdots, n, i \in I_4, \end{cases}$$

where $\bar{d}_{ij}, \bar{\alpha}_{ij}, \bar{\beta}_{ij}, i \in I_4$ are random variables expressed by

$$\bar{d}_{ij} = d_{ij}^1 + \bar{t}_i d_{ij}^2,\ \bar{\alpha}_{ij} = \alpha_{ij}^1 + \bar{t}_i\alpha_{ij}^2,\ \bar{\beta}_{ij} = \beta_{ij}^1 + \bar{t}_i\beta_{ij}^2, i \in I_4.$$

Here, $\bar{t}_i, i \in I_4$ is a random variable whose distribution function is denoted by $T_i(\cdot), i \in I_4$, which is strictly increasing and continuous, and d_{ij}^1, d_{ij}^2, $\alpha_{ij}^1, \alpha_{ij}^2, \beta_{ij}^1, \beta_{ij}^2, i \in I_4$ are constants. According to operations of LR fuzzy numbers based on the extension principle [18], each objective function $\tilde{\bar{\mathbf{c}}}_i(\omega)\mathbf{x}, i \in I_4$ can also be expressed by an LR fuzzy number whose membership function is defined [18] as

$$\mu_{\tilde{\bar{\mathbf{c}}}_i(\omega)\mathbf{x}}(y) = \begin{cases} L\left(\dfrac{\mathbf{d}_i(\omega)\mathbf{x} - y}{\alpha_i(\omega)\mathbf{x}}\right), & y \leq \mathbf{d}_i(\omega)\mathbf{x}, i \in I_4 \\ R\left(\dfrac{y - \mathbf{d}_i(\omega)\mathbf{x}}{\beta_i(\omega)\mathbf{x}}\right), & y > \mathbf{d}_i(\omega)\mathbf{x}, i \in I_4. \end{cases}$$

Considering the imprecise nature of the decision maker's judgment, we assume that the decision maker has fuzzy goals $\tilde{G}_i, i \in I_1 \cup I_2 \cup I_3 \cup I_4$ [71] for objective functions in MOP6.1. Such a fuzzy goal can be quantified by eliciting the corresponding membership function. Let us denote membership functions of the objective functions as $\mu_{\tilde{G}_i}(y), i \in I_1 \cup I_2 \cup I_3 \cup I_4$. Then, MOP6.1 can be formally transformed into the following problem.

MOP 6.2

$$\begin{aligned} &\max_{\mathbf{x} \in X} \quad \mu_{\tilde{G}_i}(z_i(\mathbf{x})), i \in I_1 \\ &\max_{\mathbf{x} \in X} \quad \mu_{\tilde{G}_i}(\tilde{z}_i(\mathbf{x})), i \in I_2 \\ &\max_{\mathbf{x} \in X} \quad \mu_{\tilde{G}_i}(\bar{z}_i(\mathbf{x})), i \in I_3 \\ &\max_{\mathbf{x} \in X} \quad \mu_{\tilde{G}_i}(\tilde{\bar{z}}_i(\mathbf{x})), i \in I_4 \end{aligned}$$

In MOP6.2, membership function $\mu_{\tilde{G}_i}(z_i(\mathbf{x})), i \in I_1$ is well-defined, but the other membership functions $\mu_{\tilde{G}_i}(\tilde{z}_i(\mathbf{x})), i \in I_2$, $\mu_{\tilde{G}_i}(\bar{z}_i(\mathbf{x})), i \in I_3$, and $\mu_{\tilde{G}_i}(\tilde{\bar{z}}_i(\mathbf{x})), i \in I_4$ are ill-defined.

To handle membership function $\mu_{\tilde{G}_i}(\tilde{z}_i(\mathbf{x})), i \in I_2$, using the concept of a possibility measure [18], we define the value of membership function $\mu_{\tilde{G}_i}(\tilde{z}_i(\mathbf{x})), i \in I_2$ as

$$\mu_{\tilde{G}_i}(\tilde{z}_i(\mathbf{x})) \stackrel{\text{def}}{=} \Pi_{\tilde{z}_i(\mathbf{x})}(\tilde{G}_i) = \sup_y \min\{\mu_{\tilde{\mathbf{c}}_i\mathbf{x}}(y), \mu_{\tilde{G}_i}(y)\}, i \in I_2. \tag{6.1}$$

Assuming that $h_i \stackrel{\text{def}}{=} \Pi_{\tilde{z}_i(\mathbf{x})}(\tilde{G}_i), i \in I_2$, then the value of membership function $\mu_{\tilde{G}_i}(\tilde{z}_i(\mathbf{x})), i \in I_2$ can be obtained as

$$\mu_{\tilde{G}_i}(\tilde{z}_i(\mathbf{x})) \stackrel{\text{def}}{=} \{\mu_{\tilde{G}_i}(f_i(\mathbf{x}, h_i)) \mid \mu_{\tilde{G}_i}^{-1}(h_i) = f_i(\mathbf{x}, h_i)\}, i \in I_2, \tag{6.2}$$

where

$$f_i(\mathbf{x}, h_i) \stackrel{\text{def}}{=} \mathbf{d}_i\mathbf{x} - L^{-1}(h_i)\alpha_i\mathbf{x}, i \in I_2 \tag{6.3}$$

For membership function $\mu_{\tilde{G}_i}(\bar{z}_i(\mathbf{x})), i \in I_3$, using a fractile optimization model [87], we define the value of membership function $\mu_{\tilde{G}_i}(\bar{z}_i(\mathbf{x})), i \in I_3$ as shown below. First, let us consider the probability with which inequality $\mu_{\tilde{G}_i}(\bar{z}_i(\mathbf{x})) \geq h_i, i \in I_3$ is satisfied, where h_i represents a target value for membership function $\mu_{\tilde{G}_i}(\bar{z}_i(\mathbf{x}))$. Then, the following relation holds.

$$\begin{aligned}
p_i(\mathbf{x}, h_i) &\stackrel{\text{def}}{=} \Pr(\omega \mid \mu_{\tilde{G}_i}(\mathbf{c}_i(\omega)\mathbf{x} + \alpha_i(\omega)) \geq h_i) \\
&= \Pr(\omega \mid (\mathbf{c}_i^2\mathbf{x} + \alpha_i^2)t_i(\omega) + (\mathbf{c}_i^1\mathbf{x} + \alpha_i^1) \leq \mu_{\tilde{G}_i}^{-1}(h_i)) \\
&= \Pr\left(\omega \mid t_i(\omega) \leq \frac{\mu_{\tilde{G}_i}^{-1}(h_i) - (\mathbf{c}_i^1\mathbf{x} + \alpha_i^1)}{\mathbf{c}_i^2\mathbf{x} + \alpha_i^2}\right) \\
&= T_i\left(\frac{\mu_{\tilde{G}_i}^{-1}(h_i) - (\mathbf{c}_i^1\mathbf{x} + \alpha_i^1)}{\mathbf{c}_i^2\mathbf{x} + \alpha_i^2}\right), i \in I_3 \qquad (6.4)
\end{aligned}$$

Here, $\mathbf{c}_i^2\mathbf{x} + \alpha_i^2 > 0, i \in I_3$, and $\Pr(\cdot)$ represents a probability measure. According to the fractile optimization model [87], the decision maker subjectively specifies permissible probability level $\hat{p}_i$ for $p_i(\mathbf{x}, h_i)$ in advance. Then, inequality relation $p_i(\mathbf{x}, h_i) \geq \hat{p}_i$ can be transformed into

$$\begin{aligned}
& p_i(\mathbf{x}, h_i) = T_i\left(\frac{\mu_{\tilde{G}_i}^{-1}(h_i) - (\mathbf{c}_i^1\mathbf{x} + \alpha_i^1)}{\mathbf{c}_i^2\mathbf{x} + \alpha_i^2}\right) \geq \hat{p}_i \\
\Leftrightarrow\ & T_i^{-1}(\hat{p}_i) \cdot (\mathbf{c}_i^2\mathbf{x} + \alpha_i^2) + (\mathbf{c}_i^1\mathbf{x} + \alpha_i^1) \leq \mu_{\tilde{G}_i}^{-1}(h_i) \\
\Leftrightarrow\ & \mu_{\tilde{G}_i}(f_i(\mathbf{x}, \hat{p}_i)) \geq h_i,
\end{aligned}$$

where

$$f_i(\mathbf{x}, \hat{p}_i) \stackrel{\text{def}}{=} T_i^{-1}(\hat{p}_i) \cdot (\mathbf{c}_i^2\mathbf{x} + \alpha_i^2) + (\mathbf{c}_i^1\mathbf{x} + \alpha_i^1), i \in I_3. \qquad (6.5)$$

As a result, we can define the value of membership function $\mu_{\tilde{G}_i}(\bar{z}_i(\mathbf{x})), i \in I_3$ based on the fractile optimization model as

$$\mu_{\tilde{G}_i}(\bar{z}_i(\mathbf{x})) \stackrel{\text{def}}{=} \mu_{\tilde{G}_i}(f_i(\mathbf{x}, \hat{p}_i))\ i \in I_3. \qquad (6.6)$$

In membership function (6.6), the decision maker specifies permissible probability levels $\hat{p}_i, i \in I_3$ in advance; however, in practice, it proves difficult for the decision maker to specify such parameters, because of the conflict between the larger value of h_i and the larger value of $\hat{p}_i$. From such a perspective, Yano [111, 116, 117] proposed fuzzy approaches in conjunction with a fractile optimization model [87] in which the decision maker is required to specify not permissible probability levels $\hat{p}_i, i \in I_3$, but rather the corresponding membership functions. Below, we assume that the decision maker has fuzzy goals for permissible probability levels $\hat{p}_i, i \in I_3$ whose membership functions are denoted as $\mu_{\hat{p}_i}(\hat{p}_i), i \in I_3$. We can then define the value of membership function $\mu_{\widetilde{G}_i}(\bar{z}_i(\mathbf{x})), i \in I_3$ as

$$\mu_{\widetilde{G}_i}(\bar{z}_i(\mathbf{x})) \stackrel{\text{def}}{=} \min\{\mu_{\hat{p}_i}(\hat{p}_i), \mu_{\widetilde{G}_i}(f_i(\mathbf{x}, \hat{p}_i))\}, i \in I_3. \tag{6.7}$$

To handle membership function $\mu_{\widetilde{G}_i}(\widetilde{\bar{z}}_i(\mathbf{x})), i \in I_4$, using the concept of a possibility measure [18], we define the value of membership function $\mu_{\widetilde{G}_i}(\widetilde{\bar{z}}_i(\mathbf{x})), i \in I_4$, as

$$\mu_{\widetilde{G}_i}(\widetilde{\bar{z}}_i(\mathbf{x})) \stackrel{\text{def}}{=} \Pi_{\widetilde{\bar{z}}_i(\mathbf{x})}(\widetilde{G}_i) = \sup_y \min\{\mu_{\widetilde{\bar{z}}_i(\mathbf{x})}(y), \mu_{\widetilde{G}_i}(y)\}, i \in I_4. \tag{6.8}$$

Using a fractile optimization model [87] for (6.8), let us consider a probability that inequality $\mu_{\widetilde{G}_i}(\widetilde{\bar{z}}_i(\mathbf{x})) \geq h_i, i \in I_4$ is satisfied, where h_i represents a target value for membership function $\mu_{\widetilde{G}_i}(\widetilde{\bar{z}}_i(\mathbf{x})), i \in I_4$. Then, the following relations hold.

$$\begin{aligned}
& \Pr(\omega \mid \Pi_{\widetilde{\bar{\mathbf{c}}}_i(\omega)\mathbf{x}}(\widetilde{G}_i) \geq h_i) \\
= \;& \Pr\left(\omega \mid (\mathbf{d}_i(\omega) - L^{-1}(h_i)\alpha_i(\omega))\mathbf{x} \leq \mu_{\widetilde{G}_i}^{-1}(h_i)\right) \\
= \;& \Pr\left(\omega \mid (\mathbf{d}_i^1 + t_i(\omega)\mathbf{d}_i^2)\mathbf{x} - L^{-1}(h_i)(\alpha_i^1 + t_i(\omega)\alpha_i^2)\mathbf{x} \leq \mu_{\widetilde{G}_i}^{-1}(h_i)\right) \\
= \;& \Pr\left(\omega \mid (\mathbf{d}_i^1\mathbf{x} - L^{-1}(h_i)\alpha_i^1\mathbf{x}) + t_i(\omega)(\mathbf{d}_i^2\mathbf{x} - L^{-1}(h_i)\alpha_i^2\mathbf{x}) \leq \mu_{\widetilde{G}_i}^{-1}(h_i)\right) \\
= \;& \Pr\left(\omega \mid t_i(\omega) \leq \frac{\mu_{\widetilde{G}_i}^{-1}(h_i) - (\mathbf{d}_i^1\mathbf{x} - L^{-1}(h_i)\alpha_i^1\mathbf{x})}{\mathbf{d}_i^2\mathbf{x} - L^{-1}(h_i)\alpha_i^2\mathbf{x}}\right) \\
= \;& T_i\left(\frac{\mu_{\widetilde{G}_i}^{-1}(h_i) - (\mathbf{d}_i^1\mathbf{x} - L^{-1}(h_i)\alpha_i^1\mathbf{x})}{\mathbf{d}_i^2\mathbf{x} - L^{-1}(h_i)\alpha_i^2\mathbf{x}}\right) \stackrel{\text{def}}{=} p_i(\mathbf{x}, h_i), i \in I_4
\end{aligned} \tag{6.9}$$

According to a fractile optimization model, the decision maker subjectively specifies permissible probability level $\hat{p}_i$ for $p_i(\mathbf{x}, h_i), i \in I_4$ in advance. Then, inequality relation $p_i(\mathbf{x}, h_i) \geq \hat{p}_i, i \in I_4$ can be transformed into

$$\begin{aligned}
& \hat{p}_i \leq p_i(\mathbf{x}, h_i) = T_i\left(\frac{\mu_{\widetilde{G}_i}^{-1}(h_i) - (\mathbf{d}_i^1\mathbf{x} - L^{-1}(h_i)\alpha_i^1\mathbf{x})}{\mathbf{d}_i^2\mathbf{x} - L^{-1}(h_i)\alpha_i^2\mathbf{x}}\right) \\
\Leftrightarrow \;& h_i \leq \mu_{\widetilde{G}_i}(f_i(\mathbf{x}, h_i, \hat{p}_i)), i \in I_4,
\end{aligned} \tag{6.10}$$

where

$$f_i(\mathbf{x}, h_i, \hat{p}_i) \stackrel{\text{def}}{=} (\mathbf{d}_i^1\mathbf{x} - L^{-1}(h_i)\alpha_i^1\mathbf{x}) + T_i^{-1}(\hat{p}_i) \cdot (\mathbf{d}_i^2\mathbf{x} - L^{-1}(h_i)\alpha_i^2\mathbf{x}), i \in I_4. \tag{6.11}$$

As a result, we can define the value of membership function $\mu_{\widetilde{G}_i}(\widetilde{\bar{z}}_i(\mathbf{x})), i \in I_4$ through a fractile optimization model [87] as

$$\mu_{\widetilde{G}_i}(\widetilde{\bar{z}}_i(\mathbf{x})) \stackrel{\text{def}}{=} \{\mu_{\widetilde{G}_i}(f_i(\mathbf{x}, h_i, \hat{p}_i)) \mid \mu_{\widetilde{G}_i}(f_i(\mathbf{x}, h_i, \hat{p}_i)) = h_i\}, i \in I_4. \tag{6.12}$$

From similar discussions presented by Yano and Sakawa [132, 133], let us assume that the decision maker specifies not permissible probability levels $\hat{p}_i, i \in I_4$, but rather corresponding membership functions $\mu_{\hat{p}_i}(\hat{p}_i), i \in I_4$. Then, instead of (6.12), we can adopt membership function

$$\mu_{\widetilde{G}_i}(\widetilde{\widetilde{z}}_i(\mathbf{x})) \stackrel{\text{def}}{=} \left\{ \min\{\mu_{\hat{p}_i}(\hat{p}_i), \mu_{\widetilde{G}_i}(f_i(\mathbf{x}, h_i, \hat{p}_i))\} \mid \mu_{\widetilde{G}_i}(f_i(\mathbf{x}, h_i, \hat{p}_i)) = h_i \right\}, i \in I_4. \tag{6.13}$$

Using definitions (6.2), (6.7), and (6.13), MOP6.2 can be transformed into the usual multiobjective programming problem shown below.

MOP 6.3

$$\max_{\mathbf{x}\in X} \mu_{D_i}(\mathbf{x}) \stackrel{\text{def}}{=} \mu_{\widetilde{G}_i}(\mathbf{c}_i\mathbf{x}), i \in I_1 \tag{6.14}$$

$$\max_{\mathbf{x}\in X, h_i\in[0,1]} \mu_{D_i}(\mathbf{x}, h_i) \stackrel{\text{def}}{=} \mu_{\widetilde{G}_i}(f_i(\mathbf{x}, h_i)), i \in I_2 \tag{6.15}$$

$$\max_{\mathbf{x}\in X, \hat{p}_i\in(0,1)} \mu_{D_i}(\mathbf{x}, \hat{p}_i) \stackrel{\text{def}}{=} \min\{\mu_{\hat{p}_i}(\hat{p}_i), \mu_{\widetilde{G}_i}(f_i(\mathbf{x}, \hat{p}_i))\}, i \in I_3 \tag{6.16}$$

$$\max_{\mathbf{x}\in X, h_i\in[0,1], \hat{p}_i\in(0,1)} \mu_{D_i}(\mathbf{x}, h_i, \hat{p}_i) \stackrel{\text{def}}{=} \min\{\mu_{\hat{p}_i}(\hat{p}_i), \mu_{\widetilde{G}_i}(f_i(\mathbf{x}, h_i, \hat{p}_i))\}, i \in I_4 \tag{6.17}$$

subject to

$$\mu_{\widetilde{G}_i}(f_i(\mathbf{x}, h_i)) = h_i, i \in I_2 \tag{6.18}$$

$$\mu_{\widetilde{G}_i}(f_i(\mathbf{x}, h_i, \hat{p}_i)) = h_i, i \in I_4 \tag{6.19}$$

Throughout this subsection, we make the following assumptions with respect to membership functions $\mu_{\widetilde{G}_i}(\cdot), i \in I_1 \cup I_2 \cup I_3 \cup I_4$, $\mu_{\hat{p}_i}(\cdot), i \in I_3 \cup I_4$.

Assumption 6.1
$\mu_{\widetilde{G}_i}(\mathbf{c}_i\mathbf{x}), i \in I_1, \mu_{\widetilde{G}_i}(f_i(\mathbf{x}, h_i)), i \in I_2, \mu_{\widetilde{G}_i}(f_i(\mathbf{x}, \hat{p}_i)), i \in I_3, \mu_{\widetilde{G}_i}(f_i(\mathbf{x}, h_i, \hat{p}_i)), i \in I_4$ are strictly decreasing and continuous with respect to $\mathbf{c}_i\mathbf{x}, i \in I_1, f_i(\mathbf{x}, h_i), i \in I_2, f_i(\mathbf{x}, \hat{p}_i), i \in I_3, f_i(\mathbf{x}, h_i, \hat{p}_i), i \in I_4$ respectively, which are defined on interval $[f_{i\min}, f_{i\max}]$, $i \in I_1 \cup I_2 \cup I_3 \cup I_4$, where $\mu_{\widetilde{G}_i}(\mathbf{c}_i\mathbf{x}) = 0$, if $\mathbf{c}_i\mathbf{x} \geq f_{i\max}$, $\mu_{\widetilde{G}_i}(\mathbf{c}_i\mathbf{x}) = 1$, if $\mathbf{c}_i\mathbf{x} \leq f_{i\min}, i \in I_1$, $\mu_{\widetilde{G}_i}(f_i(\mathbf{x}, h_i)) = 0$, if $f_i(\mathbf{x}, h_i) \geq f_{i\max}$, $\mu_{\widetilde{G}_i}(f_i(\mathbf{x}, h_i)) = 1$, if $f_i(\mathbf{x}, h_i) \leq f_{i\min}, i \in I_2$, $\mu_{\widetilde{G}_i}(f_i(\mathbf{x}, \hat{p}_i)) = 0$, if $f_i(\mathbf{x}, \hat{p}_i) \geq f_{i\max}$, $\mu_{\widetilde{G}_i}(f_i(\mathbf{x}, \hat{p}_i)) = 1$, if $f_i(\mathbf{x}, \hat{p}_i) \leq f_{i\min}$, $i \in I_3$, and $\mu_{\widetilde{G}_i}(f_i(\mathbf{x}, h_i, \hat{p}_i)) = 0$, if $f_i(\mathbf{x}, h_i, \hat{p}_i) \geq f_{i\max}$, $\mu_{\widetilde{G}_i}(f_i(\mathbf{x}, h_i, \hat{p}_i)) = 1$, if $f_i(\mathbf{x}, h_i, \hat{p}_i) \leq f_{i\min}$, $i \in I_4$.

Assumption 6.2
$\mu_{\hat{p}_i}(\hat{p}_i), i \in I_3 \cup I_4$ are strictly increasing and continuous with respect to $\hat{p}_i$, which are defined on interval $P_i \stackrel{\text{def}}{=} [p_{i\min}, p_{i\max}] \subset (0,1)$, where $\mu_{\hat{p}_i}(\hat{p}_i) = 0$ if

$\hat{p}_i \leq p_{i\min}, i \in I_3 \cup I_4$, $\mu_{\hat{p}_i}(\hat{p}_i) = 1$ if $p_{i\max} \leq \hat{p}_i, i \in I_3 \cup I_4$.

To determine these membership functions appropriately, as an example, the decision maker can set $f_{i\min}$ and $f_{i\max}, i \in I_1 \cup I_2 \cup I_3 \cup I_4$, $p_{i\min}$, $p_{i\max}, i \in I_3 \cup I_4$ according to the following procedure.

Algorithm 6.1

Step 1: *For $i \in I_1$, $f_{i\min}$ can be obtained by solving linear programming problem*

$$f_{i\min} \stackrel{\text{def}}{=} \min_{\mathbf{x} \in X} \mathbf{c}_i \mathbf{x}, i \in I_1. \tag{6.20}$$

Step 2: *For $i \in I_2$, $f_{i\min}$ can be obtained by solving optimization problem*

$$f_{i\min} \stackrel{\text{def}}{=} \min_{\mathbf{x} \in X, h_i \in [0,1]} f_i(\mathbf{x}, h_i) \text{ subject to } \mu_{\tilde{G}_i}^{-1}(h_i) = f_i(\mathbf{x}, h_i), i \in I_2. \tag{6.21}$$

Step 3: *For $i \in I_3$, the decision maker subjectively specifies $p_{i\min}$ and $p_{i\max}$ according to Assumption 6.2 and sets interval $P_i = [p_{i\min}, p_{i\max}], i \in I_3$. Corresponding to interval $P_i, i \in I_3$, $f_{i\min}, i \in I_3$ can be obtained by solving linear programming problem*

$$f_{i\min} \stackrel{\text{def}}{=} \min_{\mathbf{x} \in X} f_i(\mathbf{x}, p_{i\min}), i \in I_3. \tag{6.22}$$

Step 4: *For $i \in I_4$, the decision maker subjectively specifies $p_{i\min}$ and $p_{i\max}$ according to Assumption 6.2, and sets interval $P_i = [p_{i\min}, p_{i\max}], i \in I_4$. Corresponding to interval $P_i, i \in I_4$, $f_{i\min}, i \in I_3$ can be obtained by solving optimization problem*

$$f_{i\min} \stackrel{\text{def}}{=} \min_{\mathbf{x} \in X, h_i \in [0,1]} f_i(\mathbf{x}, h_i, p_{i\min}) \text{ subject to } \mu_{\tilde{G}_i}^{-1}(h_i) = f_i(\mathbf{x}, h_i, \hat{p}_{i\min}), i \in I_4. \tag{6.23}$$

Step 5: *Let $\mathbf{x}_i \in X, i \in I_1 \cup I_2 \cup I_3 \cup I_4$ be the optimal solutions of optimization problems (6.20), (6.21), (6.22), and (6.23). Using optimal solutions $\mathbf{x}_i \in X, i \in I_1 \cup I_2 \cup I_3 \cup I_4$, $f_{i\max}, i \in I_1 \cup I_2 \cup I_3 \cup I_4$ can be obtained as follows [139].*

$$f_{i\max} \stackrel{\text{def}}{=} \max_{\ell \in I_1 \cup I_2 \cup I_3 \cup I_4, \ell \neq i} \mathbf{c}_i \mathbf{x}_\ell, i \in I_1 \tag{6.24}$$

$$f_{i\max} \stackrel{\text{def}}{=} \max_{\ell \in I_1 \cup I_2 \cup I_3 \cup I_4, \ell \neq i, h_i \in [0,1]} f_i(\mathbf{x}_\ell, h_i)$$
$$\text{subject to } \mu_{\tilde{G}_i}^{-1}(h_i) = f_i(\mathbf{x}_\ell, h_i), i \in I_2 \tag{6.25}$$

$$f_{i\max} \stackrel{\text{def}}{=} \max_{\ell \in I_1 \cup I_2 \cup I_3 \cup I_4, \ell \neq i} f_i(\mathbf{x}_\ell, p_{i\max}), i \in I_3 \tag{6.26}$$

$$f_{i\max} \stackrel{\text{def}}{=} \max_{\ell \in I_1 \cup I_2 \cup I_3 \cup I_4, \ell \neq i, h_i \in [0,1]} f_i(\mathbf{x}_\ell, h_i, \hat{p}_{i\max})$$
$$\text{subject to } \mu_{\tilde{G}_i}^{-1}(h_i) = f_i(\mathbf{x}_\ell, h_i, \hat{p}_{i\max}), i \in I_4 \tag{6.27}$$

To handle MOP6.3, we introduce a generalized Pareto optimal solution.

Definition 6.1 $(\mathbf{x}^*, \mathbf{h}^*, \hat{\mathbf{p}}^*) \in X_c$ is a generalized Pareto optimal solution to MOP6.3 if and only if there does not exist another $(\mathbf{x}, \mathbf{h}, \hat{\mathbf{p}}) \in X_c$ such that $\mu_{D_i}(\mathbf{x}) \geq \mu_{D_i}(\mathbf{x}^*), i \in I_1$, $\mu_{D_i}(\mathbf{x}, h_i) \geq \mu_{D_i}(\mathbf{x}^*, h_i^*), i \in I_2$ $\mu_{D_i}(\mathbf{x}, \hat{p}_i) \geq \mu_{D_i}(\mathbf{x}^*, \hat{p}_i^*), i \in I_3$, $\mu_{D_i}(\mathbf{x}, h_i, \hat{p}_i) \geq \mu_{D_i}(\mathbf{x}^*, h_i^*, \hat{p}_i^*), i \in I_4$ with strict inequality holding for at least one $i \in I_1 \cup I_2 \cup I_3 \cup I_4$, where $\mathbf{h}$ and $\hat{\mathbf{p}}$ are vectors of $h_i, i \in I_2 \cup I_4$ and $\hat{p}_i, i \in I_3 \cup I_4$, respectively, and constraint set X_c is defined as

$$\begin{aligned} X_c \stackrel{\text{def}}{=} & \{(\mathbf{x}, \mathbf{h}, \hat{\mathbf{p}}) \mid \mathbf{x} \in X, h_i \in [0,1], i \in I_2 \cup I_4, \hat{p}_i \in P_i, i \in I_3 \cup I_4, \\ & \mu_{\tilde{G}_i}^{-1}(h_i) = f_i(\mathbf{x}, h_i), i \in I_2, \mu_{\tilde{G}_i}^{-1}(h_i) = f_i(\mathbf{x}, h_i, \hat{p}_i), i \in I_4\}. \end{aligned} \tag{6.28}$$

In Definition 6.1, note that in addition to $\mathbf{x}$, $\mathbf{h}$ and $\hat{\mathbf{p}}$ are also decision variables.

To generate a candidate for a satisfactory solution from a generalized Pareto optimal solution set, the decision maker specifies reference membership values [71]. Once reference membership values $\hat{\mu}_i, i \in I_1 \cup I_2 \cup I_3 \cup I_4$ are specified, the corresponding generalized Pareto optimal solution is obtained by solving the following minmax problem.

MINMAX 6.1 ($\hat{\mu}$)

$$\min_{(\mathbf{x}, \mathbf{h}, \hat{\mathbf{p}}) \in X_c, \lambda \in \Lambda} \lambda \tag{6.29}$$

subject to

$$\hat{\mu}_i - \mu_{D_i}(\mathbf{x}) \leq \lambda, i \in I_1 \tag{6.30}$$

$$\hat{\mu}_i - \mu_{D_i}(\mathbf{x}, h_i) \leq \lambda, i \in I_2 \tag{6.31}$$

$$\hat{\mu}_i - \mu_{D_i}(\mathbf{x}, \hat{p}_i) \leq \lambda, i \in I_3 \tag{6.32}$$

$$\hat{\mu}_i - \mu_{D_i}(\mathbf{x}, h_i, \hat{p}_i) \leq \lambda, i \in I_4 \tag{6.33}$$

where

$$\Lambda \stackrel{\text{def}}{=} [\max_{i=1,\cdots,k} \hat{\mu}_i - 1, \max_{i=1,\cdots,k} \hat{\mu}_i]. \tag{6.34}$$

Constraints (6.28) and (6.31) for $i \in I_2$ can be transformed as follows.

$$\hat{\mu}_i - \lambda \leq h_i, i \in I_2 \tag{6.35}$$

$$\mathbf{d}_i \mathbf{x} - L^{-1}(h_i) \alpha_i \mathbf{x} = \mu_{\tilde{G}_i}^{-1}(h_i), i \in I_2 \tag{6.36}$$

From (6.35) and Assumption 6.1, the following relations hold.

$$L^{-1}(\hat{\mu}_i - \lambda) \geq L^{-1}(h_i), i \in I_2 \tag{6.37}$$

$$\mu_{\tilde{G}_i}^{-1}(\hat{\mu}_i - \lambda) \geq \mu_{\tilde{G}_i}^{-1}(h_i), i \in I_2 \tag{6.38}$$

Using (6.35), (6.36), (6.37), and (6.38), we can reduce (6.31) to the following constraint.

$$\begin{aligned} \mathbf{d}_i \mathbf{x} - L^{-1}(\hat{\mu}_i - \lambda) \alpha_i \mathbf{x} & \leq \mathbf{d}_i \mathbf{x} - L^{-1}(h_i) \alpha_i \mathbf{x} = \mu_{\tilde{G}_i}^{-1}(h_i) \\ & \leq \mu_{\tilde{G}_i}^{-1}(\hat{\mu}_i - \lambda), i \in I_2 \end{aligned} \tag{6.39}$$

From (6.7) and (6.32) for $i \in I_3$, the following inequalities hold.

$$\hat{\mu}_i - \lambda \leq \mu_{\tilde{G}_i}(f_i(\mathbf{x}, \hat{p}_i)), i \in I_3 \quad (6.40)$$
$$\hat{\mu}_i - \lambda \leq \mu_{\hat{p}_i}(\hat{p}_i), i \in I_3 \quad (6.41)$$

Further, (6.40) can be equivalently transformed into

$$(\mathbf{c}_i^1\mathbf{x} + \alpha_i^1) + T_i^{-1}(\hat{p}_i) \cdot (\mathbf{c}_i^2\mathbf{x} + \alpha_i^2) \leq \mu_{\tilde{G}_i}^{-1}(\hat{\mu}_i - \lambda).$$

As a result, (6.32) can be expressed as

$$(\mathbf{c}_i^1\mathbf{x} + \alpha_i^1) + T_i^{-1}(\mu_{\hat{p}_i}^{-1}(\hat{\mu}_i - \lambda)) \cdot (\mathbf{c}_i^2\mathbf{x} + \alpha_i^2) \leq \mu_{\tilde{G}_i}^{-1}(\hat{\mu}_i - \lambda), i \in I_3. \quad (6.42)$$

Similarly, from (6.28) and (6.33) for $i \in I_4$, it holds that

$$\begin{aligned} & h_i = \mu_{\tilde{G}_i}(f_i(\mathbf{x}, h_i, \hat{p}_i)) \geq \hat{\mu}_i - \lambda, \\ \Leftrightarrow \quad & \mu_{\tilde{G}_i}^{-1}(h_i) = f_i(\mathbf{x}, h_i, \hat{p}_i) \leq \mu_{\tilde{G}_i}^{-1}(\hat{\mu}_i - \lambda) \\ \Leftrightarrow \quad & \mu_{\tilde{G}_i}^{-1}(h_i) = (\mathbf{d}_i^1\mathbf{x} - L^{-1}(h_i)\alpha_i^1\mathbf{x}) + T_i^{-1}(\hat{p}_i) \cdot (\mathbf{d}_i^2\mathbf{x} - L^{-1}(h_i)\alpha_i^2\mathbf{x}) \\ & \leq \mu_{\tilde{G}_i}^{-1}(\hat{\mu}_i - \lambda), i \in I_4 \end{aligned} \quad (6.43)$$

On the left-hand-side of (6.43), because of $L^{-1}(h_i) \leq L^{-1}(\hat{\mu}_i - \lambda)$ and $\alpha_i^1\mathbf{x} + T_i^{-1}(\hat{p}_i)\alpha_i^2\mathbf{x} > 0$, it holds that

$$\begin{aligned} & (\mathbf{d}_i^1\mathbf{x} - L^{-1}(h_i)\alpha_i^1\mathbf{x}) + T_i^{-1}(\hat{p}_i) \cdot (\mathbf{d}_i^2\mathbf{x} - L^{-1}(h_i)\alpha_i^2\mathbf{x}) \\ = \quad & (\mathbf{d}_i^1\mathbf{x} + T_i^{-1}(\hat{p}_i)\mathbf{d}_i^2\mathbf{x}) - L^{-1}(h_i)\left(\alpha_i^1\mathbf{x} + T_i^{-1}(\hat{p}_i)\alpha_i^2\mathbf{x}\right) \\ \geq \quad & (\mathbf{d}_i^1\mathbf{x} + T_i^{-1}(\hat{p}_i)\mathbf{d}_i^2\mathbf{x}) - L^{-1}(\hat{\mu}_i - \lambda)\left(\alpha_i^1\mathbf{x} + T_i^{-1}(\hat{p}_i)\alpha_i^2\mathbf{x}\right), \\ & i \in I_4. \end{aligned} \quad (6.44)$$

Using (6.43), (6.44), and $\hat{\mu}_i - \mu_{\hat{p}_i}(\hat{p}_i) \leq \lambda$, $i \in I_4$, it holds that

$$\begin{aligned} & \mu_{\tilde{G}_i}^{-1}(\hat{\mu}_i - \lambda) \\ \geq \quad & (\mathbf{d}_i^1\mathbf{x} + T_i^{-1}(\hat{p}_i)\mathbf{d}_i^2\mathbf{x}) - L^{-1}(\hat{\mu}_i - \lambda)\left(\alpha_i^1\mathbf{x} + T_i^{-1}(\hat{p}_i)\alpha_i^2\mathbf{x}\right) \\ = \quad & (\mathbf{d}_i^1\mathbf{x} - L^{-1}(\hat{\mu}_i - \lambda)\alpha_i^1\mathbf{x}) + T_i^{-1}(\hat{p}_i) \cdot (\mathbf{d}_i^2\mathbf{x} - L^{-1}(\hat{\mu}_i - \lambda)\alpha_i^2\mathbf{x}) \\ \geq \quad & (\mathbf{d}_i^1\mathbf{x} - L^{-1}(\hat{\mu}_i - \lambda)\alpha_i^1\mathbf{x}) + T_i^{-1}(\mu_{\hat{p}_i}^{-1}(\hat{\mu}_i - \lambda)) \cdot (\mathbf{d}_i^2\mathbf{x} - L^{-1}(\hat{\mu}_i - \lambda)\alpha_i^2\mathbf{x}), \\ & i \in I_4. \end{aligned} \quad (6.45)$$

From (6.39), (6.42), and (6.45), MINMAX6.1($\hat{\mu}$) can be reduced to the following optimization problem, in which the decision variables $h_i, i \in I_2 \cup I_4$ and $\hat{p}_i, i \in I_3 \cup I_4$ in MINMAX6.1($\hat{\mu}$) have disappeared.

MINMAX 6.2 ($\hat{\mu}$)

$$\min_{\mathbf{x}\in X,\lambda\in\Lambda} \lambda$$

subject to

$$\mathbf{c}_i\mathbf{x} \leq \mu_{\tilde{G}_i}^{-1}(\hat{\mu}_i - \lambda), i \in I_1 \tag{6.46}$$

$$\mathbf{d}_i\mathbf{x} - L^{-1}(\hat{\mu}_i - \lambda)\alpha_i\mathbf{x} \leq \mu_{\tilde{G}_i}^{-1}(\hat{\mu}_i - \lambda), i \in I_2 \tag{6.47}$$

$$(\mathbf{c}_i^1\mathbf{x} + \alpha_i^1) + T_i^{-1}(\mu_{\hat{p}_i}^{-1}(\hat{\mu}_i - \lambda)) \cdot (\mathbf{c}_i^2\mathbf{x} + \alpha_i^2) \leq \mu_{\tilde{G}_i}^{-1}(\hat{\mu}_i - \lambda), i \in I_3 \tag{6.48}$$

$$(\mathbf{d}_i^1\mathbf{x} - L^{-1}(\hat{\mu}_i - \lambda)\alpha_i^1\mathbf{x}) + T_i^{-1}(\mu_{\hat{p}_i}^{-1}(\hat{\mu}_i - \lambda)) \cdot (\mathbf{d}_i^2\mathbf{x} - L^{-1}(\hat{\mu}_i - \lambda)\alpha_i^2\mathbf{x})$$
$$\leq \mu_{\tilde{G}_i}^{-1}(\hat{\mu}_i - \lambda), i \in I_4 \tag{6.49}$$

Note that constraints (6.46), (6.47), (6.48), and (6.49) can be reduced to a set of linear inequalities for some fixed value $\lambda \in \Lambda$, which means that optimal solution $\mathbf{x}^* \in X, \lambda^* \in \Lambda$ of MINMAX6.2($\hat{\mu}$) is obtained via a combination of the bisection method with respect to λ and the first phase of the two-phase simplex method of linear programming.

The relationships between optimal solution $\mathbf{x}^* \in X, \lambda^* \in \Lambda$ of MINMAX6.2 ($\hat{\mu}$) and generalized Pareto optimal solutions are characterized by the following theorem.

Theorem 6.1

(1) If $\mathbf{x}^* \in X, \lambda^* \in \Lambda$ is a unique optimal solution of MINMAX6.2($\hat{\mu}$), then $\mathbf{x}^* \in X$, $h_i^* \stackrel{\text{def}}{=} \hat{\mu}_i - \lambda^*, i \in I_2 \cup I_4$, $\hat{p}_i^* \stackrel{\text{def}}{=} \mu_{\hat{p}_i}^{-1}(\hat{\mu}_i - \lambda^*), i \in I_3 \cup I_4$ is a generalized Pareto optimal solution to MOP6.3.

(2) If $\mathbf{x}^* \in X$, $h_i^*, i \in I_2 \cup I_4$ $\hat{p}_i^*, i \in I_3 \cup I_4$ is a generalized Pareto optimal solution to MOP6.3, then $\mathbf{x}^* \in X$, $\lambda^* \in \Lambda$ is an optimal solution of MINMAX6.2($\hat{\mu}$) for some reference membership values $\hat{\mu}_i, i \in I_1 \cup I_2 \cup I_3 \cup I_4$, where $\lambda^* \stackrel{\text{def}}{=} \hat{\mu}_i - \mu_{\tilde{G}_i}(\mathbf{c}_i\mathbf{x}^*)$, $i \in I_1$, $\lambda^* \stackrel{\text{def}}{=} \hat{\mu}_i - \mu_{\tilde{G}_i}(f_i(\mathbf{x}^*, h_i^*)) = \hat{\mu}_i - h_i^*$, $i \in I_2$, $\lambda^* \stackrel{\text{def}}{=} \hat{\mu}_i - \mu_{\hat{p}_i}(\hat{p}_i^*) = \hat{\mu}_i - \mu_{\tilde{G}_i}(f_i(\mathbf{x}^*, \hat{p}_i^*)), i \in I_3$, $\lambda^* \stackrel{\text{def}}{=} \hat{\mu}_i - \mu_{\hat{p}_i}(\hat{p}_i^*) = \hat{\mu}_i - \mu_{\tilde{G}_i}(f_i(\mathbf{x}^*, h_i^*, \hat{p}_i^*)) = \hat{\mu}_i - h_i^*, i \in I_4$.

(Proof)
(1) Since $\mathbf{x}^* \in X, \lambda^* \in \Lambda$ is an optimal solution of MINMAX6.2($\hat{\mu}$), the following inequalities hold.

$$
\begin{aligned}
\hat{\mu}_i - \lambda^* &\leq \mu_{\tilde{G}_i}(\mathbf{c}_i\mathbf{x}^*), i \in I_1 \\
\hat{\mu}_i - \lambda^* &\leq \mu_{\tilde{G}_i}(f_i(\mathbf{x}^*, h_i^*)), i \in I_2 \\
\hat{\mu}_i - \lambda^* &\leq \mu_{\tilde{G}_i}(f_i(\mathbf{x}^*, \hat{p}_i^*)), i \in I_3 \\
\hat{\mu}_i - \lambda^* &= \mu_{\hat{p}_i}(\mu_{\hat{p}_i}^{-1}(\hat{\mu}_i - \lambda^*)) = \mu_{\hat{p}_i}(\hat{p}_i^*), i \in I_3 \\
\hat{\mu}_i - \lambda^* &\leq \mu_{\tilde{G}_i}(f_i(\mathbf{x}^*, h_i^*, \hat{p}_i^*)), i \in I_4 \\
\hat{\mu}_i - \lambda^* &= \mu_{\hat{p}_i}(\mu_{\hat{p}_i}^{-1}(\hat{\mu}_i - \lambda^*)) = \mu_{\hat{p}_i}(\hat{p}_i^*), i \in I_4
\end{aligned}
$$

Assume that $\mathbf{x}^* \in X$, $h_i^* \stackrel{\text{def}}{=} \hat{\mu}_i - \lambda^*, i \in I_2 \cup I_4$, $\hat{p}_i^* \stackrel{\text{def}}{=} \mu_{\hat{p}_i}^{-1}(\hat{\mu}_i - \lambda^*), i \in I_3 \cup I_4$ is not a generalized Pareto optimal solution to MOP6.3. Then, there exists $\mathbf{x} \in X, h_i \in I_2 \cup I_4, \hat{p}_i, i \in I_3 \cup I_4$ such that

$$
\begin{aligned}
&\mu_{\tilde{G}_i}(\mathbf{c}_i\mathbf{x}) \geq \mu_{\tilde{G}_i}(\mathbf{c}_i\mathbf{x}^*) \geq \hat{\mu}_i - \lambda^*, i \in I_1, && (6.50)\\
&\mu_{\tilde{G}_i}(f_i(\mathbf{x}, h_i)) \geq \mu_{\tilde{G}_i}(f_i(\mathbf{x}^*, h_i^*)) \geq \hat{\mu}_i - \lambda^*, i \in I_2, && (6.51)\\
&h_i = \mu_{\tilde{G}_i}(f_i(\mathbf{x}, h_i)), i \in I_2, && (6.52)\\
&\min\{\mu_{\hat{p}_i}(\hat{p}_i), \mu_{\tilde{G}_i}(f_i(\mathbf{x}, \hat{p}_i))\} \geq \min\{\mu_{\hat{p}_i}(\hat{p}_i^*), \mu_{\tilde{G}_i}(f_i(\mathbf{x}^*, \hat{p}_i^*))\} \\
&= \hat{\mu}_i - \lambda^*, i \in I_3, && (6.53)\\
&\min\{\mu_{\hat{p}_i}(\hat{p}_i), \mu_{\tilde{G}_i}(f_i(\mathbf{x}, h_i, \hat{p}_i))\} \geq \min\{\mu_{\hat{p}_i}(\hat{p}_i^*), \mu_{\tilde{G}_i}(f_i(\mathbf{x}^*, h_i^*, \hat{p}_i^*))\} \\
&= \hat{\mu}_i - \lambda^*, i \in I_4, && (6.54)\\
&h_i = \mu_{\tilde{G}_i}(f_i(\mathbf{x}, h_i, \hat{p}_i)), i \in I_4. && (6.55)
\end{aligned}
$$

From (6.50) and Assumption 6.1, it holds that

$$\mathbf{c}_i\mathbf{x} \leq \mu_{\tilde{G}_i}^{-1}(\hat{\mu}_i - \lambda^*), i \in I_1. \tag{6.56}$$

From (6.51), (6.52) and Assumption 6.1, the following relations hold for $i \in I_2$.

$$\mathbf{d}_i\mathbf{x} - L^{-1}(\hat{\mu}_i - \lambda^*)\alpha_i\mathbf{x} \leq \mathbf{d}_i\mathbf{x} - L^{-1}(h_i)\alpha_i\mathbf{x} \leq \mu_{\tilde{G}_i}^{-1}(\hat{\mu}_i - \lambda^*), i \in I_2 \tag{6.57}$$

From (6.53), the following two inequalities hold for $i \in I_3$.

$$
\begin{aligned}
\mu_{\hat{p}_i}(\hat{p}_i) &\geq \hat{\mu}_i - \lambda^*, i \in I_3 && (6.58)\\
\mu_{\tilde{G}_i}(f_i(\mathbf{x}, \hat{p}_i)) &\geq \hat{\mu}_i - \lambda^*, i \in I_3 && (6.59)
\end{aligned}
$$

Because of (6.5) and Assumptions 6.1 and 6.2, (6.58) and (6.59) are equivalently transformed as follows.

$$
\begin{aligned}
\hat{p}_i &\geq \mu_{\hat{p}_i}^{-1}(\hat{\mu}_i - \lambda^*), i \in I_3 \\
\hat{p}_i &\leq T_i\left(\frac{\mu_{\tilde{G}_i}^{-1}(\hat{\mu}_i - \lambda^*) - (\mathbf{c}_i^1\mathbf{x} + \alpha_i^1)}{(\mathbf{c}_i^2\mathbf{x} + \alpha_i^2)}\right), i \in I_3
\end{aligned}
$$

As a result, the above two inequalities can be replaced by the following inequality for $i \in I_3$.

$$\mu_{\tilde{G}_i}^{-1}(\hat{\mu}_i - \lambda^*) - (\mathbf{c}_i^1\mathbf{x} + \alpha_i^1) \geq T_i^{-1}(\mu_{\hat{p}_i}^{-1}(\hat{\mu}_i - \lambda^*)) \cdot (\mathbf{c}_i^2\mathbf{x} + \alpha_i^2), i \in I_3 \tag{6.60}$$

From (6.54), the following inequality holds for $i \in I_4$.

$$\min\{\mu_{\hat{p}_i}(\hat{p}_i), \mu_{\tilde{G}_i}(f_i(\mathbf{x}, h_i, \hat{p}_i))\} \geq \hat{\mu}_i - \lambda^*, i \in I_4$$

Then, it holds that

$$\begin{aligned} \mu_{\hat{p}_i}(\hat{p}_i) &\geq \hat{\mu}_i - \lambda^*, i \in I_4 &(6.61)\\ \mu_{\tilde{G}_i}(f_i(\mathbf{x}, h_i, \hat{p}_i)) &\geq \hat{\mu}_i - \lambda^*, i \in I_4. &(6.62) \end{aligned}$$

From (6.11) and Assumptions 6.1 and 6.2, (6.61) and (6.62) can be transformed as follows.

$$\begin{aligned} \hat{p}_i &\geq \mu_{\hat{p}_i}^{-1}(\hat{\mu}_i - \lambda^*), i \in I_4\\ \hat{p}_i &\leq T_i\left(\frac{\mu_{\tilde{G}_i}^{-1}(\hat{\mu}_i - \lambda^*) - (\mathbf{d}_i^1\mathbf{x} - L^{-1}(h_i)\alpha_i^1\mathbf{x})}{\mathbf{d}_i^2\mathbf{x} - L^{-1}(h_i)\alpha_i^2\mathbf{x}}\right), i \in I_4 \end{aligned}$$

Because of (6.55), $L^{-1}(h_i) \leq L^{-1}(\hat{\mu}_i - \lambda^*), i \in I_4$, there exists $\mathbf{x} \in X$ such that

$$\begin{aligned} &\mu_{\tilde{G}_i}^{-1}(\hat{\mu}_i - \lambda^*) - (\mathbf{d}_i^1\mathbf{x} - L^{-1}(h_i)\alpha_i^1\mathbf{x})\\ &\geq T_i^{-1}(\mu_{\hat{p}_i}^{-1}(\hat{\mu}_i - \lambda^*)) \cdot (\mathbf{d}_i^2\mathbf{x} - L^{-1}(h_i)\alpha_i^2\mathbf{x})\\ \Leftrightarrow\ &\mu_{\tilde{G}_i}^{-1}(\hat{\mu}_i - \lambda^*) \geq (\mathbf{d}_i^1\mathbf{x} + T_i^{-1}(\mu_{\hat{p}_i}^{-1}(\hat{\mu}_i - \lambda^*)) \cdot \mathbf{d}_i^2\mathbf{x})\\ &-L^{-1}(h_i)(\alpha_i^1\mathbf{x} + T_i^{-1}(\mu_{\hat{p}_i}^{-1}(\hat{\mu}_i - \lambda^*)) \cdot \alpha_i^2\mathbf{x})\\ \Leftrightarrow\ &\mu_{\tilde{G}_i}^{-1}(\hat{\mu}_i - \lambda^*) \geq (\mathbf{d}_i^1\mathbf{x} + T_i^{-1}(\mu_{\hat{p}_i}^{-1}(\hat{\mu}_i - \lambda^*)) \cdot \mathbf{d}_i^2\mathbf{x})\\ &-L^{-1}(\hat{\mu}_i - \lambda^*)(\alpha_i^1\mathbf{x} + T_i^{-1}(\mu_{\hat{p}_i}^{-1}(\hat{\mu}_i - \lambda^*)) \cdot \alpha_i^2\mathbf{x}), i \in I_4. \quad (6.63) \end{aligned}$$

Inequalities (6.56), (6.57), (6.60), and (6.63) contradict the fact that $\mathbf{x}^* \in X, \lambda^* \in \Lambda$ is a unique optimal solution of MINMAX6.2($\hat{\mu}$).

(2) Assume that $\mathbf{x}^* \in X, \lambda^* \in \Lambda$ is not an optimal solution of MINMAX6.2($\hat{\mu}$) for any reference membership values $\hat{\mu}_i, i \in I_1 \cup I_2 \cup I_3 \cup I_4$ that satisfy equalities

$$\begin{aligned} \hat{\mu}_i - \lambda^* &= \mu_{\tilde{G}_i}(\mathbf{c}_i\mathbf{x}^*), i \in I_1 &(6.64)\\ \hat{\mu}_i - \lambda^* &= \mu_{\tilde{G}_i}(f_i(\mathbf{x}^*, h_i^*)) = h_i^*, i \in I_2 &(6.65)\\ \hat{\mu}_i - \lambda^* &= \mu_{\hat{p}_i}(\hat{p}_i^*) = \mu_{\tilde{G}_i}(f_i(\mathbf{x}^*, \hat{p}_i^*)), i \in I_3 &(6.66)\\ \hat{\mu}_i - \lambda^* &= \mu_{\hat{p}_i}(\hat{p}_i^*) = \mu_{\tilde{G}_i}(f_i(\mathbf{x}^*, h_i^*, \hat{p}_i^*)), i \in I_4. &(6.67) \end{aligned}$$

Then, there exists some $\mathbf{x} \in X, \lambda < \lambda^*$ such that

$$\mathbf{c}_i\mathbf{x} \leq \mu_{\tilde{G}_i}^{-1}(\hat{\mu}_i - \lambda), i \in I_1 \tag{6.68}$$

$$\mathbf{d}_i\mathbf{x} - L^{-1}(\hat{\mu}_i - \lambda)\alpha_i\mathbf{x} \leq \mu_{\tilde{G}_i}^{-1}(\hat{\mu}_i - \lambda), i \in I_2 \tag{6.69}$$

$$(\mathbf{c}_i^1\mathbf{x} + \alpha_i^1) + T_i^{-1}(\mu_{\hat{p}_i}^{-1}(\hat{\mu}_i - \lambda)) \cdot (\mathbf{c}_i^2\mathbf{x} + \alpha_i^2)$$
$$\leq \mu_{\tilde{G}_i}^{-1}(\hat{\mu}_i - \lambda), i \in I_3 \tag{6.70}$$

$$\mu_{\tilde{G}_i}^{-1}(\hat{\mu}_i - \lambda) - (\mathbf{d}_i^1\mathbf{x} - L^{-1}(\hat{\mu}_i - \lambda)\alpha_i^1\mathbf{x})$$
$$\geq T_i^{-1}(\mu_{\hat{p}_i}^{-1}(\hat{\mu}_i - \lambda)) \cdot (\mathbf{d}_i^2\mathbf{x} - L^{-1}(\hat{\mu}_i - \lambda)\alpha_i^2\mathbf{x}), i \in I_4. \tag{6.71}$$

From Assumption 6.1, (6.64), (6.68), and $\hat{\mu}_i - \lambda > \hat{\mu}_i - \lambda^*$, the following relation holds for $i \in I_1$.

$$\mu_{\tilde{G}_i}(\mathbf{c}_i\mathbf{x}) \geq \hat{\mu}_i - \lambda > \hat{\mu}_i - \lambda^* = \mu_{\tilde{G}_i}(\mathbf{c}_i\mathbf{x}^*), i \in I_1 \tag{6.72}$$

From Assumption 6.1, (6.65), (6.69), and $\hat{\mu}_i - \lambda > \hat{\mu}_i - \lambda^*$, the following relations hold for $i \in I_2$.

$$\mu_{\tilde{G}_i}(\mathbf{d}_i\mathbf{x} - L^{-1}(\hat{\mu}_i - \lambda)\alpha_i\mathbf{x}) \geq \hat{\mu}_i - \lambda$$
$$> \hat{\mu}_i - \lambda^* = \mu_{\tilde{G}_i}(\mathbf{d}_i\mathbf{x}^* - L^{-1}(\hat{\mu}_i - \lambda^*)\alpha_i\mathbf{x}^*), i \in I_2 \tag{6.73}$$

Because of Assumption 6.1, (6.66), (6.70), and $\hat{\mu}_i - \lambda > \hat{\mu}_i - \lambda^*, i \in I_3$, the following two inequalities hold for $i \in I_3$.

$$\mu_{\hat{p}_i}(\hat{p}_i) > \mu_{\hat{p}_i}(\hat{p}_i^*), i \in I_3 \tag{6.74}$$

$$\mu_{\tilde{G}_i}(f_i(\mathbf{x}, \hat{p}_i)) > \mu_{\tilde{G}_i}(f_i(\mathbf{x}^*, \hat{p}_i^*)), i \in I_3, \tag{6.75}$$

where $\hat{p}_i \stackrel{\text{def}}{=} \mu_{\hat{p}_i}^{-1}(\mu_i - \lambda), i \in I_3$. Moreover, because of Assumption 6.1, Assumption 6.2, (6.67), (6.71) and $\hat{\mu}_i - \lambda > \hat{\mu}_i - \lambda^*, i \in I_4$, the following inequalities hold.

$$\mu_{\hat{p}_i}(\hat{p}_i) \quad > \quad \mu_{\hat{p}_i}(\hat{p}_i^*), i \in I_4 \tag{6.76}$$

$$\mu_{\tilde{G}_i}(f_i(\mathbf{x}, \hat{h}_i, \hat{p}_i)) \quad > \quad \mu_{\tilde{G}_i}(f_i(\mathbf{x}^*, h_i^*, \hat{p}_i^*)), i \in I_4, \tag{6.77}$$

where $\hat{p}_i \stackrel{\text{def}}{=} \mu_{\hat{p}_i}^{-1}(\mu_i - \lambda), \hat{h}_i \stackrel{\text{def}}{=} \hat{\mu}_i - \lambda, i \in I_4$. Inequalities (6.72)–(6.77) contradict the fact that $\mathbf{x}^* \in X$, $h_i^* \in I_2 \cup I_4$, $\hat{p}_i^*, i \in I_3 \cup I_4$ is a generalized Pareto optimal solution to MOP6.3. □

6.1.2 An interactive linear programming algorithm

In this subsection, we propose an interactive algorithm for obtaining a satisfactory solution from a generalized Pareto optimal solution set. Unfortunately, from Theorem 6.1, there is no guarantee that optimal solution $\mathbf{x}^* \in X, \lambda^* \in \Lambda$ of MINMAX6.2($\hat{\mu}$) is generalized Pareto optimal if $\mathbf{x}^* \in X, \lambda^* \in \Lambda$ is not unique. To guarantee generalized Pareto optimality, we first assume that constraints (6.46)–(6.49) of MINMAX6.2($\hat{\mu}$) are active at optimal solution $\mathbf{x}^* \in X, \lambda^* \in \Lambda$, i.e.,

$$\hat{\mu}_i - \lambda^* = \mu_{\tilde{G}_i}(\mathbf{c}_i\mathbf{x}^*), i \in I_1 \tag{6.78}$$

$$\hat{\mu}_i - \lambda^* = \mu_{\tilde{G}_i}(f_i(\mathbf{x}^*, \hat{\mu}_i - \lambda^*)), i \in I_2 \tag{6.79}$$

$$\hat{\mu}_i - \lambda^* = \mu_{\tilde{G}_i}(f_i(\mathbf{x}^*, \mu_{\tilde{p}_i}^{-1}(\hat{\mu}_i - \lambda^*))), i \in I_3 \tag{6.80}$$

$$\hat{\mu}_i - \lambda^* = \mu_{\tilde{G}_i}(f_i(\mathbf{x}^*, \hat{\mu}_i - \lambda^*, \mu_{\tilde{p}_i}^{-1}(\hat{\mu}_i - \lambda^*))), i \in I_4. \tag{6.81}$$

For optimal solution $\mathbf{x}^* \in X, \lambda^* \in \Lambda$ of MINMAX6.2($\hat{\mu}$), where active conditions (6.78)–(6.81) are satisfied, we solve the generalized Pareto optimality test problem defined as follows.

Test 6.1

$$w \stackrel{\text{def}}{=} \max_{\mathbf{x} \in X, \varepsilon_i \geq 0, i \in I_1 \cup I_2 \cup I_3 \cup I_4} \sum_{i \in I_1 \cup I_2 \cup I_3 \cup I_4} \varepsilon_i$$

subject to

$$\mathbf{c}_i\mathbf{x} + \varepsilon_i = \mu_{\tilde{G}_i}^{-1}(\hat{\mu}_i - \lambda^*), i \in I_1$$

$$\mathbf{d}_i\mathbf{x} - L^{-1}(\hat{\mu}_i - \lambda^*)\alpha_i\mathbf{x} + \varepsilon_i = \mu_{\tilde{G}_i}^{-1}(\hat{\mu}_i - \lambda^*), i \in I_2$$

$$(\mathbf{c}_i^1\mathbf{x} + \alpha_i^1) + T_i^{-1}(\mu_{\tilde{p}_i}^{-1}(\hat{\mu}_i - \lambda^*)) \cdot (\mathbf{c}_i^2\mathbf{x} + \alpha_i^2) + \varepsilon_i = \mu_{\tilde{G}_i}^{-1}(\hat{\mu}_i - \lambda^*), i \in I_3$$

$$T_i^{-1}(\mu_{\tilde{p}_i}^{-1}(\hat{\mu}_i - \lambda^*)) \cdot (\mathbf{d}_i^2\mathbf{x} - L^{-1}(\hat{\mu}_i - \lambda^*)\alpha_i^2\mathbf{x})$$
$$+(\mathbf{d}_i^1\mathbf{x} - L^{-1}(\hat{\mu}_i - \lambda^*)\alpha_i^1\mathbf{x}) + \varepsilon_i = \mu_{\tilde{G}_i}^{-1}(\hat{\mu}_i - \lambda^*), i \in I_4$$

For the optimal solution of TEST6.1, the following theorem holds.

Theorem 6.2

For optimal solution $\check{\mathbf{x}} \in X, \check{\varepsilon}_i \geq 0, i \in I_1 \cup I_2 \cup I_3 \cup I_4$ of TEST6.1, if $w = 0$ (equivalently, $\check{\varepsilon}_i = 0, i \in I_1 \cup I_2 \cup I_3 \cup I_4$), then $\mathbf{x}^* \in X, h_i^* \stackrel{\text{def}}{=} \hat{\mu}_i - \lambda^*, i \in I_2 \cup I_4$, $\hat{p}_i^* \stackrel{\text{def}}{=} \mu_{\tilde{p}_i}^{-1}(\hat{\mu}_i - \lambda^*), i \in I_3 \cup I_4$ is a generalized Pareto optimal solution to MOP6.3.

(Proof)
From active conditions (6.78)–(6.81), it holds that

$$\begin{aligned}\hat{\mu}_i-\lambda^* &= \mu_{\tilde{G}_i}(\mathbf{c}_i\mathbf{x}^*), i\in I_1\\ \hat{\mu}_i-\lambda^* &= \mu_{\tilde{G}_i}(f_i(\mathbf{x}^*,\hat{\mu}_i-\lambda^*)), i\in I_2\\ \hat{\mu}_i-\lambda^* &= \mu_{\tilde{G}_i}(f_i(\mathbf{x}^*,\mu_{\hat{p}_i}^{-1}(\hat{\mu}_i-\lambda^*))), i\in I_3\\ \hat{\mu}_i-\lambda^* &= \mu_{\tilde{G}_i}(f_i(\mathbf{x}^*,\hat{\mu}_i-\lambda^*,\mu_{\hat{p}_i}^{-1}(\hat{\mu}_i-\lambda^*))), i\in I_4.\end{aligned}$$

If $\mathbf{x}^*\in X$, $h_i^* \stackrel{\text{def}}{=} \hat{\mu}_i-\lambda^*, i\in I_2\cup I_4$, $\hat{p}_i^* \stackrel{\text{def}}{=} \mu_{\hat{p}_i}^{-1}(\hat{\mu}_i-\lambda^*), i\in I_3\cup I_4$ is not a generalized Pareto optimal solution to MOP6.3, there exists some $\mathbf{x}\in X, h_i, i\in I_2\cup I_4, \hat{p}_i, i\in I_3\cup I_4$ such that

$$\mu_{\tilde{G}_i}(\mathbf{c}_i\mathbf{x}) \geq \mu_{\tilde{G}_i}(\mathbf{c}_i\mathbf{x}^*) = \hat{\mu}_i-\lambda^*, i\in I_1 \tag{6.82}$$
$$\mu_{\tilde{G}_i}(f_i(\mathbf{x},h_i)) = h_i \geq \mu_{\tilde{G}_i}(f_i(\mathbf{x}^*,\hat{\mu}_i-\lambda^*)) = \hat{\mu}_i-\lambda^*, i\in I_2 \tag{6.83}$$
$$\begin{aligned}&\mu_{D_i}(\mathbf{x},\hat{p}_i) = \min\{\mu_{\hat{p}_i}(\hat{p}_i),\mu_{\tilde{G}_i}(f_i(\mathbf{x},\hat{p}_i))\} \geq \mu_{D_i}(\mathbf{x}^*,\mu_{\hat{p}_i}^{-1}(\hat{\mu}_i-\lambda^*))\\ &= \hat{\mu}_i-\lambda^*, i\in I_3\end{aligned} \tag{6.84}$$
$$\begin{aligned}&\mu_{D_{G_i}}(\mathbf{x},h_i,\hat{p}_i) = \min\{\mu_{\hat{p}_i}(\hat{p}_i),\mu_{\tilde{G}_i}(f_i(\mathbf{x},h_i,\hat{p}_i))\}\\ &\geq \mu_{D_{G_i}}(\mathbf{x}^*,\hat{\mu}_i-\lambda^*,\mu_{\hat{p}_i}^{-1}(\hat{\mu}_i-\lambda^*)) = \hat{\mu}_i-\lambda^*, i\in I_4\end{aligned} \tag{6.85}$$
$$\mu_{\tilde{G}_i}(f_i(\mathbf{x},h_i,\hat{p}_i)) = h_i, i\in I_4, \tag{6.86}$$

with strict inequality holding for at least one $i\in I_1\cup I_2\cup I_3\cup I_4$. From (6.82) and Assumption 6.1, it holds that

$$\mathbf{c}_i\mathbf{x} \leq \mu_{\tilde{G}_i}^{-1}(\hat{\mu}_i-\lambda^*), i\in I_1. \tag{6.87}$$

From (6.83), Assumption 6.1 and $L^{-1}(h_i)\leq L^{-1}(\hat{\mu}_i-\lambda^*)$, the following inequality holds.

$$\mathbf{d}_i\mathbf{x}-L^{-1}(\hat{\mu}_i-\lambda^*)\alpha_i\mathbf{x} \leq \mu_{\tilde{G}_i}^{-1}(\hat{\mu}_i-\lambda^*), i\in I_2 \tag{6.88}$$

From (6.84), the following inequalities hold.

$$\mu_{\hat{p}_i}(\hat{p}_i) \geq \hat{\mu}_i-\lambda^*, i\in I_3 \tag{6.89}$$
$$\mu_{\tilde{G}_i}(f_i(\mathbf{x},\hat{p}_i)) \geq \hat{\mu}_i-\lambda^*, i\in I_3 \tag{6.90}$$

From (6.5) and Assumptions 6.1 and 6.2, two inequalities (6.89) and (6.90) can be transformed into single inequality.

$$\mu_{\tilde{G}_i}^{-1}(\hat{\mu}_i-\lambda^*)-(\mathbf{c}_i^1\mathbf{x}+\alpha_i^1) \geq T_i^{-1}(\mu_{\hat{p}_i}^{-1}(\hat{\mu}_i-\lambda^*))\cdot(\mathbf{c}_i^2\mathbf{x}+\alpha_i^2), i\in I_3 \tag{6.91}$$

From (6.85), it holds that

$$\mu_{\hat{p}_i}(\hat{p}_i) \geq \hat{\mu}_i-\lambda^*, i\in I_4, \tag{6.92}$$
$$\mu_{\tilde{G}_i}(f_i(\mathbf{x},h_i,\hat{p}_i)) \geq \hat{\mu}_i-\lambda^*, i\in I_4. \tag{6.93}$$

Because of (6.86), Assumptions 6.1 and 6.2, and $L^{-1}(h_i) \leq L^{-1}(\hat{\mu}_i - \lambda^*)$, (6.92) and (6.93) can be transformed into

$$\hat{p}_i \geq \mu_{\tilde{p}_i}^{-1}(\hat{\mu}_i - \lambda^*), i \in I_4 \tag{6.94}$$

$$\begin{aligned} \hat{p}_i &\leq T_i\left(\frac{\mu_{\tilde{G}_i}^{-1}(\hat{\mu}_i - \lambda^*) - (\mathbf{d}_i^1\mathbf{x} - L^{-1}(h_i)\alpha_i^1\mathbf{x})}{\mathbf{d}_i^2\mathbf{x} - L^{-1}(h_i)\alpha_i^2\mathbf{x}}\right) \\ &\leq T_i\left(\frac{\mu_{\tilde{G}_i}^{-1}(\hat{\mu}_i - \lambda^*) - (\mathbf{d}_i^1\mathbf{x} - L^{-1}(\hat{\mu}_i - \lambda^*)\alpha_i^1\mathbf{x})}{\mathbf{d}_i^2\mathbf{x} - L^{-1}(\hat{\mu}_i - \lambda^*)\alpha_i^2\mathbf{x}}\right), i \in I_4. \end{aligned} \tag{6.95}$$

From (6.94) and (6.95), there exists some $\mathbf{x} \in X$ such that

$$\begin{aligned} &\mu_{\tilde{G}_i}^{-1}(\hat{\mu}_i - \lambda^*) - (\mathbf{d}_i^1\mathbf{x} - L^{-1}(\hat{\mu}_i - \lambda^*)\alpha_i^1\mathbf{x}) \\ &\geq T_i^{-1}(\mu_{\tilde{p}_i}^{-1}(\hat{\mu}_i - \lambda^*)) \cdot (\mathbf{d}_i^2\mathbf{x} - L^{-1}(\hat{\mu}_i - \lambda^*)\alpha_i^2\mathbf{x}), i \in I_4 \end{aligned} \tag{6.96}$$

Inequalities (6.87), (6.88), (6.91), and (6.96) indicate that $\check{\varepsilon}_i > 0$ for some $i \in I_1 \cup I_2 \cup I_3 \cup I_4$, and $w > 0$. □

Given the above, we next present our interactive algorithm for deriving a satisfactory solution from a generalized Pareto optimal solution set.

Algorithm 6.2

Step 1: *The decision maker sets his or her membership functions $\mu_{\tilde{G}_i}(\cdot), i \in I_1 \cup I_2 \cup I_3 \cup I_4$ for the objective functions of MOP6.1 according to Assumption 6.1.*
Step 2: *The decision maker sets his or her membership functions $\mu_{\tilde{p}_i}(\hat{p}_i), i \in I_3 \cup I_4$ for permissible probability levels $\hat{p}_i, i \in I_3 \cup I_4$ according to Assumption 6.2.*
Step 3: *Set initial reference membership values to $\hat{\mu}_i = 1, i \in I_1 \cup I_2 \cup I_3 \cup I_4$.*
Step 4: *Solve MINMAX6.2($\hat{\mu}$) and TEST6.1 to obtain the corresponding generalized Pareto optimal solution to MOP6.3.*
Step 5: *If the decision maker is satisfied with the current values of generalized Pareto optimal solution $\mu_{D_i}(\mathbf{x}^*), i \in I_1$, $\mu_{D_i}(\mathbf{x}^*, h_i^*), i \in I_2$ $\mu_{D_i}(\mathbf{x}^*, \hat{p}_i^*), i \in I_3$, $\mu_{D_i}(\mathbf{x}^*, h_i^*, \hat{p}_i^*), i \in I_4$, where $h_i^* \stackrel{\text{def}}{=} \hat{\mu}_i - \lambda^*, i \in I_2 \cup I_4$, $\hat{p}_i^* \stackrel{\text{def}}{=} \mu_{\tilde{p}_i}^{-1}(\hat{\mu}_i - \lambda^*), i \in I_3 \cup I_4$, then stop. Otherwise, the decision maker updates his or her reference membership values $\hat{\mu}_i, i \in I_1 \cup I_2 \cup I_3 \cup I_4$, and returns to Step 4.*

6.1.3 A numerical example

In this subsection, to demonstrate the feasibility of our interactive algorithm, we present the following three objective programming problem whose objective functions are comprised of fuzzy, random variable, and fuzzy random variable coefficients.

Table 6.1: Parameters of the objective functions and constraints in MOP6.4

j	x_1	x_2	x_3	x_4	x_5	x_6	x_7	x_8	x_9	x_{10}
d_{1j}	31	56	29	14	34	47	50	19	40	41
c^1_{2j}	12	–46	–23	–38	–33	–48	12	8	19	20
c^2_{2j}	1	2	4	2	2	1	2	1	2	1
d^1_{3j}	–18	–26	–22	–28	–15	–29	–10	–19	–17	–28
d^2_{3j}	2	1	3	2	1	2	3	3	2	1
α_{1j}	0.51	0.54	0.53	0.55	0.48	0.57	0.47	0.52	0.5	0.53
α_{3j}	0.59	0.56	0.58	0.62	0.6	0.52	0.65	0.57	0.64	0.63
β_{1j}	0.51	0.54	0.53	0.55	0.48	0.57	0.47	0.52	0.5	0.53
β_{3j}	0.59	0.56	0.58	0.62	0.6	0.52	0.65	0.57	0.64	0.63
a_{1j}	12	–2	4	–7	13	–1	–6	6	11	–8
a_{2j}	–2	5	3	16	6	–12	12	4	–7	–10
a_{3j}	3	–16	–4	–8	–8	2	–12	–12	4	–3
a_{4j}	–11	6	–5	9	–1	8	–4	6	–9	6
a_{5j}	–4	7	–6	–5	13	6	–2	–5	14	–6
a_{6j}	5	–3	14	–3	–9	–7	4	–4	–5	9
a_{7j}	–3	–4	–6	9	6	18	11	–9	–4	7

MOP 6.4

$$\begin{aligned} \min_{\mathbf{x}\in X} \widetilde{\mathbf{c}}_1\mathbf{x} &= \sum_{j=1}^{10} \widetilde{c}_{1j}x_j, \ I_2 = \{1\} \\ \min_{\mathbf{x}\in X} \overline{\mathbf{c}}_2\mathbf{x} &= \sum_{j=1}^{10} \overline{c}_{2j}x_j, \ I_3 = \{2\} \\ \min_{\mathbf{x}\in X} \widetilde{\overline{\mathbf{c}}}_3\mathbf{x} &= \sum_{j=1}^{10} \widetilde{\overline{c}}_{3j}x_j, \ I_4 = \{3\} \end{aligned}$$

subject to

$$\mathbf{x} \in X \stackrel{\text{def}}{=} \left\{ (x_1, \cdots, x_{10})^T \geq \mathbf{0} \mid \mathbf{a}_\ell \mathbf{x} \leq b_\ell, \ell = 1, \cdots, 7 \right\}$$

where $\mathbf{b} = (164, -190, -184, 99, -150, 154, 142)$, and $\mathbf{a}_\ell, \ell = 1, \cdots, 7$ are shown in Table 6.1. Further, $\widetilde{c}_{1j}, j = 1, \cdots, 10$ are LR fuzzy numbers whose membership functions are defined as

$$\mu_{\widetilde{c}_{1j}}(s) = \begin{cases} L\left(\frac{d_{1j}-s}{\alpha_{1j}}\right), & s \leq d_{1j}, j = 1, \cdots, 10 \\ R\left(\frac{s-d_{1j}}{\beta_{1j}}\right), & s > d_{1j}, j = 1, \cdots, 10, \end{cases}$$

where function $L(t)(=R(t)) \stackrel{\text{def}}{=} \max\{0, 1-|t|\}$, and parameters $d_{1j}, \alpha_{1j}, \beta_{1j}, j = 1, \cdots, 10$ are shown in Table 6.1. Further,

$$\bar{c}_{2j} \stackrel{\text{def}}{=} c_{2j}^1 + \bar{t}_2 c_{2j}^2, j = 1, \cdots, 10$$

are random variables whose parameters $c_{2j}^1, c_{2j}^2, j = 1, \cdots, 10$ are shown in Table 6.1, and $\bar{t}_2 \sim \mathrm{N}(3, 3^2)$. Given the occurrence of each elementary event ω, $\tilde{\bar{c}}_{3j}(\omega), j = 1, \cdots, 10$ are realizations of fuzzy random variables $\tilde{\bar{c}}_{3j}, j = 1, \cdots, 10$, which are LR fuzzy numbers whose membership functions are defined as

$$\mu_{\tilde{\bar{c}}_{3j}(\omega)}(s) = \begin{cases} L\left(\frac{d_{3\ell}(\omega)-s}{\alpha_{3j}}\right), & s \leq d_{3j}(\omega), j = 1, \cdots, 10 \\ R\left(\frac{s-d_{3j}(\omega)}{\beta_{3j}}\right), & s > d_{3j}(\omega), j = 1, \cdots, 10, \end{cases}$$

where function $L(t)(=R(t)) \stackrel{\text{def}}{=} \max\{0, 1-|t|\}$. Further, parameters $\bar{d}_{3j}, j = 1, \cdots, 10$ are random variables expressed as

$$\bar{d}_{3j} \stackrel{\text{def}}{=} d_{3j}^1 + \bar{t}_3 d_{3j}^2, j = 1, \cdots, 10,$$

where $\bar{t}_3 \sim \mathrm{N}(3, 2^2)$. d_{3j}^1, d_{3j}^2, α_{3j}, and $\beta_{3j}, j = 1, \cdots, 10$ are constants as shown in Table 6.1. Given a hypothetical decision maker, we apply Algorithm 6.2 to MOP6.4.

At Step 1, assume that the hypothetical decision maker sets his or her linear membership functions of fuzzy goals $\tilde{G}_i, i = 1, 2, 3$ as follows.

$$\begin{aligned} \mu_{\tilde{G}_1}(y_1) &= \frac{y_1 - 1800}{1700 - 1800} \\ \mu_{\tilde{G}_2}(y_2) &= \frac{y_2 - 700}{600 - 700} \\ \mu_{\tilde{G}_3}(y_3) &= \frac{y_3 - (-900)}{(-1000) - (-900)} \end{aligned}$$

At Step 2, assume that the hypothetical decision maker sets his or her membership functions corresponding to fuzzy goals for permissible probability levels $\hat{p}_i, i = 2, 3$ as

$$\begin{aligned} \mu_{p_2}(\hat{p}_2) &= \frac{\hat{p}_2 - 0.1099}{(0.9996 - 0.1099)} \\ \mu_{p_3}(\hat{p}_3) &= \frac{\hat{p}_3 - 0.0754}{(0.9906 - 0.0754)}. \end{aligned}$$

At Step 3, set the initial reference membership values to $\hat{\mu}_i = 1, i = 1, 2, 3$. At Step 4, solve MINMAX6.2($\hat{\mu}$) and TEST6.1, thus obtaining the corresponding generalized Pareto optimal solution. The interactive processes under the hypothetical decision maker are summarized in Table 6.2, where $h_i^* \stackrel{\text{def}}{=} \hat{\mu}_i - \lambda^*, i = 1, 3$, $\hat{p}_i^* \stackrel{\text{def}}{=} \mu_{\hat{p}_i}^{-1}(\hat{\mu}_i - \lambda^*), i = 2, 3$.

Table 6.2: Interactive processes in MOP6.4

	1	2	3
$\hat{\mu}_1$	1	0.43	0.43
$\hat{\mu}_2$	1	0.48	0.46
$\hat{\mu}_2$	1	0.45	0.43
$\mu_{D_1}(\mathbf{x}^*, h_1^*)$	0.43648	0.41761	0.43611
$\mu_{D_2}(\mathbf{x}^*, \hat{p}_2^*)$	0.43648	0.46761	0.46611
$\mu_{D_3}(\mathbf{x}^*, h_3^*, \hat{p}_2^*)$	0.43648	0.43761	0.43611
$f_1(\mathbf{x}^*, h_1^*)$	1756.35	1758.24	1756.39
$f_2(\mathbf{x}^*, \hat{p}_2^*)$	656.352	653.239	653.389
$f_3(\mathbf{x}^*, h_3^*, \hat{p}_3^*)$	–943.648	–943.761	–943.611
h_1^*	0.43648	0.41761	0.43611
$\hat{p}_2^*$	0.498333	0.526031	0.524695
h_3^*	0.43648	0.43761	0.43611
$\hat{p}_3^*$	0.474915	0.475951	0.474577

6.2 GMOPs with variance covariance matrices

In this section, we turn our attention to GMOPs with variance covariance matrices. In Section 6.2.1, we formulate GMOPs with variance covariance matrices as multiobjective programming problems using a fractile optimization model and a possibility measure; further, we introduce a generalized Pareto optimal solution. In Section 6.2.2, we propose an interactive convex programming algorithm for obtaining a satisfactory solution from a generalized Pareto optimal solution set.

6.2.1 A formulation using a fractile model and a possibility measure

In this subsection, we focus on a multiobjective programming problem with variance covariance matrices defined as shown below in which fuzzy coefficients, random variable coefficients, and fuzzy random variable coefficients are included in the objective functions.

MOP 6.5

$$\min_{\mathbf{x}\in X} z_i(\mathbf{x}) \stackrel{\text{def}}{=} \mathbf{c}_i\mathbf{x}, \; i \in I_1$$

$$\min_{\mathbf{x}\in X} \tilde{z}_i(\mathbf{x}) \stackrel{\text{def}}{=} \tilde{\mathbf{c}}_i\mathbf{x}, \; i \in I_2$$

$$\min_{\mathbf{x}\in X} \bar{z}_i(\mathbf{x}) \stackrel{\text{def}}{=} \bar{\mathbf{c}}_i\mathbf{x}, \; i \in I_3$$

$$\min_{\mathbf{x}\in X} \tilde{\bar{z}}_i(\mathbf{x}) \stackrel{\text{def}}{=} \tilde{\bar{\mathbf{c}}}_i\mathbf{x}, \; i \in I_4$$

Here, $\mathbf{x} = (x_1, \cdots, x_n)^T$ is an n-dimensional decision variable column vector, X is a linear constraint set, and I_1, I_2, I_3, I_4 are index sets of the objective functions. All coefficients of the objective functions, i.e., $\mathbf{c_i}, i \in I_1$, $\widetilde{\mathbf{c}}_\mathbf{i}, i \in I_2$, $\bar{\mathbf{c}}_\mathbf{i}, i \in I_3$, and $\widetilde{\bar{\mathbf{c}}}_i, i \in I_4$ are n-dimensional vectors. Further, each element of $\mathbf{c_i}, i \in I_1$, $\widetilde{\mathbf{c}}_\mathbf{i}, i \in I_2$, $\bar{\mathbf{c}}_\mathbf{i}, i \in I_3$ and $\widetilde{\bar{\mathbf{c}}}_i, i \in I_4$ is a real number, a fuzzy coefficient, a random variable coefficient, and a fuzzy random variable coefficient [46, 65, 77] respectively.

For objective function $\widetilde{z}_i(\mathbf{x}), i \in I_2$, we assume that each element of fuzzy coefficients $\widetilde{\mathbf{c}}_\mathbf{i}, i \in I_2$ is an LR fuzzy number [18] whose membership function is defined as

$$\mu_{\widetilde{c}_{ij}}(s) = \begin{cases} L\left(\frac{d_{ij}-s}{\alpha_{ij}}\right), & s \leq d_{ij}, j = 1, \cdots, n, i \in I_2 \\ R\left(\frac{s-d_{ij}}{\beta_{ij}}\right), & s > d_{ij}, j = 1, \cdots, n, i \in I_2, \end{cases}$$

where function $L(t) \stackrel{\text{def}}{=} \max\{0, l(t)\}$ is a real-valued continuous function from $[0, \infty)$ to $[0, 1]$ and $l(t)$ is a strictly decreasing continuous function satisfying $l(0) = 1$. Further, $R(t) \stackrel{\text{def}}{=} \max\{0, r(t)\}$ satisfies the same conditions and d_{ij} is the mean value, α_{ij}, β_{ij} are called left and right spreads respectively [18]. According to operations of fuzzy numbers based on the extension principle [18], each objective function $\widetilde{\mathbf{c}}_i\mathbf{x}, i \in I_2$ can also be expressed by an LR fuzzy number whose membership function is defined as

$$\mu_{\widetilde{\mathbf{c}}_i\mathbf{x}}(y) = \begin{cases} L\left(\dfrac{\mathbf{d}_i\mathbf{x} - y}{\alpha_i\mathbf{x}}\right), & y \leq \mathbf{d}_i\mathbf{x}, i \in I_2 \\ R\left(\dfrac{y - \mathbf{d}_i\mathbf{x}}{\beta_i\mathbf{x}}\right), & y > \mathbf{d}_i\mathbf{x}, i \in I_2. \end{cases} \tag{6.97}$$

where $\mathbf{d}_i = (d_{i1}, \cdots, d_{in})$, $\alpha_i = (\alpha_{i1}, \cdots, \alpha_{in}) > \mathbf{0}$, $\beta_i = (\beta_{i1}, \cdots, \beta_{in}) > \mathbf{0}$, $\mathbf{x} \geq \mathbf{0}$.

For objective function $\bar{z}_i(\mathbf{x}), i \in I_3$, we assume that each element $\bar{c}_{ij}$ of random variable coefficient vector $\bar{\mathbf{c}}_\mathbf{i}, i \in I_3$ is a Gaussian random variable, i.e.,

$$\bar{c}_{ij} \sim \mathrm{N}(E[\bar{c}_{ij}], \sigma_{ijj}), i \in I_3, j = 1, \cdots, n, \tag{6.98}$$

and variance covariance matrices between Gaussian random variables $\bar{c}_{ij}, j = 1, \cdots, n$ are given as $(n \times n)$-dimensional matrices

$$V_i = \begin{pmatrix} \sigma_{i11} & \sigma_{i12} & \cdots & \sigma_{i1n} \\ \sigma_{i21} & \sigma_{i22} & \cdots & \sigma_{i2n} \\ \vdots & \vdots & \ddots & \vdots \\ \sigma_{in1} & \sigma_{in2} & \cdots & \sigma_{inn} \end{pmatrix}, i \in I_3. \tag{6.99}$$

Let us denote the vectors of the expectation for the random variable row vectors $\bar{\mathbf{c}}_i, i \in I_3$ as

$$\mathbf{E}[\bar{\mathbf{c}}_i] = (E[\bar{c}_{i1}], \cdots, E[\bar{c}_{in}]), i \in I_3. \tag{6.100}$$

Then, using the variance covariance matrices $V_i, i \in I_3$, we can formally express the vector $\bar{\mathbf{c}}_i$ as

$$\bar{\mathbf{c}}_i \sim \mathrm{N}(\mathbf{E}[\bar{\mathbf{c}}_i], V_i), i \in I_3. \tag{6.101}$$

From the property of Gaussian random variables, the objective function $\bar{\mathbf{c}}_i\mathbf{x}, i \in I_3$ becomes Gaussian random variable.

$$\bar{\mathbf{c}}_i\mathbf{x} \sim \mathrm{N}(\mathbf{E}[\bar{\mathbf{c}}_i]\mathbf{x}, \mathbf{x}^T V_i\mathbf{x}), i \in I_3 \tag{6.102}$$

For objective function $\tilde{\bar{z}}_i(\mathbf{x}), i \in I_4$, we assume that $\tilde{\bar{c}}_{ij}, i \in I_4$ is an LR-type fuzzy random variable [36, 38, 77]. Given each elementary event ω, $\tilde{\bar{c}}_{ij}(\omega)$ is a realization of LR-type fuzzy random variable $\tilde{\bar{c}}_{ij}$, which is an LR fuzzy number [18] whose membership function is defined as

$$\mu_{\tilde{\bar{c}}_{ij}(\omega)}(s) = \begin{cases} L\left(\frac{d_{ij}(\omega)-s}{\alpha_{ij}}\right), & s \leq d_{ij}(\omega), j=1,\cdots,n, i \in I_4 \\ R\left(\frac{s-d_{ij}(\omega)}{\beta_{ij}}\right), & s > d_{ij}(\omega), j=1,\cdots,n, i \in I_4, \end{cases}$$

where function $L(t) \stackrel{\text{def}}{=} \max\{0, l(t)\}$ is a real-valued continuous function from $[0,\infty)$ to $[0,1]$, and $l(t)$ is a strictly decreasing continuous function satisfying $l(0)=1$. Further, $R(t) \stackrel{\text{def}}{=} \max\{0, r(t)\}$ satisfies the same conditions. Let us assume that $\alpha_{ij} > 0, \beta_{ij} > 0, i \in I_4, j=1,\cdots,n$, are spread parameters [18], mean value [18] $\bar{d}_{ij}, i \in I_4$ is a Gaussian random variable, i.e.,

$$\bar{d}_{ij} \sim \mathrm{N}(E[\bar{d}_{ij}], \sigma_{ijj}), i \in I_4, j=1,\cdots,n,$$

and positive-definite variance covariance matrices $V_i, i \in I_4$ between Gaussian random variables $\bar{d}_{ij}, i \in I_4, j=1,\cdots,n$ are given as

$$V_i = \begin{pmatrix} \sigma_{i11} & \sigma_{i12} & \cdots & \sigma_{i1n} \\ \sigma_{i21} & \sigma_{i22} & \cdots & \sigma_{i2n} \\ \vdots & \vdots & \ddots & \vdots \\ \sigma_{in1} & \sigma_{in2} & \cdots & \sigma_{inn} \end{pmatrix}, i \in I_4. \tag{6.103}$$

Let us also denote vectors of the expectation for random variable row vector $\bar{\mathbf{d}}_i$ as

$$\mathbf{E}[\bar{\mathbf{d}}_i] = (E[\bar{d}_{i1}],\cdots,E[\bar{d}_{in}]), i \in I_4.$$

As shown by Katagiri et al. [38], realizations $\tilde{\bar{\mathbf{c}}}_i(\omega)\mathbf{x}, i \in I_4$ become LR fuzzy numbers characterized by the following membership functions based on the extension principle [18].

$$\mu_{\tilde{\bar{\mathbf{c}}}_i(\omega)\mathbf{x}}(y) = \begin{cases} L\left(\frac{\mathbf{d}_i(\omega)\mathbf{x}-y}{\alpha_i\mathbf{x}}\right), & y \leq \mathbf{d}_i(\omega)\mathbf{x}, i \in I_4 \\ R\left(\frac{y-\mathbf{d}_i(\omega)\mathbf{x}}{\beta_i\mathbf{x}}\right), & y > \mathbf{d}_i(\omega)\mathbf{x}, i \in I_4 \end{cases}$$

Considering the imprecise nature of the decision maker's judgment, we assume that the decision maker has fuzzy goals $\widetilde{G}_i, i \in I_1 \cup I_2 \cup I_3 \cup I_4$ [71] for the objective functions of MOP6.1. Such fuzzy goals can be quantified by eliciting the corresponding membership functions. Let us denote membership functions of the objective functions as $\mu_{\widetilde{G}_i}(y), i \in I_1 \cup I_2 \cup I_3 \cup I_4$. Then, MOP6.5 can be formally transformed into the following problem.

MOP 6.6

$$\begin{array}{ll} \max\limits_{\mathbf{x} \in X} & \mu_{\widetilde{G}_i}(z_{i_i}\mathbf{x}), i \in I_1 \\ \max\limits_{\mathbf{x} \in X} & \mu_{\widetilde{G}_i}(\widetilde{z}_i\mathbf{x}), i \in I_2 \\ \max\limits_{\mathbf{x} \in X} & \mu_{\widetilde{G}_i}(\bar{z}_i\mathbf{x}), i \in I_3 \\ \max\limits_{\mathbf{x} \in X} & \mu_{\widetilde{G}_i}(\widetilde{\bar{z}}_i(\mathbf{x})), i \in I_4 \end{array}$$

In MOP6.6, membership function $\mu_{\widetilde{G}_i}(z_i(\mathbf{x})), i \in I_1$ is well-defined, but the other membership functions $\mu_{\widetilde{G}_i}(\widetilde{z}_i(\mathbf{x})), i \in I_2$, $\mu_{\widetilde{G}_i}(\bar{z}_i(\mathbf{x})), i \in I_3$, and $\mu_{\widetilde{G}_i}(\widetilde{\bar{z}}_i(\mathbf{x})), i \in I_4$ are ill-defined.

To handle membership function $\mu_{\widetilde{G}_i}(\widetilde{z}_i(\mathbf{x})), i \in I_2$, using the concept of a possibility measure [18], we define the value of membership function $\mu_{\widetilde{G}_i}(\widetilde{z}_i(\mathbf{x})), i \in I_2$ as

$$\mu_{\widetilde{G}_i}(\widetilde{z}_i(\mathbf{x})) \stackrel{\text{def}}{=} \Pi_{\widetilde{z}_i(\mathbf{x})}(\widetilde{G}_i) = \sup_y \min\{\mu_{\widetilde{z}_i(\mathbf{x})}(y), \mu_{\widetilde{G}_i}(y)\}, i \in I_2. \tag{6.104}$$

Assuming that $h_i \stackrel{\text{def}}{=} \Pi_{\widetilde{\mathbf{c}}_i(\mathbf{x})}(\widetilde{G}_i), i \in I_2$, then, the value of membership function $\mu_{\widetilde{G}_i}(\widetilde{z}_i(\mathbf{x})), i \in I_2$ can be obtained as

$$\mu_{\widetilde{G}_i}(\widetilde{z}_i(\mathbf{x})) \stackrel{\text{def}}{=} \{\mu_{\widetilde{G}_i}(f_i(\mathbf{x}, h_i)) \mid \mu_{\widetilde{G}_i}^{-1}(h_i) = f_i(\mathbf{x}, h_i)\}, i \in I_2, \tag{6.105}$$

where

$$f_i(\mathbf{x}, h_i) \stackrel{\text{def}}{=} \mathbf{d}_i\mathbf{x} - L^{-1}(h_i)\alpha_i\mathbf{x}, i \in I_2 \tag{6.106}$$

For membership function $\mu_{\widetilde{G}_i}(\bar{z}_i(\mathbf{x})), i \in I_3$, using a fractile optimization model [87], we define the value of membership function $\mu_{\widetilde{G}_i}(\bar{z}_i(\mathbf{x})), i \in I_3$ as follows. First, let us consider the probability that inequality $\mu_{\widetilde{G}_i}(\bar{z}_i(\mathbf{x})) \geq h_i, i \in I_3$ is satisfied, where h_i represents a target value for membership function $\mu_{\widetilde{G}_i}(\bar{z}_i(\mathbf{x}))$.

Then, the following relation holds.

$$\begin{aligned} p_i(\mathbf{x},h_i) &\stackrel{\text{def}}{=} \Pr(\omega \mid \mu_{\tilde{G}_i}(\bar{\mathbf{c}}_i(\omega)\mathbf{x}) \geq h_i) \\ &= \Pr(\omega \mid \mathbf{c}_i(\omega)\mathbf{x} \leq \mu_{\tilde{G}_i}^{-1}(h_i)) \\ &= \Pr\left(\omega \mid \frac{\mathbf{c}_i(\omega)\mathbf{x} - \mathbf{E}[\bar{\mathbf{c}}_i]\mathbf{x}}{\sqrt{\mathbf{x}^T V_i \mathbf{x}}} \leq \frac{\mu_{\tilde{G}_i}^{-1}(h_i) - \mathbf{E}[\mathbf{c}_i]\mathbf{x}}{\sqrt{\mathbf{x}^T V_i \mathbf{x}}}\right) \\ &= \Phi\left(\frac{\mu_{\tilde{G}_i}^{-1}(h_i) - \mathbf{E}[\bar{\mathbf{c}}_i]\mathbf{x}}{\sqrt{\mathbf{x}^T V_i \mathbf{x}}}\right), \; i \in I_3 \end{aligned} \tag{6.107}$$

Here, $\Pr(\cdot)$ represents a probability measure, and $\Phi(\cdot)$ indicates a standard normal distribution function. According to the fractile optimization model [87], the decision maker subjectively specifies permissible probability level $\hat{p}_i$ for $p_i(\mathbf{x},h_i)$ in advance. Then, inequality $p_i(\mathbf{x},h_i) \geq \hat{p}_i$ can be transformed into

$$\begin{aligned} & \hat{p}_i \leq p_i(\mathbf{x},h_i) \\ \Leftrightarrow \; & \mu_{\tilde{G}_i}^{-1}(h_i) \geq \mathbf{E}[\bar{\mathbf{c}}_i]\mathbf{x} + \Phi^{-1}(\hat{p}_i) \cdot \sqrt{\mathbf{x}^T V_i \mathbf{x}} \\ \Leftrightarrow \; & h_i \leq \mu_{\tilde{G}_i}(f_i(\mathbf{x},\hat{p}_i)), i \in I_3, \end{aligned} \tag{6.108}$$

where

$$f_i(\mathbf{x},\hat{p}_i) \stackrel{\text{def}}{=} \mathbf{E}[\bar{\mathbf{c}}_i]\mathbf{x} + \Phi^{-1}(\hat{p}_i) \cdot \sqrt{\mathbf{x}^T V_i \mathbf{x}}, i \in I_3. \tag{6.109}$$

As a result, we can define the value of the membership function $\mu_{\tilde{G}_i}(\bar{z}_i(\mathbf{x})), i \in I_3$ based on a fractile optimization model as follows.

$$\mu_{\tilde{G}_i}(\bar{z}_i(\mathbf{x})) \stackrel{\text{def}}{=} \mu_{\tilde{G}_i}(f_i(\mathbf{x},\hat{p}_i)) \; i \in I_3 \tag{6.110}$$

In membership function (6.110), the decision maker must specify permissible probability levels $\hat{p}_i, i \in I_3$ in advance; however, in practice, it seems to be difficult for decision makers to specify such parameters appropriately, because of the conflict between the larger value of h_i and the larger value of $\hat{p}_i$. From such a perspective, Yano [111, 116, 117] proposed fuzzy approaches through a fractile optimization model [87] in which the decision maker is required to specify not permissible probability levels $\hat{p}_i, i \in I_3$, but rather the corresponding membership functions. Below, we assume that the decision maker has fuzzy goals for permissible probability levels $\hat{p}_i, i \in I_3$, whose membership functions are denoted as $\mu_{\hat{p}_i}(\hat{p}_i), i \in I_3$. We can then define the value of membership function $\mu_{\tilde{G}_i}(\bar{z}_i(\mathbf{x})), i \in I_3$ as

$$\mu_{\tilde{G}_i}(\bar{z}_i(\mathbf{x})) \stackrel{\text{def}}{=} \min\{\mu_{\hat{p}_i}(\hat{p}_i), \mu_{\tilde{G}_i}(f_i(\mathbf{x},\hat{p}_i))\}, i \in I_3. \tag{6.111}$$

To handle membership function $\mu_{\widetilde{G}_i}(\widetilde{\overline{z}}_i(\mathbf{x})), i \in I_4$, using the concept of a possibility measure [18], we define the value of membership function $\mu_{\widetilde{G}_i}(\widetilde{\overline{z}}_i(\mathbf{x})), i \in I_4$, as

$$\mu_{\widetilde{G}_i}(\widetilde{\overline{z}}_i(\mathbf{x})) \stackrel{\text{def}}{=} \Pi_{\widetilde{\overline{z}}_i(\mathbf{x})}(\widetilde{G}_i) = \sup_y \min\{\mu_{\widetilde{\overline{z}}_i(\mathbf{x})}(y), \mu_{\widetilde{G}_i}(y)\}, i \in I_4. \tag{6.112}$$

Using a fractile optimization model [87] for (6.112), let us consider the probability that inequality $\mu_{\widetilde{G}_i}(\widetilde{\overline{z}}_i(\mathbf{x})) \geq h_i, i \in I_4$ is satisfied, where h_i represents a target value for membership function $\mu_{\widetilde{G}_i}(\widetilde{\overline{z}}_i(\mathbf{x})), i \in I_4$. Then, the following relations hold.

$$\begin{aligned}
& \Pr(\omega \mid \Pi_{\widetilde{\overline{\mathbf{c}}}_i(\omega)\mathbf{x}}(\widetilde{G}_i) \geq h_i) \\
= \; & \Phi\left(\frac{\mu_{\widetilde{G}_i}^{-1}(h_i) - (\mathbf{E}[\bar{\mathbf{d}}_i]\mathbf{x} - L^{-1}(h_i)\alpha_i\mathbf{x})}{\sqrt{\mathbf{x}^T V_i \mathbf{x}}}\right) \\
\stackrel{\text{def}}{=} \; & p_i(\mathbf{x}, h_i), i \in I_4
\end{aligned} \tag{6.113}$$

According to the fractile optimization model, the decision maker subjectively specifies a permissible probability level $\hat{p}_i$ for $p_i(\mathbf{x}, h_i), i \in I_4$ in advance. Then, inequalities $p_i(\mathbf{x}, h_i) \geq \hat{p}_i, i \in I_4$ for the fractile optimization model can be transformed into

$$\begin{aligned}
& \hat{p}_i \leq p_i(\mathbf{x}, h_i) \\
\Leftrightarrow \; & \mu_{\widetilde{G}_i}^{-1}(h_i) \geq (\mathbf{E}[\bar{\mathbf{d}}_i]\mathbf{x} - L^{-1}(h_i)\alpha_i\mathbf{x}) + \Phi^{-1}(\hat{p}_i) \cdot \sqrt{\mathbf{x}^T V_i \mathbf{x}} \\
\Leftrightarrow \; & h_i \leq \mu_{\widetilde{G}_i}(f_i(\mathbf{x}, h_i, \hat{p}_i)),
\end{aligned} \tag{6.114}$$

where

$$f_i(\mathbf{x}, h_i, \hat{p}_i) \stackrel{\text{def}}{=} (\mathbf{E}[\bar{\mathbf{d}}_i]\mathbf{x} - L^{-1}(h_i)\alpha_i\mathbf{x}) + \Phi^{-1}(\hat{p}_i) \cdot \sqrt{\mathbf{x}^T V_i \mathbf{x}}. \tag{6.115}$$

As a result, we can define the value of membership function $\mu_{\widetilde{G}_i}(\widetilde{\overline{z}}_i(\mathbf{x})), i \in I_4$ through the fractile optimization model [87] as

$$\mu_{\widetilde{G}_i}(\widetilde{\overline{z}}_i(\mathbf{x})) \stackrel{\text{def}}{=} \{\mu_{\widetilde{G}_i}(f_i(\mathbf{x}, h_i, \hat{p}_i)) \mid \mu_{\widetilde{G}_i}(f_i(\mathbf{x}, h_i, \hat{p}_i)) = h_i\}, i \in I_4. \tag{6.116}$$

Based on similar discussions presented by Yano [119, 120], let us assume that the decision maker specifies not permissible probability levels $\hat{p}_i, i \in I_4$, but rather corresponding membership functions $\mu_{\hat{p}_i}(\hat{p}_i), i \in I_4$. Then, instead of (6.116), we can adopt membership function as follows.

$$\mu_{\widetilde{G}_i}(\widetilde{\overline{z}}_i(\mathbf{x})) \stackrel{\text{def}}{=} \left\{\min\{\mu_{\hat{p}_i}(\hat{p}_i), \mu_{\widetilde{G}_i}(f_i(\mathbf{x}, h_i, \hat{p}_i))\} \mid \mu_{\widetilde{G}_i}(f_i(\mathbf{x}, h_i, \hat{p}_i)) = h_i\right\}, i \in I_4. \tag{6.117}$$

Using the definitions (6.105), (6.111), and (6.117), MOP6.6 can be transformed into the usual multiobjective programming problem.

MOP 6.7

$$\max_{\mathbf{x}\in X} \mu_{D_i}(\mathbf{x}) \stackrel{\text{def}}{=} \mu_{\tilde{G}_i}(\mathbf{c}_i\mathbf{x}), i \in I_1 \tag{6.118}$$

$$\max_{\mathbf{x}\in X, h_i\in[0,1]} \mu_{D_i}(\mathbf{x}, h_i) \stackrel{\text{def}}{=} \mu_{\tilde{G}_i}(f_i(\mathbf{x}, h_i)), i \in I_2 \tag{6.119}$$

$$\max_{\mathbf{x}\in X, \hat{p}_i\in(0,1)} \mu_{D_i}(\mathbf{x}, \hat{p}_i) \stackrel{\text{def}}{=} \min\{\mu_{\hat{p}_i}(\hat{p}_i), \mu_{\tilde{G}_i}(f_i(\mathbf{x}, \hat{p}_i))\}, i \in I_3 \tag{6.120}$$

$$\max_{\mathbf{x}\in X, h_i\in[0,1], \hat{p}_i\in(0,1)} \mu_{D_i}(\mathbf{x}, h_i, \hat{p}_i) \stackrel{\text{def}}{=} \min\{\mu_{\hat{p}_i}(\hat{p}_i), \mu_{\tilde{G}_i}(f_i(\mathbf{x}, h_i, \hat{p}_i))\}, i \in I_4 \tag{6.121}$$

subject to

$$\mu_{\tilde{G}_i}(f_i(\mathbf{x}, h_i)) = h_i, i \in I_2 \tag{6.122}$$

$$\mu_{\tilde{G}_i}(f_i(\mathbf{x}, h_i, \hat{p}_i)) = h_i, i \in I_4 \tag{6.123}$$

Throughout this subsection, we make the following assumptions with respect to membership functions $\mu_{\tilde{G}_i}(\cdot), i \in I_1 \cup I_2 \cup I_3 \cup I_4$, $\mu_{\hat{p}_i}(\cdot), i \in I_3 \cup I_4$.

Assumption 6.3
$\mu_{\tilde{G}_i}(\mathbf{c}_i\mathbf{x}), i \in I_1, \mu_{\tilde{G}_i}(f_i(\mathbf{x}, h_i)), i \in I_2, \mu_{\tilde{G}_i}(f_i(\mathbf{x}, \hat{p}_i)), i \in I_3, \mu_{\tilde{G}_i}(f_i(\mathbf{x}, h_i, \hat{p}_i)), i \in I_4$ are strictly decreasing and continuous with respect to $\mathbf{c}_i\mathbf{x}, i \in I_1, f_i(\mathbf{x}, h_i), i \in I_2, f_i(\mathbf{x}, \hat{p}_i), i \in I_3, f_i(\mathbf{x}, h_i, \hat{p}_i), i \in I_4$ respectively, which are defined on interval $[f_{i\min}, f_{i\max}]$, $i \in I_1 \cup I_2 \cup I_3 \cup I_4$, where $\mu_{\tilde{G}_i}(\mathbf{c}_i\mathbf{x}) = 0$, if $\mathbf{c}_i\mathbf{x} \geq f_{i\max}$, $\mu_{\tilde{G}_i}(\mathbf{c}_i\mathbf{x}) = 1$, if $\mathbf{c}_i\mathbf{x} \leq f_{i\min}, i \in I_1$, $\mu_{\tilde{G}_i}(f_i(\mathbf{x}, h_i)) = 0$, if $f_i(\mathbf{x}, h_i) \geq f_{i\max}$, $\mu_{\tilde{G}_i}(f_i(\mathbf{x}, h_i)) = 1$, if $f_i(\mathbf{x}, h_i) \leq f_{i\min}, i \in I_2$, $\mu_{\tilde{G}_i}(f_i(\mathbf{x}, \hat{p}_i)) = 0$, if $f_i(\mathbf{x}, \hat{p}_i) \geq f_{i\max}$, $\mu_{\tilde{G}_i}(f_i(\mathbf{x}, \hat{p}_i)) = 1$, if $f_i(\mathbf{x}, \hat{p}_i) \leq f_{i\min}$, $i \in I_3$, and $\mu_{\tilde{G}_i}(f_i(\mathbf{x}, h_i, \hat{p}_i)) = 0$, if $f_i(\mathbf{x}, h_i, \hat{p}_i) \geq f_{i\max}$, $\mu_{\tilde{G}_i}(f_i(\mathbf{x}, h_i, \hat{p}_i)) = 1$, if $f_i(\mathbf{x}, h_i, \hat{p}_i) \leq f_{i\min}$, $i \in I_4$.

Assumption 6.4
$\mu_{\hat{p}_i}(\hat{p}_i), i \in I_3 \cup I_4$ are strictly increasing and continuous with respect to $\hat{p}_i$, which are defined on interval $P_i \stackrel{\text{def}}{=} [p_{i\min}, p_{i\max}] \subset (0.5, 1)$, where $\mu_{\hat{p}_i}(\hat{p}_i) = 0$ if $\hat{p}_i \leq p_{i\min}, i \in I_3 \cup I_4$, $\mu_{\hat{p}_i}(\hat{p}_i) = 1$ if $p_{i\max} \leq \hat{p}_i, i \in I_3 \cup I_4$.

To handle MOP6.7, we next introduce generalized Pareto optimality.

Definition 6.2 $(\mathbf{x}^*, \mathbf{h}^*, \hat{\mathbf{p}}^*) \in X_c$ is a generalized Pareto optimal solution to MOP6.7 if and only if there does not exist another $(\mathbf{x}, \mathbf{h}, \hat{\mathbf{p}}) \in X_c$ such that $\mu_{D_i}(\mathbf{x}) \geq \mu_{D_i}(\mathbf{x}^*), i \in I_1$, $\mu_{D_i}(\mathbf{x}, h_i) \geq \mu_{D_i}(\mathbf{x}^*, h_i^*), i \in I_2$ $\mu_{D_i}(\mathbf{x}, \hat{p}_i) \geq \mu_{D_i}(\mathbf{x}^*, \hat{p}_i^*), i \in I_3$, $\mu_{D_i}(\mathbf{x}, h_i, \hat{p}_i) \geq \mu_{D_i}(\mathbf{x}^*, h_i^*, \hat{p}_i^*), i \in I_4$ with strict inequality holding for at least one $i \in I_1 \cup I_2 \cup I_3 \cup I_4$, where $\mathbf{h}$ and $\hat{\mathbf{p}}$ are vectors of $h_i, i \in I_2 \cup I_4$ and $\hat{p}_i, i \in I_3 \cup I_4$ respectively, and constraint set X_c is defined as

$$\begin{aligned} X_c \stackrel{\text{def}}{=} & \{(\mathbf{x}, \mathbf{h}, \hat{\mathbf{p}}) \mid \mathbf{x} \in X, h_i \in [0,1], i \in I_2 \cup I_4, \hat{p}_i \in P_i, i \in I_3 \cup I_4, \\ & \mu_{\tilde{G}_i}^{-1}(h_i) = f_i(\mathbf{x}, h_i), i \in I_2, \mu_{\tilde{G}_i}^{-1}(h_i) = f_i(\mathbf{x}, h_i, \hat{p}_i), i \in I_4\}. \quad (6.124) \end{aligned}$$

To generate a candidate for a satisfactory solution from a generalized Pareto optimal solution set, the decision maker specifies reference membership values [71]. Once reference membership values $\hat{\mu}_i, i \in I_1 \cup I_2 \cup I_3 \cup I_4$ are specified, the corresponding generalized Pareto optimal solution is obtained by solving the following minmax problem.

MINMAX 6.3 ($\hat{\mu}$)

$$\min_{(\mathbf{x}, \mathbf{h}, \hat{\mathbf{p}}) \in X_c, \lambda \in \Lambda} \lambda \tag{6.125}$$

subject to

$$\hat{\mu}_i - \mu_{D_i}(\mathbf{x}) \leq \lambda, i \in I_1 \tag{6.126}$$

$$\hat{\mu}_i - \mu_{D_i}(\mathbf{x}, h_i) \leq \lambda, i \in I_2 \tag{6.127}$$

$$\hat{\mu}_i - \mu_{D_i}(\mathbf{x}, \hat{p}_i) \leq \lambda, i \in I_3 \tag{6.128}$$

$$\hat{\mu}_i - \mu_{D_i}(\mathbf{x}, h_i, \hat{p}_i) \leq \lambda, i \in I_4 \tag{6.129}$$

where

$$\Lambda \stackrel{\text{def}}{=} [\lambda_{\min}, \lambda_{\max}] = [\max_{i=1,\cdots,k} \hat{\mu}_i - 1, \max_{i=1,\cdots,k} \hat{\mu}_i]. \tag{6.130}$$

Constraints (6.126) for $i \in I_1$ can be transformed into

$$\mu_{\tilde{G}_i}^{-1}(\hat{\mu}_i - \lambda) \geq \mathbf{c}_i \mathbf{x}, i \in I_1. \tag{6.131}$$

Similarly, constraints (6.124) and (6.127) for $i \in I_2$ can be transformed into

$$\hat{\mu}_i - \lambda \leq h_i, i \in I_2 \tag{6.132}$$

$$\mathbf{d}_i \mathbf{x} - L^{-1}(h_i) \alpha_i \mathbf{x} = \mu_{\tilde{G}_i}^{-1}(h_i), i \in I_2. \tag{6.133}$$

From (6.132) and Assumption 6.3, the following relations hold.

$$L^{-1}(\hat{\mu}_i - \lambda) \geq L^{-1}(h_i), i \in I_2 \tag{6.134}$$

$$\mu_{\tilde{G}_i}^{-1}(\hat{\mu}_i - \lambda) \geq \mu_{\tilde{G}_i}^{-1}(h_i), i \in I_2 \tag{6.135}$$

Using (6.132), (6.133), (6.134), and (6.135), we can reduce (6.127) to the following constraint.

$$\begin{aligned} & \mathbf{d}_i \mathbf{x} - L^{-1}(\hat{\mu}_i - \lambda) \alpha_i \mathbf{x} \\ \leq \ & \mathbf{d}_i \mathbf{x} - L^{-1}(h_i) \alpha_i \mathbf{x} = \mu_{\tilde{G}_i}^{-1}(h_i) \\ \leq \ & \mu_{\tilde{G}_i}^{-1}(\hat{\mu}_i - \lambda), i \in I_2 \end{aligned} \tag{6.136}$$

From (6.117) and (6.128) for $i \in I_3$, the following inequalities hold.

$$\hat{\mu}_i - \lambda \leq \mu_{\tilde{G}_i}(f_i(\mathbf{x}, \hat{p}_i)), i \in I_3 \tag{6.137}$$
$$\hat{\mu}_i - \lambda \leq \mu_{\hat{p}_i}(\hat{p}_i), i \in I_3 \tag{6.138}$$

Further, (6.137) can be equivalently transformed into

$$\begin{aligned} & f_i(\mathbf{x}, \hat{p}_i) \leq \mu_{\tilde{G}_i}^{-1}(\hat{\mu}_i - \lambda) \\ \Leftrightarrow \quad & \mathbf{E}[\bar{\mathbf{c}}_i]\mathbf{x} + \Phi^{-1}(\hat{p}_i) \cdot \sqrt{\mathbf{x}^T V_i \mathbf{x}} \leq \mu_{\tilde{G}_i}^{-1}(\hat{\mu}_i - \lambda), i \in I_3. \end{aligned} \tag{6.139}$$

As a result, (6.128) can be expressed as

$$\mathbf{E}[\bar{\mathbf{c}}_i]\mathbf{x} + \Phi^{-1}(\mu_{\hat{p}_i}^{-1}(\hat{\mu}_i - \lambda)) \cdot \sqrt{\mathbf{x}^T V_i \mathbf{x}} \leq \mu_{\tilde{G}_i}^{-1}(\hat{\mu}_i - \lambda), i \in I_3. \tag{6.140}$$

Similarly, from (6.124) and (6.129) for $i \in I_4$, it holds that

$$\begin{aligned} & h_i = \mu_{\tilde{G}_i}(f_i(\mathbf{x}, h_i, \hat{p}_i)) \geq \hat{\mu}_i - \lambda, \\ \Leftrightarrow \quad & \mu_{\tilde{G}_i}^{-1}(h_i) = f_i(\mathbf{x}, h_i, \hat{p}_i) \leq \mu_{\tilde{G}_i}^{-1}(\hat{\mu}_i - \lambda) \\ \Leftrightarrow \quad & \mu_{\tilde{G}_i}^{-1}(h_i) = (\mathbf{E}[\bar{\mathbf{d}}_i]\mathbf{x} - L^{-1}(h_i)\alpha_i\mathbf{x}) + \Phi^{-1}(\hat{p}_i) \cdot \sqrt{\mathbf{x}^T V_i \mathbf{x}} \\ & \leq \mu_{\tilde{G}_i}^{-1}(\hat{\mu}_i - \lambda), i \in I_4 \end{aligned} \tag{6.141}$$

Because of $L^{-1}(h_i) \leq L^{-1}(\hat{\mu}_i - \lambda)$, the following relation holds for the left-hand-side of inequality (6.141).

$$\begin{aligned} & (\mathbf{E}[\bar{\mathbf{d}}_i]\mathbf{x} - L^{-1}(h_i)\alpha_i\mathbf{x}) + \Phi^{-1}(\hat{p}_i) \cdot \sqrt{\mathbf{x}^T V_i \mathbf{x}} \\ \geq \quad & (\mathbf{E}[\bar{\mathbf{d}}_i]\mathbf{x} - L^{-1}(\hat{\mu}_i - \lambda)\alpha_i\mathbf{x}) + \Phi^{-1}(\hat{p}_i) \cdot \sqrt{\mathbf{x}^T V_i \mathbf{x}}, i \in I_4 \end{aligned} \tag{6.142}$$

From inequalities (6.141) and (6.142), the following inequality holds.

$$\mu_{\tilde{G}_i}^{-1}(\hat{\mu}_i - \lambda) \geq (\mathbf{E}[\bar{\mathbf{d}}_i]\mathbf{x} - L^{-1}(\hat{\mu}_i - \lambda)\alpha_i\mathbf{x}) + \Phi^{-1}(\hat{p}_i) \cdot \sqrt{\mathbf{x}^T V_i \mathbf{x}}, i \in I_4 \tag{6.143}$$

On the other hand, from Assumption 6.4 and (6.121), it holds that $\hat{p}_i \geq \mu_{\hat{p}_i}^{-1}(\hat{\mu}_i - \lambda), i \in I_4$. As a result, inequality (6.143) can be transformed into

$$\begin{aligned} \mu_{\tilde{G}_i}^{-1}(\hat{\mu}_i - \lambda) \quad & \geq \quad (\mathbf{E}[\bar{\mathbf{d}}_i]\mathbf{x} - L^{-1}(\hat{\mu}_i - \lambda)\alpha_i\mathbf{x}) + \Phi^{-1}(\hat{p}_i) \cdot \sqrt{\mathbf{x}^T V_i \mathbf{x}} \\ & \geq \quad (\mathbf{E}[\bar{\mathbf{d}}_i]\mathbf{x} - L^{-1}(\hat{\mu}_i - \lambda)\alpha_i\mathbf{x}) + \Phi^{-1}(\mu_{\hat{p}_i}^{-1}(\hat{\mu}_i - \lambda)) \cdot \sqrt{\mathbf{x}^T V_i \mathbf{x}}, \\ & \quad\; i \in I_4. \end{aligned}$$

Therefore, MINMAX6.3($\hat{\mu}$) can be reduced to the following simple minmax problem, where both decision variables $\hat{p}_i, i \in I_3 \cup I_4$, and $h_i, i \in I_2 \cup I_4$ have disappeared.

MINMAX 6.4 $(\hat{\mu})$

$$\min_{\mathbf{x}\in X,\lambda\in\Lambda} \lambda \tag{6.144}$$

subject to

$$\mathbf{c}_i\mathbf{x} \leq \mu_{\tilde{G}_i}^{-1}(\hat{\mu}_i-\lambda), i\in I_1 \tag{6.145}$$

$$\mathbf{d}_i\mathbf{x} - L^{-1}(\hat{\mu}_i-\lambda)\alpha_i\mathbf{x} \leq \mu_{\tilde{G}_i}^{-1}(\hat{\mu}_i-\lambda), i\in I_2 \tag{6.146}$$

$$\mathbf{E}[\bar{\mathbf{c}}_i]\mathbf{x} + \Phi^{-1}(\mu_{\hat{p}_i}^{-1}(\hat{\mu}_i-\lambda))\cdot\sqrt{\mathbf{x}^T V_i \mathbf{x}} \leq \mu_{\tilde{G}_i}^{-1}(\hat{\mu}_i-\lambda), i\in I_3 \tag{6.147}$$

$$\mu_{\tilde{G}_i}^{-1}(\hat{\mu}_i-\lambda) \geq (\mathbf{E}[\bar{\mathbf{d}}_i]\mathbf{x} - L^{-1}(\hat{\mu}_i-\lambda)\alpha_i\mathbf{x}) + \Phi^{-1}(\mu_{\hat{p}_i}^{-1}(\hat{\mu}_i-\lambda))\cdot\sqrt{\mathbf{x}^T V_i \mathbf{x}},$$
$$i\in I_4 \tag{6.148}$$

The relationships between optimal solution $\mathbf{x}^*\in X, \lambda^*\in\Lambda$ of MINMAX6.4 $(\hat{\mu})$ and generalized Pareto optimal solutions to MOP6.7 are characterized by the following theorem.

Theorem 6.3

(1) If $\mathbf{x}^*\in X, \lambda^*\in\Lambda$ is a unique optimal solution of MINMAX6.4($\hat{\mu}$), then $\mathbf{x}^*\in X$, $h_i^* \stackrel{\text{def}}{=} \hat{\mu}_i-\lambda^*, i\in I_2\cup I_4$, $\hat{p}_i^* \stackrel{\text{def}}{=} \mu_{\hat{p}_i}^{-1}(\hat{\mu}_i-\lambda^*), i\in I_3\cup I_4$ is a generalized Pareto optimal solution to MOP6.7.

(2) If $\mathbf{x}^*\in X$, $h_i^*, i\in I_2\cup I_4$ $\hat{p}_i^*, i\in I_3\cup I_4$ is a generalized Pareto optimal solution to MOP6.7, then $\mathbf{x}^*\in X$, $\lambda^*\in\Lambda$ is an optimal solution of MINMAX6.4($\hat{\mu}$) for some reference membership values $\hat{\mu}_i, i\in I_1\cup I_2\cup I_3\cup I_4$, where $\lambda^* \stackrel{\text{def}}{=} \hat{\mu}_i - \mu_{\tilde{G}_i}(\mathbf{c}_i\mathbf{x}^*)$, $i\in I_1$, $\lambda^* \stackrel{\text{def}}{=} \hat{\mu}_i - \mu_{\tilde{G}_i}(f_i(\mathbf{x}^*, h_i^*)) = \hat{\mu}_i - h_i^*$, $i\in I_2$, $\lambda^* \stackrel{\text{def}}{=} \hat{\mu}_i - \mu_{\hat{p}_i}(\hat{p}_i^*) = \hat{\mu}_i - \mu_{\tilde{G}_i}(f_i(\mathbf{x}^*, \hat{p}_i^*)), i\in I_3$, $\lambda^* \stackrel{\text{def}}{=} \hat{\mu}_i - \mu_{\hat{p}_i}(\hat{p}_i^*) = \hat{\mu}_i - \mu_{\tilde{G}_i}(f_i(\mathbf{x}^*, h_i^*, \hat{p}_i^*)) = \hat{\mu}_i - h_i^*, i\in I_4$.

(Proof)

(1) Since $\mathbf{x}^*\in X, \lambda^*\in\Lambda$ is an optimal solution of MINMAX6.4($\hat{\mu}$), the following inequalities hold.

$$\begin{aligned}
\hat{\mu}_i-\lambda^* &\leq \mu_{\tilde{G}_i}(\mathbf{c}_i\mathbf{x}^*), i\in I_1\\
\hat{\mu}_i-\lambda^* &\leq \mu_{\tilde{G}_i}(f_i(\mathbf{x}^*, h_i^*)), i\in I_2\\
\hat{\mu}_i-\lambda^* &\leq \mu_{\tilde{G}_i}(f_i(\mathbf{x}^*, \hat{p}_i^*)), i\in I_3\\
\hat{\mu}_i-\lambda^* &= \mu_{\hat{p}_i}(\mu_{\hat{p}_i}^{-1}(\hat{\mu}_i-\lambda^*)) = \mu_{\hat{p}_i}(\hat{p}_i^*), i\in I_3\\
\hat{\mu}_i-\lambda^* &\leq \mu_{\tilde{G}_i}(f_i(\mathbf{x}^*, h_i^*, \hat{p}_i^*)), i\in I_4\\
\hat{\mu}_i-\lambda^* &= \mu_{\hat{p}_i}(\mu_{\hat{p}_i}^{-1}(\hat{\mu}_i-\lambda^*)) = \mu_{\hat{p}_i}(\hat{p}_i^*), i\in I_4
\end{aligned}$$

Assume that $\mathbf{x}^*\in X$, $h_i^* \stackrel{\text{def}}{=} \hat{\mu}_i-\lambda^*, i\in I_2\cup I_4$, $\hat{p}_i^* \stackrel{\text{def}}{=} \mu_{\hat{p}_i}^{-1}(\hat{\mu}_i-\lambda^*), i\in I_3\cup I_4$ is not a generalized Pareto optimal solution to MOP6.7. Then, there exists $\mathbf{x} \in X, h_i \in I_2 \cup I_4, \hat{p}_i, i \in I_3 \cup I_4$ such that

$$\mu_{\tilde{G}_i}(\mathbf{c}_i\mathbf{x}) \geq \mu_{\tilde{G}_i}(\mathbf{c}_i\mathbf{x}^*) \geq \hat{\mu}_i - \lambda^*, i \in I_1, \tag{6.149}$$

$$\mu_{\tilde{G}_i}(f_i(\mathbf{x}, h_i)) \geq \mu_{\tilde{G}_i}(f_i(\mathbf{x}^*, h_i^*)) \geq \hat{\mu}_i - \lambda^*, i \in I_2, \tag{6.150}$$

$$h_i = \mu_{\tilde{G}_i}(f_i(\mathbf{x}, h_i)), i \in I_2, \tag{6.151}$$

$$\min\{\mu_{\hat{p}_i}(\hat{p}_i), \mu_{\tilde{G}_i}(f_i(\mathbf{x}, \hat{p}_i))\} \geq \min\{\mu_{\hat{p}_i}(\hat{p}_i^*), \mu_{\tilde{G}_i}(f_i(\mathbf{x}^*, \hat{p}_i^*))\}$$
$$= \hat{\mu}_i - \lambda^*, i \in I_3, \tag{6.152}$$

$$\min\{\mu_{\hat{p}_i}(\hat{p}_i), \mu_{\tilde{G}_i}(f_i(\mathbf{x}, h_i, \hat{p}_i))\} \geq \min\{\mu_{\hat{p}_i}(\hat{p}_i^*), \mu_{\tilde{G}_i}(f_i(\mathbf{x}^*, h_i^*, \hat{p}_i^*))\}$$
$$= \hat{\mu}_i - \lambda^*, i \in I_4, \tag{6.153}$$

$$h_i = \mu_{\tilde{G}_i}(f_i(\mathbf{x}, h_i, \hat{p}_i)), i \in I_4, \tag{6.154}$$

with strict inequality holding for at least one $i \in I_1 \cup I_2 \cup I_3 \cup I_4$. From (6.149) and Assumption 6.3, it holds that

$$\mathbf{c}_i\mathbf{x} \leq \mu_{\tilde{G}_i}^{-1}(\hat{\mu}_i - \lambda^*), i \in I_1. \tag{6.155}$$

From (6.150), (6.151), and Assumption 6.3, the following relations hold for $i \in I_2$.

$$\mathbf{d}_i\mathbf{x} - L^{-1}(\hat{\mu}_i - \lambda^*)\alpha_i\mathbf{x} \leq \mathbf{d}_i\mathbf{x} - L^{-1}(h_i)\alpha_i\mathbf{x} \leq \mu_{\tilde{G}_i}^{-1}(\hat{\mu}_i - \lambda^*), i \in I_2 \tag{6.156}$$

From (6.152), the following two inequalities hold for $i \in I_3$.

$$\mu_{\hat{p}_i}(\hat{p}_i) \geq \hat{\mu}_i - \lambda^*, i \in I_3 \tag{6.157}$$

$$\mu_{\tilde{G}_i}(f_i(\mathbf{x}, \hat{p}_i)) \geq \hat{\mu}_i - \lambda^*, i \in I_3 \tag{6.158}$$

Because of (6.109) and Assumptions 6.3 and 6.4, (6.157) and (6.158) are equivalently transformed as follows.

$$\hat{p}_i \geq \mu_{\hat{p}_i}^{-1}(\hat{\mu}_i - \lambda^*), i \in I_3$$

$$\hat{p}_i \leq \Phi\left(\frac{\mu_{\tilde{G}_i}^{-1}(\hat{\mu}_i - \lambda^*) - \mathbf{E}[\bar{\mathbf{c}}_i]\mathbf{x}}{\sqrt{\mathbf{x}^T V_i \mathbf{x}}}\right), i \in I_3$$

As a result, the above two inequalities can be replaced by the following inequality for $i \in I_3$.

$$\mu_{\tilde{G}_i}^{-1}(\hat{\mu}_i - \lambda^*) - \mathbf{E}[\bar{\mathbf{c}}_i]\mathbf{x} \geq \Phi(\mu_{\hat{p}_i}^{-1}(\hat{\mu}_i - \lambda^*)) \cdot \sqrt{\mathbf{x}^T V_i \mathbf{x}}, i \in I_3 \tag{6.159}$$

From (6.153), the following inequality holds for $i \in I_4$.

$$\min\{\mu_{\hat{p}_i}(\hat{p}_i), \mu_{\tilde{G}_i}(f_i(\mathbf{x}, h_i, \hat{p}_i))\} \geq \hat{\mu}_i - \lambda^*, i \in I_4$$

Then, it holds that

$$\begin{aligned} \mu_{\hat{p}_i}(\hat{p}_i) &\geq \hat{\mu}_i - \lambda^*, i \in I_4 & (6.160)\\ \mu_{\tilde{G}_i}(f_i(\mathbf{x}, h_i, \hat{p}_i)) &\geq \hat{\mu}_i - \lambda^*, i \in I_4. & (6.161)\end{aligned}$$

From (6.115) and Assumptions 6.3 and 6.4, we can transform (6.160) and (6.161) into

$$\begin{aligned} \hat{p}_i &\geq \mu_{\hat{p}_i}^{-1}(\hat{\mu}_i - \lambda^*), i \in I_4\\ \hat{p}_i &\leq \Phi\left(\frac{\mu_{\tilde{G}_i}^{-1}(\hat{\mu}_i - \lambda^*) - (\mathbf{E}[\bar{\mathbf{d}}_i]\mathbf{x} - L^{-1}(h_i)\alpha_i\mathbf{x})}{\sqrt{\mathbf{x}^T V_i \mathbf{x}}}\right), i \in I_4.\end{aligned}$$

From (6.154), it holds that $h_i \geq \hat{\mu}_i - \lambda^*, i \in I_4$. This means that the following inequalities hold.

$$\begin{aligned} \mu_{\tilde{G}_i}^{-1}(\hat{\mu}_i - \lambda^*) &\geq (\mathbf{E}[\bar{\mathbf{d}}_i]\mathbf{x} - L^{-1}(h_i)\alpha_i\mathbf{x}) + \Phi^{-1}(\mu_{\hat{p}_i}^{-1}(\hat{\mu}_i - \lambda^*)) \cdot \sqrt{\mathbf{x}^T V_i \mathbf{x}}\\ &\geq (\mathbf{E}[\bar{\mathbf{d}}_i]\mathbf{x} - L^{-1}(\hat{\mu}_i - \lambda^*)\alpha_i\mathbf{x})\\ &\quad + \Phi^{-1}(\mu_{\hat{p}_i}^{-1}(\hat{\mu}_i - \lambda^*)) \cdot \sqrt{\mathbf{x}^T V_i \mathbf{x}}, i \in I_4 & (6.162)\end{aligned}$$

Inequalities (6.155), (6.156), (6.159), and (6.162) contradict the fact that $\mathbf{x}^* \in X, \lambda^* \in \Lambda$ is a unique optimal solution of MINMAX6.4($\hat{\mu}$).

(2) Assume that $\mathbf{x}^* \in X, \lambda^* \in \Lambda$ is not an optimal solution of MINMAX6.4($\hat{\mu}$) for any reference membership values $\hat{\mu}_i, i \in I_1 \cup I_2 \cup I_3 \cup I_4$ that satisfy the following equalities.

$$\begin{aligned} \hat{\mu}_i - \lambda^* &= \mu_{\tilde{G}_i}(\mathbf{c}_i\mathbf{x}^*), i \in I_1 & (6.163)\\ \hat{\mu}_i - \lambda^* &= \mu_{\tilde{G}_i}(f_i(\mathbf{x}^*, h_i^*)) = h_i^*, i \in I_2 & (6.164)\\ \hat{\mu}_i - \lambda^* &= \mu_{\hat{p}_i}(\hat{p}_i^*) = \mu_{\tilde{G}_i}(f_i(\mathbf{x}^*, \hat{p}_i^*)), i \in I_3 & (6.165)\\ \hat{\mu}_i - \lambda^* &= \mu_{\hat{p}_i}(\hat{p}_i^*) = \mu_{\tilde{G}_i}(f_i(\mathbf{x}^*, h_i^*, \hat{p}_i^*)), i \in I_4 & (6.166)\end{aligned}$$

Then, there exists some $\mathbf{x} \in X, \lambda < \lambda^*$ such that

$$\mathbf{c}_i\mathbf{x} \leq \mu_{\tilde{G}_i}^{-1}(\hat{\mu}_i - \lambda), i \in I_1 \tag{6.167}$$

$$\mathbf{d}_i\mathbf{x} - L^{-1}(\hat{\mu}_i - \lambda)\alpha_i\mathbf{x} \leq \mu_{\tilde{G}_i}^{-1}(\hat{\mu}_i - \lambda), i \in I_2 \tag{6.168}$$

$$\mathbf{E}[\bar{\mathbf{c}}_i]\mathbf{x} + \Phi^{-1}(\mu_{\hat{p}_i}^{-1}(\hat{\mu}_i - \lambda)) \cdot \sqrt{\mathbf{x}^T V_i \mathbf{x}} \leq \mu_{\tilde{G}_i}^{-1}(\hat{\mu}_i - \lambda), i \in I_3 \tag{6.169}$$

$$\mu_{\tilde{G}_i}^{-1}(\hat{\mu}_i - \lambda) \geq (\mathbf{E}[\bar{\mathbf{d}}_i]\mathbf{x} - L^{-1}(\hat{\mu}_i - \lambda)\alpha_i\mathbf{x}) + \Phi^{-1}(\mu_{\hat{p}_i}^{-1}(\hat{\mu}_i - \lambda)) \cdot \sqrt{\mathbf{x}^T V_i \mathbf{x}},\ i \in I_4 \tag{6.170}$$

From Assumption 6.3, (6.163), (6.167), and $\hat{\mu}_i - \lambda > \hat{\mu}_i - \lambda^*$, the following relation holds for $i \in I_1$.

$$\mu_{\tilde{G}_i}(\mathbf{c}_i\mathbf{x}) \geq \hat{\mu}_i - \lambda > \hat{\mu}_i - \lambda^* = \mu_{\tilde{G}_i}(\mathbf{c}_i\mathbf{x}^*), i \in I_1 \tag{6.171}$$

From Assumption 6.3, (6.164), (6.168), and $\hat{\mu}_i - \lambda > \hat{\mu}_i - \lambda^*$, the following relations hold for $i \in I_2$.

$$\begin{aligned} &\mu_{\tilde{G}_i}(\mathbf{d}_i\mathbf{x} - L^{-1}(\hat{\mu}_i - \lambda)\alpha_i\mathbf{x}) \geq \hat{\mu}_i - \lambda \\ &> \hat{\mu}_i - \lambda^* = \mu_{\tilde{G}_i}(\mathbf{d}_i\mathbf{x}^* - L^{-1}(\hat{\mu}_i - \lambda^*)\alpha_i\mathbf{x}^*), i \in I_2 \end{aligned} \tag{6.172}$$

Because of Assumption 6.3, (6.165), (6.169), and $\hat{\mu}_i - \lambda > \hat{\mu}_i - \lambda^*, i \in I_3$, the following two inequalities hold for $i \in I_3$.

$$\mu_{\hat{p}_i}(\hat{p}_i) > \mu_{\hat{p}_i}(\hat{p}_i^*), i \in I_3 \tag{6.173}$$

$$\mu_{\tilde{G}_i}(f_i(\mathbf{x}, \hat{p}_i)) > \mu_{\tilde{G}_i}(f_i(\mathbf{x}^*, \hat{p}_i^*)), i \in I_3 \tag{6.174}$$

Here, $\hat{p}_i \stackrel{\text{def}}{=} \mu_{\hat{p}_i}^{-1}(\mu_i - \lambda), i \in I_3$. Moreover, because of Assumptions 6.3 and 6.4, (6.166), (6.170), and $\hat{\mu}_i - \lambda > \hat{\mu}_i - \lambda^*, i \in I_4$, the following inequalities hold.

$$\mu_{\hat{p}_i}(\hat{p}_i) \; > \; \mu_{\hat{p}_i}(\hat{p}_i^*), i \in I_4 \tag{6.175}$$

$$\mu_{\tilde{G}_i}(f_i(\mathbf{x}, \hat{h}_i, \hat{p}_i)) \; > \; \mu_{\tilde{G}_i}(f_i(\mathbf{x}^*, h_i^*, \hat{p}_i^*)), i \in I_4 \tag{6.176}$$

Here, $\hat{p}_i \stackrel{\text{def}}{=} \mu_{\hat{p}_i}^{-1}(\mu_i - \lambda)$, $\hat{h}_i \stackrel{\text{def}}{=} \hat{\mu}_i - \lambda, i \in I_4$. Inequalities (6.171)–(6.176) contradict the fact that $\mathbf{x}^* \in X$, $h_i^* \in I_2 \cup I_4$, $\hat{p}_i^*, i \in I_3 \cup I_4$ is a generalized Pareto optimal solution to MOP6.7. □

It is very difficult to solve MINMAX6.4($\hat{\mu}$) directly, because the constraints (6.147) and (6.148) are nonlinear. To address this problem, we first define the following functions for constraints (6.145), (6.146), (6.147), and (6.148), respectively.

$$g_i(\mathbf{x}, \lambda) \stackrel{\text{def}}{=} \mu_{\tilde{G}_i}^{-1}(\hat{\mu}_i - \lambda) - \mathbf{c}_i\mathbf{x}, i \in I_1 \tag{6.177}$$

$$g_i(\mathbf{x}, \lambda) \stackrel{\text{def}}{=} \mu_{\tilde{G}_i}^{-1}(\hat{\mu}_i - \lambda) - \mathbf{d}_i\mathbf{x} + L^{-1}(\hat{\mu}_i - \lambda)\alpha_i\mathbf{x}, i \in I_2 \tag{6.178}$$

$$\begin{aligned} g_i(\mathbf{x}, \lambda) \stackrel{\text{def}}{=} \; & \mu_{\tilde{G}_i}^{-1}(\hat{\mu}_i - \lambda) - \mathbf{E}[\bar{\mathbf{c}}_i]\mathbf{x} - \Phi^{-1}(\mu_{\hat{p}_i}^{-1}(\hat{\mu}_i - \lambda)) \cdot \sqrt{\mathbf{x}^T V_i \mathbf{x}}, \\ & i \in I_3 \end{aligned} \tag{6.179}$$

$$\begin{aligned} g_i(\mathbf{x}, \lambda) \stackrel{\text{def}}{=} \; & \mu_{\tilde{G}_i}^{-1}(\hat{\mu}_i - \lambda) - (\mathbf{E}[\bar{\mathbf{d}}_i]\mathbf{x} - L^{-1}(\hat{\mu}_i - \lambda)\alpha_i\mathbf{x}) \\ & -\Phi^{-1}(\mu_{\hat{p}_i}^{-1}(\hat{\mu}_i - \lambda)) \cdot \sqrt{\mathbf{x}^T V_i \mathbf{x}}, i \in I_4 \end{aligned} \tag{6.180}$$

From Assumption 6.3, it holds that $\Phi^{-1}(\mu_{\hat{p}_i}^{-1}(\hat{\mu}_i - \lambda)) > 0, i \in I_3 \cup I_4$. Therefore, constraint set

$$G(\lambda) \stackrel{\text{def}}{=} \{\mathbf{x} \in X \mid g_i(\mathbf{x}, \lambda) \geq 0, i \in I_1 \cup I_2 \cup I_3 \cup I_4\} \tag{6.181}$$

is convex with respect to the fixed value of $\lambda \in \Lambda$. Thus, MINMAX6.4($\hat{\mu}$) can be solved by applying a convex programming technique and the bisection method.

6.2.2 An interactive convex programming algorithm

In this subsection, we propose an interactive algorithm for obtaining a satisfactory solution from a generalized Pareto optimal solution set. Unfortunately, from Theorem 6.3, there is no guarantee that optimal solution $\mathbf{x}^* \in X, \lambda^* \in \Lambda$ of MINMAX6.4($\hat{\mu}$) is generalized Pareto optimal if $\mathbf{x}^* \in X, \lambda^* \in \Lambda$ is not unique. To guarantee generalized Pareto optimality, we first assume that constraints (6.145)–(6.148) of MINMAX6.4($\hat{\mu}$) are active at optimal solution $\mathbf{x}^* \in X, \lambda^* \in \Lambda$, i.e.,

$$\mathbf{c}_i\mathbf{x}^* = \mu_{\tilde{G}_i}^{-1}(\hat{\mu}_i - \lambda^*), i \in I_1 \tag{6.182}$$

$$\mathbf{d}_i\mathbf{x}^* - L^{-1}(\hat{\mu}_i - \lambda^*)\alpha_i\mathbf{x}^* = \mu_{\tilde{G}_i}^{-1}(\hat{\mu}_i - \lambda^*), i \in I_2 \tag{6.183}$$

$$\mathbf{E}[\bar{\mathbf{c}}_i]\mathbf{x}^* + \Phi^{-1}(\mu_{\tilde{p}_i}^{-1}(\hat{\mu}_i - \lambda^*)) \cdot \sqrt{\mathbf{x}^{*T}V_i\mathbf{x}^*} = \mu_{\tilde{G}_i}^{-1}(\hat{\mu}_i - \lambda^*),$$
$$i \in I_3 \tag{6.184}$$

$$\mu_{\tilde{G}_i}^{-1}(\hat{\mu}_i - \lambda^*) = (\mathbf{E}[\bar{\mathbf{d}}_i]\mathbf{x}^* - L^{-1}(\hat{\mu}_i - \lambda)\alpha_i\mathbf{x}^*)$$
$$+\Phi^{-1}(\mu_{\tilde{p}_i}^{-1}(\hat{\mu}_i - \lambda^*)) \cdot \sqrt{\mathbf{x}^{*T}V_i\mathbf{x}^*}, i \in I_4 \tag{6.185}$$

For optimal solution $\mathbf{x}^* \in X, \lambda^* \in \Lambda$ of MINMAX6.4($\hat{\mu}$), where active conditions (6.182)–(6.185) are satisfied, we solve the generalized Pareto optimality test problem defined as follows.

Test 6.2

$$w \stackrel{\text{def}}{=} \max_{\mathbf{x} \in X, \varepsilon_i \geq 0, i \in I_1 \cup I_2 \cup I_3 \cup I_4} \sum_{i \in I_1 \cup I_2 \cup I_3 \cup I_4} \varepsilon_i$$

subject to

$$\mathbf{c}_i\mathbf{x} + \varepsilon_i = \mu_{\tilde{G}_i}^{-1}(\hat{\mu}_i - \lambda^*), i \in I_1$$

$$\mathbf{d}_i\mathbf{x} - L^{-1}(\hat{\mu}_i - \lambda^*)\alpha_i\mathbf{x} + \varepsilon_i = \mu_{\tilde{G}_i}^{-1}(\hat{\mu}_i - \lambda^*), i \in I_2$$

$$\mathbf{E}[\bar{\mathbf{c}}_i]\mathbf{x} + \Phi^{-1}(\mu_{\tilde{p}_i}^{-1}(\hat{\mu}_i - \lambda^*)) \cdot \sqrt{\mathbf{x}^T V_i\mathbf{x}} + \varepsilon_i = \mu_{\tilde{G}_i}^{-1}(\hat{\mu}_i - \lambda^*), i \in I_3$$

$$(\mathbf{E}[\bar{\mathbf{d}}_i]\mathbf{x} - L^{-1}(\hat{\mu}_i - \lambda^*)\alpha_i\mathbf{x})$$
$$+\Phi^{-1}(\mu_{\tilde{p}_i}^{-1}(\hat{\mu}_i - \lambda^*)) \cdot \sqrt{\mathbf{x}^T V_i\mathbf{x}} + \varepsilon_i = \mu_{\tilde{G}_i}^{-1}(\hat{\mu}_i - \lambda^*), i \in I_4$$

For the optimal solution of TEST6.2, the following theorem holds.

Theorem 6.4

For optimal solution $\check{\mathbf{x}} \in X, \check{\varepsilon}_i \geq 0, i \in I_1 \cup I_2 \cup I_3 \cup I_4$ of TEST6.2, if $w = 0$ (equivalently, $\check{\varepsilon}_i = 0, i \in I_1 \cup I_2 \cup I_3 \cup I_4$), then $\mathbf{x}^* \in X, h_i^* \stackrel{\text{def}}{=} \hat{\mu}_i - \lambda^*, i \in I_2 \cup I_4$, $\hat{p}_i^* \stackrel{\text{def}}{=} \mu_{\tilde{p}_i}^{-1}(\hat{\mu}_i - \lambda^*), i \in I_3 \cup I_4$ is a generalized Pareto optimal solution to MOP6.7.

Given the above, we next present an interactive algorithm for MOP6.7 for deriving a satisfactory solution from a generalized Pareto optimal solution set.

Algorithm 6.3

Step 1: *The decision maker sets his or her membership functions* $\mu_{\tilde{G}_i}(\cdot), i \in I_1 \cup I_2 \cup I_3 \cup I_4$ *for the objective functions in MOP6.5 according to Assumption 6.3.*
Step 2: *The decision maker sets his or her membership functions* $\mu_{\hat{p}_i}(\hat{p}_i), i \in I_3 \cup I_4$ *for permissible probability levels* $\hat{p}_i, i \in I_3 \cup I_4$ *according to Assumption 6.4.*
Step 3: *Set initial reference membership values to* $\hat{\mu}_i = 1, i \in I_1 \cup I_2 \cup I_3 \cup I_4$.
Step 4: *Solve MINMAX6.4(*$\hat{\mu}$*) and TEST6.2 to obtain corresponding generalized Pareto optimal solution* $\mathbf{x}^* \in X, \lambda^* \in \Lambda$.
Step 5: *If the decision maker is satisfied with the current values of the generalized Pareto optimal solution* $\mu_{D_i}(\mathbf{x}^*), i \in I_1,\ \mu_{D_i}(\mathbf{x}^*, h_i^*), i \in I_2\ \mu_{D_i}(\mathbf{x}^*, \hat{p}_i^*), i \in I_3,$ $\mu_{D_i}(\mathbf{x}^*, h_i^*, \hat{p}_i^*), i \in I_4$*, where* $h_i^* \stackrel{\text{def}}{=} \hat{\mu}_i - \lambda^*, i \in I_2 \cup I_4$*,* $\hat{p}_i^* \stackrel{\text{def}}{=} \mu_{\hat{p}_i}^{-1}(\hat{\mu}_i - \lambda^*), i \in I_3 \cup I_4$*, then stop. Otherwise, the decision maker updates his or her reference membership values* $\hat{\mu}_i, i \in I_1 \cup I_2 \cup I_3 \cup I_4$*, and returns to Step 4.*

Chapter 7

Applications in Farm Planning

CONTENTS

In this chapter, we formulate and solve four applications in farm planning by applying the interactive algorithms discussed in the previous chapters above with a hypothetical decision maker. In Section 7.1, a farmer must decide the planting ratio for ten crops on his or her farmland to maximize total income, maximize total leisure hours for one year, and maximize minimum leisure hours for one period [21]. To handle the uncertainty of profit coefficients of the ten crops, we formulate the corresponding farm planning problem as a GMOP with variance covariance matrices and apply Algorithm 6.3 to obtain a satisfactory solution from a generalized Pareto optimal solution set. In Section 7.2, a farmer must decide the planting ratio for seven crops on his or her farmland to maximize total income for one year and minimize total work hours for one year [136]. Similar to the previous section, to handle the uncertainty of profit coefficients of the seven crops, we formulate the corresponding farm planning problem as a GMOP with variance covariance matrices and apply Algorithm 6.3 to obtain a satisfactory solution from a generalized Pareto optimal solution set. In Section 7.3, we formulate a farm planning problem with simple recourse as a two objective fuzzy random simple recourse programming problem in which the parameters involving water availability constraints are defined as a single fuzzy random variable. We use Algorithm 3.7 to obtain a satisfactory solution from a γ-Pareto optimal solution set. In Section 7.4, we formulate a vegetable shipment problem as a two objective two-person zero-sum game in which Player 1 is a farmer and Player 2 is Nature; further, the two payoff matrices are price coefficients of two vegetables. The farmer must decide his or her mixed strategy corresponding to each month and each vegetable. We apply Algorithm 5.1 to obtain a pessimistic compromise solution from a pessimistic Pareto optimal solution set.

7.1 A farm planning problem in Japan

In this subsection, to show the efficiency of Algorithm 6.3, we consider a farm planning problem in Japan, a problem originally formulated by Hayashi [21].

In the model farm, a family engages in agriculture, with a labor force consisting of one couple and their grandfather (i.e., 2.5 people). A fixed-term contract employee (i.e., one person) is hired from April to November. They make land available for growing in a greenhouse (i.e., from 10 to 20 acres), for outdoor cultivation (i.e., from 30 to 60 acres), and for rice paddy fields (i.e., 30 acres). In the greenhouse, tomatoes and spinach (x_1), as well as strawberries (x_2) are grown; for outdoor cultivation, cucumber (x_3), eggplant (x_4), cabbage (x_5), lettuce1 (x_6), Chinese cabbage (x_7), lettuce2 (x_8), radish (x_9), and rice (x_{10}) are grown, where x_j indicates the cultivation area with one unit equal to 10 acres for each crop $j = 1, \cdots, 10$. We assume that the farmer has 2.5 people available as family labor and that he or she must decide the planting ratio for the ten crops ($x_j, j = 1, \cdots, 10$) on his or her farmland to maximize total income, maximize total leisure hours for one year, and maximize minimum leisure hours for one period (i.e., 10 days).

Tables 7.1 and 7.2 show profit coefficients c_{tj} of ten crops j ($j = 1, \cdots, 10$) in each year t ($t = 1, \cdots, 10$) [21]. Below, we assume that the profit coefficients for the ten crops are normal random variables, i.e.,

$$\bar{c}_j \sim \mathrm{N}(E[\bar{c}_j], \sigma_{jj}), j = 1, \cdots, 10,$$

and also assume that expected values $E[\bar{c}_j], j = 1, \cdots, 10$ and each element of variance covariance matrix V between $\bar{c}_{j_1}$ and $\bar{c}_{j_2}$, $j_1, j_2 = 1, \cdots, 10$ are defined as

$$
\begin{aligned}
E[\bar{c}_j] &\stackrel{\text{def}}{=} \frac{1}{10}\sum_{t=1}^{10} c_{tj}, j = 1,\cdots,10\\
v_{jj} &\stackrel{\text{def}}{=} \frac{1}{9}\sum_{t=1}^{10} (c_{tj} - E[\bar{c}_j])^2, j = 1,\cdots,10\\
v_{j_1 j_2} &\stackrel{\text{def}}{=} \frac{1}{9}\sum_{t=1}^{10} (c_{tj_1} - E[\bar{c}_{j_1}])(c_{tj_2} - E[\bar{c}_{j_2}]), j_1, j_2 = 1,\cdots,10, j_1 \neq j_2.
\end{aligned}
$$

Then, from Tables 7.1 and 7.2, the vector of expected values $\mathbf{E}[\bar{\mathbf{c}}] \stackrel{\text{def}}{=} (E[\bar{c}_1],\cdots,E[\bar{c}_{10}])$ and variance covariance matrix V are computed as follows.

$$\mathbf{E}[\bar{\mathbf{c}}] = (120.0, 113.9, 24.64, 50.41, 19.39, 11.98, 20.19, 35.93, 5.811, 12.12)$$

$$
V_1 = \begin{pmatrix}
762.19 & 251.97 & -199.03 & 21.539 & 185.20\\
251.97 & 310.83 & -167.79 & -74.733 & 89.404\\
-199.03 & -167.79 & 264.51 & 157.62 & -31.907\\
21.539 & -74.733 & 157.62 & 416.15 & 118.72\\
185.20 & 89.404 & -31.907 & 118.72 & 350.18\\
49.273 & 57.254 & 8.0999 & -16.177 & 113.98\\
60.540 & 56.517 & -20.129 & 61.555 & 55.325\\
130.605 & 5.9142 & 87.730 & 148.45 & 136.67\\
25.228 & 1.9699 & 13.703 & 27.049 & -2.5597\\
11.262 & -2.1582 & -9.8473 & -22.923 & -2.9920
\end{pmatrix}
$$

$$
V_2 = \begin{pmatrix}
49.273 & 60.540 & 130.61 & 25.228 & 11.262\\
57.254 & 56.517 & 5.9142 & 1.9699 & -2.1582\\
8.0999 & -20.129 & 87.730 & 13.703 & -9.8473\\
-16.177 & 61.555 & 148.45 & 27.049 & -22.923\\
113.98 & 55.325 & 136.66 & -2.5597 & -2.9920\\
70.884 & 23.193 & 43.044 & -3.5996 & 0.84808\\
23.193 & 86.748 & 4.8858 & 3.7239 & -0.13111\\
43.044 & 4.8858 & 227.60 & 30.978 & -8.4568\\
-3.5996 & 3.7239 & 30.978 & 8.4808 & -2.6816\\
0.84808 & -0.13111 & -8.4568 & -2.6816 & 3.6589
\end{pmatrix}
$$

$$V = (V_1 \mid V_2)$$

The first objective function $\bar{z}_1(\mathbf{x})$ (total profit, unit: 10000 Japanese yen) can then be defined as

$$\bar{z}_1(\mathbf{x}) \stackrel{\text{def}}{=} \sum_{j=1}^{10} \bar{c}_j x_j = \bar{\mathbf{c}}\mathbf{x} \sim \mathrm{N}(\mathbf{E}[\bar{\mathbf{c}}]\mathbf{x}, \mathbf{x}^T V \mathbf{x}).$$

Table 7.1: Profit coefficients in MOP7.1 (No.1)

	j	1	2	3	4	5
t	c_{tj}	tomato+	strawberry	cucumber	eggplant	cabbage
1	10 years ago	136.91	101.41	30.707	42.873	8.899
2	9 years ago	136.75	134.29	4.145	52.643	33.04
3	8 years ago	102.85	95.411	34.606	87.47	12.188
4	7 years ago	111.53	140.13	1.075	16.394	11.825
5	6 years ago	46.728	90.361	50.135	41.921	6.578
6	5 years ago	126.71	134.63	30.237	63.095	9.984
7	4 years ago	99.907	102.57	9.732	47.333	15.802
8	3 years ago	132.37	119.02	29.741	66.615	68.377
9	2 years ago	134.46	106.32	15.428	26.397	10.998
10	1 year ago	121.36	114.23	40.599	58.858	16.154

Table 7.2: Profit coefficients in MOP7.1 (No.2)

	j	6	7	8	9	10
t	c_{tj}	lettuce1	Chinese cabbage	lettuce2	radish	rice
1	10 years ago	9.787	17.349	55.482	8.831	14.996
2	9 years ago	19.973	40.834	28.096	4.081	12.916
3	8 years ago	0.061	18.828	47.4	10.336	8.194
4	7 years ago	13.021	9.766	33.413	5.421	11.462
5	6 years ago	15.229	20.234	26.61	3.587	11.981
6	5 years ago	8.469	30.156	37.2	9.157	11.141
7	4 years ago	0.513	14.897	16.252	0.99	14.338
8	3 years ago	27.913	22.495	61.956	6.549	11.387
9	2 years ago	10.178	16.717	18.268	5.14	13.301
10	1 year ago	15.823	10.637	34.58	4.02	11.529

The second objective function is total leisure hours for one year. Tables 7.3, 7.4, 7.5, and 7.6 show required working hours $L_{\ell j}$, supply possible working hours TF_ℓ of his or her family, TA_ℓ of a fixed-term contract employee for each crop j and each period ℓ [21]. Given this, the second objective function $z_2(\mathbf{x})$ (unit: 1 hour) can be expressed as

$$z_2(\mathbf{x}) \stackrel{\text{def}}{=} \sum_{\ell=1}^{31} \left((2.5 \cdot TF_\ell + 1 \cdot TA_\ell) - \sum_{j=1}^{10} L_{\ell j} x_j \right).$$

Table 7.3: Supply possible working hours and required working hours in MOP7.1 (No.1)

ℓ	period	TF_ℓ	TA_ℓ	$L_{\ell 1}$	$L_{\ell 2}$	$L_{\ell 3}$	$L_{\ell 4}$
1	Feb-2	76.8					
2	Feb-3	63.1		13			
3	Mar-1	81		10			
4	Mar-2	83.4		10			7
5	Mar-3	94.7		10			3
6	Apr-1	88.8	64	15			12
7	Apr-2	91.3	64	10.5			3
8	Apr-3	93.7	64	79.5	74		13
9	May-1	95.9	64	141.5	98.5		45
10	May-2	97.8	64	24.5	151.5	4	31
11	May-3	109.3	72	32.5	65	12	16
12	Jun-1	100.5	64	44	10	30	21
13	Jun-2	101	64	44	17	71	27
14	Jun-3	101.1	64	43.5		24	17
15	Jul-1	100.6	64	98.5		23	33

The third objective function $z_3(\mathbf{x})$ (unit: 1 hour) is the minimum leisure hours for one period, which can be defined as

$$z_3(\mathbf{x}) \stackrel{\text{def}}{=} \min_{\ell=1,\cdots,31} \left((2.5 \cdot TF_\ell + 1 \cdot TA_\ell) - \sum_{j=1}^{10} L_{\ell j} x_j \right).$$

As land area constraints (unit: 10 acres) for the greenhouse,

$$1 \leq x_1 + x_2 \leq 2$$

must be satisfied. Similarly, as land area constraints (unit: 10 acres) for outdoor cultivation of double cropping, the following constraints must be satisfied.

$$3 \leq x_3 + x_4 + x_5 + x_6 \leq 6$$
$$3 \leq x_7 + x_8 + x_9 \leq 6$$

Then, we formulate the three objective programming problem involving random variable coefficients as follows (see MOP6.5).

Table 7.4: Supply possible working hours and required working hours in MOP7.1 (No.2)

ℓ	period	TF_ℓ	TA_ℓ	$L_{\ell 1}$	$L_{\ell 2}$	$L_{\ell 3}$	$L_{\ell 4}$
16	Jul-2	99.7	64	159		53	45
17	Jul-3	108	72	104		67	65
18	Aug-1	96.4	64	46.5		60.5	92
19	Aug-2	94.3	64	56.5		38	69
20	Aug-3	101.1	72	40	131	20	65
21	Sep-1	89.4	64	24			52
22	Sep-2	86.9	64	86			43
23	Sep-3	84.4	64	85			30
24	Oct-1	81.9	64	83	8		8
25	Oct-2	79.5	64	83	3		
26	Oct-3	84.8	72	83	36		
27	Nov-1	74.8	64	71	38		
28	Nov-2	72.9	64	71	10		
29	Nov-3	71.2	64	69			
30	Dec-1	70		63			
31	Dec-2	69.3		8			

MOP 7.1

$$\max \quad \bar{z}_1(\mathbf{x}) \stackrel{\text{def}}{=} \sum_{j=1}^{10} \bar{c}_j \cdot x_j \sim \mathrm{N}(\mathbf{E}[\bar{\mathbf{c}}]\mathbf{x}, \mathbf{x}^T V \mathbf{x})$$

$$\max \quad z_2(\mathbf{x}) \stackrel{\text{def}}{=} \sum_{\ell=1}^{31} \left((2.5 \cdot TF_\ell + 1 \cdot TA_\ell) - \sum_{j=1}^{10} L_{\ell j} x_j \right)$$

$$\max \quad z_3(\mathbf{x}) \stackrel{\text{def}}{=} x_{11}$$

subject to

$$(2.5 \cdot TF_\ell + 1 \cdot TA_\ell) - \sum_{j=1}^{10} L_{\ell j} x_j \geq x_{11}, \ell = 1, \cdots, 31$$

$$1 \leq x_1 + x_2 \leq 2$$

$$3 \leq x_3 + x_4 + x_5 + x_6 \leq 6$$

$$3 \leq x_7 + x_8 + x_9 \leq 6$$

$$x_{10} = 3, x_i \geq 0, i = 1, \cdots, 11$$

Table 7.5: Required working hours in MOP7.1 (No.3)

ℓ	period	$L_{\ell 5}$	$L_{\ell 6}$	$L_{\ell 7}$	$L_{\ell 8}$	$L_{\ell 9}$	$L_{\ell 10}$
1	Feb-2		8.5				
2	Feb-3		14				
3	Mar-1		5				
4	Mar-2	3	4				
5	Mar-3	3	25				4.4
6	Apr-1	2	23				3.4
7	Apr-2	15	5				3.3
8	Apr-3	15					4
9	May-1		3				7.7
10	May-2	4	2				2.7
11	May-3	1	65				0.4
12	Jun-1	1	8				0.6
13	Jun-2						0.8
14	Jun-3	19					2.9
15	Jul-1	26					0.6

We apply Algorithm 6.3 to MOP7.1 with a hypothetical decision maker. At Step 1, the hypothetical decision maker sets his or her membership functions corresponding to fuzzy goals $\widetilde{G}_i, i = 1,2,3$ for objective functions $\bar{z}_1(\mathbf{x}), z_2(\mathbf{x})$, $z_3(\mathbf{x})$ as

$$\begin{aligned}
\mu_{\widetilde{G}_1}(\bar{z}_1(\mathbf{x})) &= \frac{\bar{z}_1(\mathbf{x}) - 226}{455 - 226} \\
\mu_{\widetilde{G}_2}(z_2(\mathbf{x})) &= \frac{z_2(\mathbf{x}) - 3340}{7057 - 3340} \\
\mu_{\widetilde{G}_3}(z_3(\mathbf{x})) &= \frac{z_3(\mathbf{x}) - 0}{156 - 0}.
\end{aligned}$$

At Step 2, the hypothetical decision maker sets his or her membership function corresponding to the fuzzy goal for permissible probability level $\hat{p}_1$ for objective function $\bar{z}_1(\mathbf{x})$ as

$$\mu_{\hat{p}_1}(\hat{p}_1) = \frac{\hat{p}_1 - 0.7}{0.9 - 0.7}.$$

Then, the corresponding three objective programming problem is formulated as follows (see MOP6.7).

Table 7.6: Required working hours in MOP7.1 (No.4)

ℓ	period	$L_{\ell 5}$	$L_{\ell 6}$	$L_{\ell 7}$	$L_{\ell 8}$	$L_{\ell 9}$	$L_{\ell 10}$
16	Jul-2	13					2.7
17	Jul-3	5					1.7
18	Aug-1			4	8.5		1.6
19	Aug-2			20	14		
20	Aug-3			16	23	7	
21	Sep-1				23		3.7
22	Sep-2			3	5	4	16
23	Sep-3			3	3	1	2.3
24	Oct-1				65		
25	Oct-2			16	8	1	
26	Oct-3			42		20	
27	Nov-1			4		40	
28	Nov-2					4	
29	Nov-3						
30	Dec-1						
31	Dec-2						

MOP 7.2

$$\begin{aligned}
\max \mu_{D_1}(\mathbf{x}, \hat{p}_1) &\stackrel{\text{def}}{=} \min\left\{\mu_{\hat{p}_1}(\hat{p}_1), \mu_{\tilde{G}_1}(\mathbf{E}[\bar{\mathbf{c}}]\mathbf{x} + \Phi^{-1}(1-\hat{p}_1)\cdot\sqrt{\mathbf{x}^T V \mathbf{x}})\right\} \\
\max \mu_{D_2}(\mathbf{x}) &\stackrel{\text{def}}{=} \mu_{\tilde{G}_2}\left(\sum_{\ell=1}^{31}((2.5\cdot TF_\ell + 1\cdot TA_\ell) - \sum_{j=1}^{10} L_{\ell j}x_j)\right) \\
\max \mu_{D_3}(\mathbf{x}) &\stackrel{\text{def}}{=} \mu_{\tilde{G}_2}(x_{11}) \\
\text{subject to} &
\end{aligned}$$

$$(2.5\cdot TF_\ell + 1\cdot TA_\ell) - \sum_{j=1}^{10} L_{\ell j}x_j \geq x_{11}, \ell = 1, \cdots, 31$$

$$1 \leq x_1 + x_2 \leq 2$$

$$3 \leq x_3 + x_4 + x_5 + x_6 \leq 6$$

$$3 \leq x_7 + x_8 + x_9 \leq 6$$

$$x_{10} = 3, x_i \geq 0, i = 1, \cdots, 11, 0.5 < \hat{p}_1 < 1$$

To obtain a generalized Pareto optimal solution defined by Definition 6.2, we formulate the minmax problem as follows (see MINMAX6.4($\hat{\mu}$)).

MINMAX 7.1 ($\hat{\mu}$)

$$\min_{\mathbf{x}, \lambda \in \Lambda} \lambda$$

subject to

$$\mathbf{E}[\bar{\mathbf{c}}]\mathbf{x}+\Phi^{-1}(1-\mu_{\hat{p}_1}^{-1}(\hat{\mu}_1-\lambda))\cdot\sqrt{\mathbf{x}^T V\mathbf{x}} \geq \mu_{\tilde{G}_1}^{-1}(\hat{\mu}_1-\lambda)$$

$$\sum_{\ell=1}^{31}\left((2.5\cdot TF_\ell+1\cdot TA_\ell)-\sum_{j=1}^{10}L_{\ell j}x_j\right)\geq \mu_{\tilde{G}_2}^{-1}(\hat{\mu}_2-\lambda)$$

$$(2.5\cdot TF_\ell+1\cdot TA_\ell)-\sum_{j=1}^{10}L_{\ell j}x_j\geq \mu_{\tilde{G}_3}^{-1}(\hat{\mu}_3-\lambda),\ell=1,\cdots,31$$

$$1\leq x_1+x_2\leq 2$$

$$3\leq x_3+x_4+x_5+x_6\leq 6$$

$$3\leq x_7+x_8+x_9\leq 6$$

$$x_{10}=3,x_i\geq 0,i=1,\cdots,10$$

Note that an inequality sign is reversed in comparison with MINMAX6.4($\hat{\mu}$), because the three objective functions are not minimized, but rather maximized.

At Step 3, we set initial reference membership values to $\hat{\mu}_i=1,i=1,2,3$. At Step 4, by solving MINMAX7.1($\hat{\mu}$) and TEST6.2, the corresponding generalized Pareto optimal solution is obtained as follows, where $\hat{p}^* \stackrel{\text{def}}{=} \mu_{\hat{p}_1}^{-1}(\hat{\mu}_1-\lambda^*)$.

$$(\mu_{D_1}(\mathbf{x}^*,\hat{p}^*),\mu_{D_2}(\mathbf{x}^*),\mu_{D_3}(\mathbf{x}^*))=(0.50240,0.50240,0.50240)$$

The hypothetical decision maker is not satisfied with the current values of the membership functions. Therefore, he or she updates his or her reference membership values to $(\hat{\mu}_1,\hat{\mu}_2,\hat{\mu}_3)=(0.55,0.51,0.48)$ to improve the first membership function $\mu_{D_1}(\mathbf{x},\hat{p}_1)$ at the expense of the third one $\mu_{D_3}(\mathbf{x})$ (i.e., Step 5). Then, the corresponding generalized Pareto optimal solution is obtained as (i.e., Step 4)

$$(\mu_{D_1}(\mathbf{x}^*,\hat{p}^*),\mu_{D_2}(\mathbf{x}^*),\mu_{D_3}(\mathbf{x}^*))=(0.52664,0.48664,0.45664).$$

The hypothetical decision maker is again not satisfied with the current values of the membership functions. Therefore, he or she updates his or her reference membership values to $(\hat{\mu}_1,\hat{\mu}_2,\hat{\mu}_3)=(0.55,0.48,0.45)$ to further improve the first membership function $\mu_{D_1}(\mathbf{x},\hat{p}_1)$ at the expense of both the second one $\mu_{D_2}(\mathbf{x})$ and the third one $\mu_{D_3}(\mathbf{x})$ (i.e., Step 5). Then, the corresponding generalized Pareto optimal solution is obtained as (i.e., Step 4)

$$(\mu_{D_1}(\mathbf{x}^*,\hat{p}^*),\mu_{D_2}(\mathbf{x}^*),\mu_{D_3}(\mathbf{x}^*))=(0.53993,0.46993,0.43993).$$

The interactive processes involved with this hypothetical decision maker are summarized in Table 7.7. Here, a satisfactory solution is obtained at the third iteration.

To investigate the influence on the generalized Pareto optimal solutions from the reference membership values, the generalized Pareto optimal solution for reference membership values $(\hat{\mu}_1,\hat{\mu}_2,\hat{\mu}_3)=(0.48,0.58,0.53)$ is shown at the fourth

Table 7.7: Interactive processes in MOP7.1

	1	2	3	4
$\hat{\mu}_1$	1.	0.55	0.55	0.5
$\hat{\mu}_2$	1.	0.51	0.48	0.54
$\hat{\mu}_3$	1.	0.48	0.45	0.54
$\mu_{D_1}(\mathbf{x}^*, \hat{p}_1^*)$	0.50240	0.52664	0.53993	0.48465
$\mu_{D_2}(\mathbf{x}^*)$	0.50240	0.48664	0.46993	0.52465
$\mu_{D_3}(\mathbf{x}^*)$	0.50240	0.45664	0.43993	0.52465
$\hat{p}_1^*$	0.80048	0.80533	0.80799	0.79693
$f_1(\mathbf{x}^*, p_1^*)$	341.05	346.60	349.64	336.98
$E[\bar{z}_1(\mathbf{x}^*)]$	452.94	460.36	462.94	450.17
$z_2(\mathbf{x}^*)$	5207.4	5148.8	5086.7	5290.1
$z_3(\mathbf{x}^*)$	78.375	71.236	68.629	81.845
tomato (x_1^*)	0.61798	0.62013	0.64355	0.58965
strawberry (x_2^*)	0.99295	1.0099	1.0202	0.97972
cucumber (x_3^*)	0.0	0.0	0.0	0.0
eggplant (x_4^*)	0.0	0.0	0.0	0.0
cabbage (x_5^*)	4.1746	3.9579	3.7167	4.5673
lettuce1 (x_6^*)	1.8254	2.0421	2.2833	1.4327
Chinese cabbage (x_7^*)	2.8235	2.9747	2.9816	2.8082
lettuce2 (x_8^*)	2.0175	2.1225	2.1315	2.0019
radish (x_9^*)	0.000	0.000	0.000	0.000
rice (x_{10}^*)	3.000	3.000	3.000	3.000

iteration of Table 7.7. Let us compare two generalized Pareto optimal solutions at the third and fourth iterations. If the decision maker gives priority to total profit at the third iteration, tomato (x_1^*) and lettuce1(x_6^*) drastically increase. Conversely, cabbage (x_5^*) drastically decreases in comparison with the results at the fourth iteration. Such differences seem to result from required working hours $L_{\ell j}$ for each crop and each period, expected values of profit coefficient $\bar{c}_i$ for each crop, and variance covariance matrix V.

7.2 A farm planning problem in the Philippines

In this subsection, we formulate a second crop planning problem of paddy fields in the Philippines as a multiobjective programming problem involving random variable coefficients; this problem was originally formulated by Yokoyama et al. [136]. In the model farm, in the wet season between May and October, only rice (x_1) is grown; in the dry season between November and April, tobacco (x_2), tomatoes (x_3), garlic (x_4), mung beans (x_5), corn (x_6), and sweet peppers (x_7) are grown, where x_j indicates the cultivation area (unit: 1 hectare) for each crop $j = 1, \cdots, 7$. We assume that the farmer has two people available in terms of family labor, but does not have access to hired labor; further, he or she must decide

Table 7.8: Profit coefficients in MOP7.3

	j	1	2	3	4	5	6	7
t	year	c_{t1}	c_{t2}	c_{t3}	c_{t4}	c_{t5}	c_{t6}	c_{t7}
1	1989	4.5	32.6	4.0	72.6	7.3	2.7	10.9
2	1990	5.7	22.7	29.5	13.6	6.3	4.3	24.5
3	1991	3.8	26.3	42.3	42.9	5.1	1.2	13.3
4	1992	3.5	21.3	20.1	35.7	5.2	2.5	26.1
5	1993	4.4	26.2	39.3	22.5	8.4	2.2	26.6

the planting ratio for seven crops (i.e., $x_j, j = 1, \cdots, 7$) on his or her farmland to maximize total income for one year and minimize total work hours for one year. Table 7.8 shows profit coefficients c_{tj} of the seven crops j ($j = 1, \cdots, 7$) in each year t from 1989 to 1993 ($t = 1, \cdots, 5$) [136]. From Table 7.8, the vector of expected values $\mathbf{E}[\bar{\mathbf{c}}] \stackrel{\text{def}}{=} (E[\bar{c}_1], \cdots, E[\bar{c}_7])$ and variance covariance matrix V are computed as follows.

$$\mathbf{E}[\bar{\mathbf{c}}] = (4.38, 25.82, 27.04, 37.46, 6.46, 2.58, 20.28),$$

$$V = \begin{pmatrix} 0.717 & 0.1005 & -0.504 & -7.296 & 0.4565 & 0.787 & 0.8745 \\ 0.1005 & 19.13 & -30.13 & 79.39 & 2.994 & -1.250 & -26.01 \\ -0.504 & -30.13 & 242.06 & -239.13 & -1.993 & -5.924 & 39.27 \\ -7.296 & 79.39 & -239.13 & 515.2 & -0.217 & -9.626 & -143.3 \\ 0.4565 & 2.994 & -1.993 & -0.217 & 1.983 & 0.2665 & 1.467 \\ 0.787 & -1.250 & -5.924 & -9.626 & 0.2665 & 1.257 & 3.225 \\ 0.8745 & -26.015 & 39.27 & -143.3 & 1.467 & 3.225 & 57.08 \end{pmatrix}$$

Below, we assume that profit coefficients for the seven crops are normal random variables, i.e.,

$$\bar{c}_j \sim \mathrm{N}(E[\bar{c}_j], \sigma_{jj}), j = 1, \cdots, 7.$$

Given the above, the first objective function $\bar{z}_1(\mathbf{x})$ (total profit, unit: 1000 pesos) can be defined as

$$\bar{z}_1(\mathbf{x}) \stackrel{\text{def}}{=} \sum_{j=1}^{7} \bar{c}_j x_j = \bar{\mathbf{c}}\mathbf{x} \sim \mathrm{N}(\mathbf{E}[\bar{\mathbf{c}}]\mathbf{x}, \mathbf{x}^T V \mathbf{x}),$$

where $\mathbf{x} \stackrel{\text{def}}{=} (x_1, \cdots, x_7)$. The second objective function is total working hours. Table 7.9 shows required working hours $L_{\ell j}$ for each crop ($j = 1, \cdots, 7$) and each period ($\ell = 1, \cdots, 27$) [136]. Then, the second objective function $z_2(\mathbf{x})$ (i.e., total number of working hours, unit: 1 hour) can be expressed as

$$z_2(\mathbf{x}) \stackrel{\text{def}}{=} \sum_{\ell=1}^{27} \sum_{j=1}^{7} L_{\ell j} x_j.$$

Table 7.9: Required working hours in MOP7.3

ℓ	period	$L_{\ell 1}$	$L_{\ell 2}$	$L_{\ell 3}$	$L_{\ell 4}$	$L_{\ell 5}$	$L_{\ell 6}$	$L_{\ell 7}$
1	May-2	26						
2	Jun-1	16						
3	Jun-2	160						
4	Jul-1	16						
5	Jul-2	6						
6	Aug-2	8						
7	Sep-3	140						
8	Oct-1	32	8					
9	Oct-3		46			6		
10	Nov-1		36		174		8	
11	Nov-2		100	10	44	12	54	50
12	Nov-3			22	16	12	8	20
13	Dec-1		8	38	16	10	16	108
14	Dec-2		16	94	16			72
15	Dec-3		8	32	16			24
16	Jan-1		8	14				64
17	Jan-2		36	14		12		16
18	Jan-3		70	6				
19	Feb-1		70	6	180			48
20	Feb-2		36	14			60	56
21	Feb-3		36	6				48
22	Mar-1			36				56
23	Mar-2			30		32		
24	Mar-3			30				
25	Apr-1			38		56		
26	Apr-2			30				
27	Apr-3			26				

Since the upper limit on working hours for each period ($\ell = 1, \cdots, 27$) can be computed as 8 (hours) $\times$ 2 (persons) $\times$ 10 (days) $=$ 160 (hours), the following constraints must be satisfied.

$$\sum_{j=1}^{7} L_{\ell j} x_j \leq 160, \ \ell = 1, \cdots, 27$$

As two land area constraints (unit: 1 hectare) for the wet and dry season,

$$x_1 \leq 1, \sum_{j=2}^{7} x_j \leq 1, x_j \geq 0, j = 1, \cdots, 7$$

must be satisfied. We assume that the water availability constraint in the dry season is expressed as

$$\sum_{j=1}^{7} w_j x_j \leq 400, \tag{7.1}$$

where water demand coefficients w_j for crops $(j = 2, \cdots, 7)$ are set to [136]

$$(w_2, w_3, w_4, w_5, w_6, w_7) = (264.6, 232.3, 352.8, 88.2, 44.1, 220.5).$$

Given the above, we formulate the two objective programming problem involving random variable coefficients as follows (see MOP6.5).

MOP 7.3

$$\max_{\mathbf{x}\in X} \quad \bar{z}_1(\mathbf{x}) \stackrel{\text{def}}{=} \sum_{j=1}^{7} \bar{c}_j x_j \sim \mathrm{N}(\mathbf{E}[\bar{\mathbf{c}}]\mathbf{x}, \mathbf{x}^T V \mathbf{x})$$

$$\min_{\mathbf{x}\in X} \quad z_2(\mathbf{x}) \stackrel{\text{def}}{=} \sum_{\ell=1}^{27} \sum_{j=1}^{7} L_{\ell j} x_j$$

where

$$X \stackrel{\text{def}}{=} \{\mathbf{x} \in \mathrm{R}^7 \mid \sum_{j=1}^{7} L_{\ell j} x_j \leq 160,\ \ell = 1, \cdots, 27,$$
$$\sum_{j=1}^{7} w_j x_j \leq 400, x_1 \leq 1, \sum_{j=2}^{7} x_j \leq 1, x_j \geq 0, j = 1, \cdots, 7\}.$$

We apply Algorithm 6.3 to MOP7.3 with a hypothetical decision maker. At Step 1, the hypothetical decision maker sets his or her membership functions corresponding to fuzzy goals $\widetilde{G}_i, i = 1, 2$ for objective functions $\bar{z}_1(\mathbf{x}), z_2(\mathbf{x})$ as

$$\mu_{\widetilde{G}_1}(\bar{z}_1(\mathbf{x})) = \frac{\bar{z}_1(\mathbf{x}) - 26}{31 - 26}$$
$$\mu_{\widetilde{G}_2}(z_2(\mathbf{x})) = \frac{z_2(\mathbf{x}) - 900}{300 - 900}.$$

At Step 2, the hypothetical decision maker sets his or her membership function corresponding to the fuzzy goal for permissible probability level $\hat{p}_1$ for objective function $\bar{z}_1(\mathbf{x})$ as

$$\mu_{\hat{p}_1}(\hat{p}_1) = \frac{\hat{p}_1 - 0.7}{0.9 - 0.7}.$$

Then, MOP7.3 can be transformed into the following two objective programming problem (see MOP6.7).

MOP 7.4

$$\max_{\mathbf{x}\in X} \mu_{D_1}(\mathbf{x}, \hat{p}_1) \stackrel{\text{def}}{=} \min\left\{\mu_{\hat{p}_1}(\hat{p}_1), \mu_{\tilde{G}_1}(\mathbf{E}[\bar{\mathbf{c}}]\mathbf{x} + \Phi^{-1}(1-\hat{p}_1)\cdot\sqrt{\mathbf{x}^T V\mathbf{x}})\right\}$$

$$\max_{\mathbf{x}\in X} \mu_{D_2}(\mathbf{x}) \stackrel{\text{def}}{=} \mu_{\tilde{G}_2}\left(\sum_{\ell=1}^{27}\sum_{j=1}^{7} L_{\ell j}x_j\right)$$

We next formulate the corresponding minmax problem to obtain generalized Pareto optimal solutions as follows (see MINMAX6.4($\hat{\mu}$)).

MINMAX 7.2 ($\hat{\mu}$)

$$\min_{\mathbf{x}\in X, \lambda\in\Lambda} \lambda$$

subject to

$$\mathbf{E}[\bar{\mathbf{c}}]\mathbf{x} + \Phi^{-1}(1-\mu_{\hat{p}_1}^{-1}(\hat{\mu}_1-\lambda))\cdot\sqrt{\mathbf{x}^T V\mathbf{x}} \geq \mu_{\tilde{G}_1}^{-1}(\hat{\mu}_1-\lambda)$$

$$\sum_{\ell=1}^{27}\sum_{j=1}^{7} L_{\ell j}x_j \leq \mu_{\tilde{G}_2}^{-1}(\hat{\mu}_2-\lambda)$$

At Step 3, we set initial reference membership values to $\hat{\mu}_i = 1, i = 1,2$. Next, at Step 4, by solving MINMAX7.2($\hat{\mu}$) and TEST6.2, we obtain the corresponding generalized Pareto optimal solution as shown below, where $\hat{p}^* \stackrel{\text{def}}{=} \mu_{\hat{p}_1}^{-1}(\hat{\mu}_1 - \lambda^*)$.

$$(\mu_{D_1}(\mathbf{x}^*, \hat{p}^*), \mu_{D_2}(\mathbf{x}^*)) = (0.414066, 0.414066)$$

The hypothetical decision maker is not satisfied with the current values of the membership functions. Therefore, he or she updates his or her reference membership values to $(\hat{\mu}_1, \hat{\mu}_2) = (0.45, 0.4)$ to improve second membership function $\mu_{D_2}(\cdot)$ at the expense of the first one $\mu_{D_1}(\cdot)$. Then, the corresponding generalized Pareto optimal solution is obtained as

$$(\mu_{D_1}(\mathbf{x}^*, \hat{p}^*), \mu_{D_2}(\mathbf{x}^*)) = (0.434333, 0.384333).$$

The interactive processes under the hypothetical decision maker are summarized in Table 7.10. From the table, note that we obtain a satisfactory solution at the third iteration. The satisfactory solution reflects not only the preference the decision maker has for total profit and total working hours but also the effect the variance covariance matrix V has between the profit coefficients of the seven crops.

7.3 A farm planning problem with simple recourse in the Philippines

Similar to Section 7.2 above, in this section, we consider a second crop planning problem of paddy fields in the Philippines [136] as a multiobjective fuzzy ran-

Table 7.10: Interactive processes in MOP7.3

	1	2	3
$\hat{\mu}_1$	1.00	0.45	0.45
$\hat{\mu}_2$	1.00	0.4	0.43
$\mu_{D_1}(\mathbf{x}^*, \hat{p}_1^*)$	0.414066	0.434333	0.422176
$\mu_{D_2}(\mathbf{x}^*)$	0.414066	0.384333	0.402176
$\hat{p}_1^*$	0.782813	0.786867	0.784435
$f_1(\mathbf{x}^*, p_1^*)$	28.0703	28.1717	28.1109
$E[\bar{\mathbf{c}}\mathbf{x}^*]$	34.1283	34.3053	34.1990
$z_2(\mathbf{x}^*)$	651.560	669.400	658.694
rice (x_1^*)	0.490029	0.534251	0.507714
tobacco (x_2^*)	0.000	0.000	0.000
tomatoes (x_3^*)	0.525720	0.527326	0.526370
garlic (x_4^*)	0.474280	0.472674	0.473630
mung beans (x_5^*)	0.000	0.000	0.000
corn (x_6^*)	0.000	0.000	0.000
sweet peppers (x_7^*)	0.000	0.000	0.000

dom simple recourse programming problem; much of this problem has already been described in Section 7.2 above as a two objective stochastic programming problem. Except for the right-hand-side parameter of water availability constraint (7.1) in Section 7.2, all parameters shown in Tables 7.8 and 7.9 are used again in this section. To show the feasibility of Algorithm 3.7, we formulate the following two objective fuzzy random simple recourse programming problem with a hypothetical decision maker [135].

First, we assume that profit coefficients $c_j, j = 1, \cdots, 7$ for the seven crops are set to mean values $E[\bar{c}_j], j = 1, \cdots, 7$ shown below and from Table 7.8.

$$\begin{aligned} \mathbf{E}[\bar{\mathbf{c}}] &\stackrel{\text{def}}{=} (E[\bar{c}_1], E[\bar{c}_2], E[\bar{c}_3], E[\bar{c}_4], E[\bar{c}_5], E[\bar{c}_6], E[\bar{c}_7]) \\ &= (4.38, 25.82, 27.04, 37.46, 6.46, 2.58, 20.28). \end{aligned}$$

We also assume that, in the right-hand-side parameter of water availability constraint (7.1), water supply possible amount is defined as an LR-type fuzzy random variable $\tilde{\bar{d}}$ (unit : 1000 gallons), i.e.,

$$\mu_{\tilde{\bar{d}}(\omega)}(s) = \begin{cases} L\left(\frac{b(\omega)-s}{\alpha}\right), & s \le b(\omega) \\ R\left(\frac{s-b(\omega)}{\beta}\right), & s > b(\omega) \end{cases}$$

where

$$\bar{b} \sim \mathrm{N}(300, 5^2), \alpha = \beta = 30, L(t) = R(t) = 1 - t, 0 \le t \le 1.$$

Then, the corresponding two objective fuzzy random programming problem is formulated as follows (see MOP3.41).

MOP 7.5

$$\max_{\mathbf{x}\in X} \quad z_1(\mathbf{x}) \stackrel{\text{def}}{=} \sum_{j=1}^{7} E[\bar{c}_j]x_j$$

$$\min_{\mathbf{x}\in X} \quad z_2(\mathbf{x}) \stackrel{\text{def}}{=} \sum_{\ell=1}^{27}\sum_{j=1}^{7} L_{\ell j}x_j$$

subject to

$$\sum_{j=1}^{7} w_j x_j = \tilde{\bar{d}}, \tag{7.2}$$

where

$$X \stackrel{\text{def}}{=} \{\mathbf{x}\in \mathrm{R}^7 \mid \sum_{j=1} L_{\ell j}x_j \le 160,\ \ell = 1,\cdots,27,$$
$$x_1 \le 1, \sum_{j=2}^{7} x_j \le 1, x_j \ge 0, j = 1,\cdots,7\}.$$

Here, water demand coefficients w_j [136] for crops ($j = 2,\cdots,7$) are set to

$$(w_2,w_3,w_4,w_5,w_6,w_7) = (264.6, 232.3, 352.8, 88.2, 44.1, 220.5),$$

and $L_{\ell j}, \ell = 1,\cdots,27, j = 1,\cdots,7$, as shown in Table 7.9.

For equality constraint (7.2), we introduce permissible possibility level $\gamma(0 \le \gamma \le 1)$, shortage y^+, and excess y^-, which have already discussed in Section 3.4.1 above. We then transform MOP7.5 into the following two objective fuzzy random simple recourse programming problem (see MOP3.42(γ)).

MOP 7.6 (γ)

$$\min_{\mathbf{x}\in X} -\sum_{j=1}^{7} \mathbf{E}[\bar{c}_j]x_j + E\left[\min_{y^+,y^-}\left(q^+y^+ + q^-y^-\right)\right]$$

$$\min_{\mathbf{x}\in X} \sum_{\ell=1}^{27}\sum_{j=1}^{7} L_{\ell j}x_j$$

subject to

$$\sum_{j=1}^{7} w_j x_j + y^+ \ge b(\omega) - L^{-1}(\gamma)\alpha$$

$$\sum_{j=1}^{7} w_j x_j - y^- \le b(\omega) + R^{-1}(\gamma)\beta$$

$$\mathbf{x}\in X, y^+ \ge 0, y^- \ge 0$$

where q^+ and q^- are weighting parameters for y^+ and y^-, respectively.

Since a penalty cost arises only for a shortage of water resource, we set weighting parameters to $q^+ = 0$ and $q^- = 10$. We can then apply Algorithm 3.7 to MOP7.6(γ) with a hypothetical decision maker. For reference objective value $\hat{z}_\ell, \ell = 1,2$ specified by the decision maker, the corresponding γ-Pareto optimal solution is obtained by solving the following minmax problem (see MINMAX3.10($\hat{\mathbf{z}}, \gamma$)).

MINMAX 7.3 ($\hat{\mathbf{z}}, \gamma$)

$$\min_{\mathbf{x} \in X, \lambda \in \mathrm{R}^1} \lambda$$

subject to

$$-\sum_{j=1}^{7} E[\bar{c}_j] x_j + d(\mathbf{x}, \gamma) - \hat{z}_1 \le \lambda$$

$$\sum_{\ell=1}^{27} \sum_{j=1}^{7} L_{\ell j} x_j - \hat{z}_2 \le \lambda$$

where, according to the definition of $d(\mathbf{x}, \gamma)$, given by (3.171), $d(\mathbf{x}, \gamma)$ is defined as

$$\begin{aligned} d(\mathbf{x}, \gamma) \stackrel{\text{def}}{=} q_1^- \Bigg\{ & (\sum_{j=2}^{7} w_j x_j - R^{-1}(\gamma)\beta) \cdot \Phi(\sum_{j=2}^{7} w_j x_j - R^{-1}(\gamma)\beta) \\ & - \int_{-\infty}^{\sum_{j=2}^{7} w_j x_j - R^{-1}(\gamma)\beta} b\phi(b) db \Bigg\}, \end{aligned}$$

where $\phi(\cdot)$ and $\Phi(\cdot)$ represent the probability density function and distribution function for $\mathrm{N}(300, 5^2)$, respectively. Here, MINMAX7.3($\hat{\mathbf{z}}, \gamma$) can be solved using Mathematica (Wolfram Research, Inc.). To investigate the influence of permissible possibility level γ on the corresponding γ-Pareto optimal solution, we show two interactive processes with a hypothetical decision maker in Table 7.11 for $\gamma = 1$ and in Table 7.12 for $\gamma = 0.5$. In each table, a satisfactory solution is obtained at the third iteration for the same reference objective values $(\hat{z}_1, \hat{z}_2)$. By comparing the results of Table 7.11 with those of Table 7.12, it is clear that any γ-Pareto optimal solution for $\gamma = 0.5$ is superior to any γ-Pareto optimal solution for $\gamma = 1$, because of the definition of possibility measure (3.167). In any γ-Pareto optimal solution of Tables 7.11 and 7.12, only tomatoes (x_3) and garlic (x_4) in the dry season and rice (x_1) in the wet season are cultivated. The larger the permissible possibility level γ, the larger the planting ratio of tomatoes (x_3) and the smaller the planting ratio of garlic (x_4), because of the differences in the water demand coefficients of tomatoes (x_3) and garlic (x_4). At the fourth iterations of

Table 7.11: Interactive processes in MOP7.6(γ) for $\gamma = 1$

	1	2	3	4
μ	300	300	300	300
σ	5	5	5	10
$\hat{z}_1$	–33	–33	–30	–30
$\hat{z}_2$	680	620	620	620
$z_1(\mathbf{x}^*)$	–34.1648	–33.5212	–33.4891	–32.3451
$z_2(\mathbf{x}^*)$	678.835	619.479	616.511	617.655
rice (x_1^*)	0.557996	0.411074	0.403728	0.410482
tobacco (x_2^*)	0.000	0.000	0.000	0.000
tomatoes (x_3^*)	0.537202	0.537202	0.537202	0.636229
garlic (x_4^*)	0.462798	0.462798	0.462798	0.363771
mung beans (x_5^*)	0.000	0.000	0.000	0.000
corn (x_6^*)	0.000	0.000	0.000	0.000
sweet peppers (x_7^*)	0.000	0.000	0.000	0.000

Tables 7.11 and 7.12, γ-Pareto optimal solutions are shown under the assumption that σ of random variable $\bar{b}$ is changed from 5 to 10. Objective function $z_1(\mathbf{x}^*)$ at the fourth iteration worsens versus objective function $z_1(\mathbf{x}^*)$ at the third iteration, because of corresponding penalty cost $d(\mathbf{x}^*, \gamma)$.

7.4 A vegetable shipment problem in Japan

In this section, we apply Algorithm 5.1 to multi-variety vegetable shipment planning problems. We assume that a farmer (i.e., Player 1) must decide the ratio of shipment amounts between tomatoes and cucumbers. Tables 7.13 and 7.14 show price lists B^1 and B^2 (unit : Japanese yen/kg) of tomatoes and cucumbers in the Nagoya Central Wholesale Market in Japan for each period from January 2009 to December 2013 [63], respectively. We assume that some data of price lists B^1 and B^2 occur in the future (i.e., Nature or Player 2 selects a year in the range 2009 through 2013) [34]. We also assume that miscellaneous costs required to cultivate vegetables with manure can be ignored. Utilizing (12×5)-dimensional matrices $B^k, k = 1,2$ of the price lists of tomatoes and cucumbers, we define (24×5)-dimensional profit matrices $A^k, k = 1,2$ as

$$A^1 = (a_{ij}^1) \stackrel{\text{def}}{=} \begin{pmatrix} B^1 \\ \mathbf{0} \end{pmatrix}, A^2 = (a_{ij}^2) \stackrel{\text{def}}{=} \begin{pmatrix} \mathbf{0} \\ B^2 \end{pmatrix},$$

where $\mathbf{0}$ represents a (12×5)-dimensional zero matrix. Next, we formulate such a shipment planning problem as a two-person zero-sum matrix game. Let

$$\mathbf{x} = (x_1, \cdots, x_{12}, x_{13}, \cdots, x_{24})$$

Table 7.12: Interactive processes in MOP7.6(γ) for $\gamma = 0.5$

	1	2	3	4
μ	300	300	300	300
σ^2	5	5	5	10
$\hat{z}_1$	–33	–33	–30	–30
$\hat{z}_2$	680	620	620	620
$z_1(\mathbf{x}^*)$	–35.4266	–34.7831	–34.7509	–33.6070
$z_2(\mathbf{x}^*)$	677.574	618.217	615.249	616.393
rice (x_1^*)	0.549943	0.403021	0.395675	0.402428
tobacco (x_2^*)	0.000	0.000	0.000	0.000
tomatoes (x_3^*)	0.41272	0.41272	0.41272	0.511748
garlic (x_4^*)	0.58728	0.58728	0.58728	0.488252
mung beans (x_5^*)	0.000	0.000	0.000	0.000
corn (x_6^*)	0.000	0.000	0.000	0.000
sweet peppers (x_7^*)	0.000	0.000	0.000	0.000

be a mixed strategy of Player 1 (i.e., the farmer), where $(x_1, \cdots, x_{12})$ corresponds to tomato and $(x_{13}, \cdots, x_{24})$ to cucumber. Also, let $\mathbf{y} = (y_1, \cdots, y_5)$ be a mixed strategy of Player 2 (i.e., Nature). As an example, if $y_j = 1$, it follows that Nature selects the j-th year, $2008 + j$. This model implies that the farmer wishes to maximize expected income taking into account the worst-cost scenario. At Step 1 of Algorithm 5.1, suppose that Player 1 sets his or her membership functions for expected profits $\mathbf{x}A^k\mathbf{y}^T, k = 1, 2$ as

$$\mu_1^1\left(\min_j(\sum_{i=1}^{24} x_i a_{ij}^1)\right) = \left(\frac{\min_j(\sum_{i=1}^{24} x_i a_{ij}^1) - 183}{555 - 183}\right)^{\frac{1}{3}}$$

$$\mu_1^2\left(\min_j(\sum_{i=1}^{24} x_i a_{ij}^2)\right) = \left(\frac{\min_j(\sum_{i=1}^{24} x_i a_{ij}^2) - 136}{510 - 136}\right)^{\frac{1}{3}}.$$

Given the assumption that Player 1 supposes Player 2 adopts the most disadvantageous strategy, we formulate the following two objective programming problem for Player 1 (see MOP5.1).

Table 7.13: Price list B^1 of tomato in the Nagoya Central Wholesale Market in Japan (yen/kg)

year	2009	2010	2011	2012	2013
January	323	306	293	371	317
February	316	349	285	444	361
March	423	385	296	500	383
April	377	415	268	448	356
May	281	249	183	329	217
June	216	226	259	263	221
July	225	226	305	272	314
August	303	302	364	247	277
September	377	555	424	415	412
October	278	458	446	555	440
November	212	433	383	518	455
December	259	277	413	389	402

MOP 7.7

$$\max_{\mathbf{x}\in X}\left(\min_{\mathbf{y}\in Y}\mu_{\widetilde{G}_{11}}(\mathbf{x}A^1\mathbf{y}^T),\min_{\mathbf{y}\in Y}\mu_{\widetilde{G}_{12}}(\mathbf{x}A^2\mathbf{y}^T)\right)$$

where

$$X \stackrel{\text{def}}{=} \left\{\mathbf{x}=(x_1,\ldots,x_{24}) \mid \sum_{i=1}^{24} x_i = 1, x_i \geq 0, i=1,\ldots,24\right\}$$

$$Y \stackrel{\text{def}}{=} \left\{\mathbf{y}=(y_1,\ldots,y_5) \mid \sum_{j=1}^{5} y_j = 1, y_j \geq 0, j=1,\ldots,5\right\}.$$

At Step 2, we set initial reference membership values to $\hat{\mu}_1 = (\hat{\mu}_1^1, \hat{\mu}_1^2) = (1,1)$, and we then obtain the corresponding pessimistic Pareto optimal solution to MOP7.7 by solving the following minmax problem (see MINMAX5.2).

MINMAX 7.4 ($\hat{\mu}_1$)

$$\min_{\mathbf{x}\in X,\lambda\in\Lambda} \lambda$$

subject to

$$\sum_{i=1}^{24} x_i a_{ij}^1 \geq (\hat{\mu}_1^1 - \lambda)^3 \cdot (555-183) + 183,\ j=1,\cdots,5$$

$$\sum_{i=1}^{24} x_i a_{ij}^2 \geq (\hat{\mu}_1^2 - \lambda)^3 \cdot (510-136) + 136,\ j=1,\cdots,5$$

Table 7.14: Price list B^2 of cucumber in the Nagoya Central Wholesale Market in Japan (yen/kg)

year	2009	2010	2011	2012	2013
January	340	320	293	423	437
February	318	371	285	421	292
March	368	375	296	412	208
April	206	295	268	232	222
May	156	172	183	206	145
June	162	196	259	169	220
July	149	165	305	168	205
August	234	195	364	136	165
September	160	307	424	194	370
October	221	289	446	256	313
November	355	331	383	335	423
December	371	326	413	510	360

Table 7.15: Interactive processes of Player 1 in MOP7.7

	1	2	3
$\hat{\mu}_1^1$	1.	0.45	0.5
$\hat{\mu}_1^2$	1.	0.4	0.35
$\mu_1^1\left(\min_j\left(\sum_{i=1}^{12} x_i^* a_{ij}^1\right)\right)$	0.40911	0.43406	0.47474
$\mu_1^2\left(\min_j\left(\sum_{i=1}^{12} x_i^* a_{ij}^2\right)\right)$	0.40911	0.38406	0.32474
$\min_j\left(\sum_{i=1}^{12} x_i^* a_{ij}^1\right)$	208.47	213.42	222.80
$\min_j\left(\sum_{i=1}^{12} x_i^* a_{ij}^2\right)$	161.61	157.19	148.81
tomato(March) x_3^*	0.14460	0.14803	0.15454
tomato(September) x_9^*	0.39073	0.40001	0.41759
cucumber(February) x_{14}^*	0.04932	0.04797	0.04541
cucumber(March) x_{15}^*	0.13249	0.12887	0.12200
cucumber(November) x_{23}^*	0.28286	0.27512	0.26045

At Step 3, we solve MINMAX7.4($\hat{\mu}_1$) via a combination of the bisection method and the first phase of the two-phase simplex method of linear programming. According to Algorithm 5.1, Player 1 updates his or her reference membership values to obtain a candidate of the pessimistic compromise solution from a pessimistic Pareto optimal solution set. Interactive processes for Player 1 are summarized in Table 7.15.

References

[1] Maria Joao Alves and João Clímaco. A review of interactive methods for multiobjective integer and mixed-integer programming. *European Journal of Operational Research*, 180(1):99–115, 2007.

[2] G Anandalingam. A mathematical programming model of decentralized multi-level systems. *Journal of the Operational Research Society*, 39(11):1021–1033, 1988.

[3] Evelyn ML Beale. On minimizing a convex function subject to linear inequalities. *Journal of the Royal Statistical Society. Series B (Methodological)*, 173–184, 1955.

[4] CR Bector, Suresh Chandra et al. *Fuzzy mathematical programming and fuzzy matrix games*, volume 169. Springer, 2005.

[5] CR Bector, Suresh Chandra, and Vidyottama Vijay. Duality in linear programming with fuzzy parameters and matrix games with fuzzy pay-offs. *Fuzzy Sets and Systems*, 146(2):253–269, 2004.

[6] Richard E Bellman and Lotfi Asker Zadeh. Decision-making in a fuzzy environment. *Management Science*, 17(4):B–141, 1970.

[7] John R Birge and Francois Louveaux. *Introduction to stochastic programming*. Springer Science & Business Media, New York, 2011.

[8] Lourdes Campos. Fuzzy linear programming models to solve fuzzy matrix games. *Fuzzy Sets and Systems*, 32(3):275–289, 1989.

[9] Vira Chankong and Yacov Y Haimes. *Multiobjective decision making: theory and methodology*. Number 8. North-Holland, New York, 1983.

[10] Abraham Charnes and William W Cooper. Management models and industrial applications of linear programming. *Management Science*, 4(1):38–91, 1957.

[11] Abraham Charnes and William W Cooper. Chance-constrained programming. *Management Science*, 6(1):73–79, 1959.

[12] Abraham Charnes and William W Cooper. Programming with linear fractional functionals. *Naval Research Logistics Quarterly*, 9(3-4):181–186, 1962.

[13] Abraham Charnes and William W Cooper. Deterministic equivalents for optimizing and satisficing under chance constraints. *Operations Research*, 11(1):18–39, 1963.

[14] Bruno Contini. A stochastic approach to goal programming. *Operations Research*, 16(3):576–586, 1968.

[15] Wade D Cook. Zero-sum games with multiple goals. *Naval Research Logistics Quarterly*, 23(4):615–621, 1976.

[16] George B Dantzig. Linear programming under uncertainty. *Management Science*, 1(3-4):197–206, 1955.

[17] Werner Dinkelbach. On nonlinear fractional programming. *Management Science*, 13(7):492–498, 1967.

[18] Didier Dubois and Henri Prade. *Fuzzy sets and systems: theory and applications*, volume 144. Academic Press, 1980.

[19] James S Dyer. Interactive goal programming. *Management Science*, 19(1):62–70, 1972.

[20] Franz Eisenführ, Thomas Langer, Martin Weber, Thomas Langer, and Martin Weber. *Rational decision making*. Springer, 2010.

[21] Japan International Research Center for Agricultural Sciences. *Farm management design and its analysis manual based on linear programming (in Japanese)*. Agriculture, Forestry and Statistics Association, 1999.

[22] Arthur M Geoffrion, James S Dyer, and A Feinberg. An interactive approach for multi-criterion optimization, with an application to the operation of an academic department. *Management Science*, 19(4-part-1):357–368, 1972.

[23] D Ghose and UR Prasad. Solution concepts in two-person multicriteria games. *Journal of Optimization Theory and Applications*, 63(2):167–189, 1989.

[24] John J Glen. Mathematical models in farm planning: a survey. *Operations Research*, 35(5):641–666, 1987.

[25] Ambrose Goicoechea, Don R Hansen, and Lucien Duckstein. *Multiobjective decision analysis with engineering and business applications.* John Wiley & Sons, 1982.

[26] Kiyotada Hayashi. Multicriteria analysis for agricultural resource management: a critical survey and future perspectives. *European Journal of Operational Research*, 122(2):486–500, 2000.

[27] GH Huang and DP Louck. An inexact two-stage stochastic programming model for water resources management under uncertainty. *Civil Engineering Environmental Systems*, 17:95–118, 2000.

[28] Suwarna Hulsurkar, Mahendra P Biswal, and Surabhi B Sinha. Fuzzy programming approach to multi-objective stochastic linear programming problems. *Fuzzy Sets and Systems*, 88(2):173–181, 1997.

[29] James P Ignizio. *Goal programming and extensions.* Lexington Books, 1976.

[30] James P Ignizio. *Linear programming in single- & multiple-objective systems.* Prentice Hall, 1982.

[31] Masahiro Inuiguchi and Masatoshi Sakawa. A possibilistic linear program is equivalent to a stochastic linear program in a special case. *Fuzzy Sets and Systems*, 76(3):309–317, 1995.

[32] Takeshi Itoh, Hiroaki Ishii, and Teruaki Nanseki. A model of crop planning under uncertainty in agricultural management. *International Journal of Production Economics*, 81:555–558, 2003.

[33] Peter Kall and Janos Mayer. *Stochastic linear programming: models, theory and computation. International series in operations research and management science.* Springer, 2005.

[34] K Kasahara, J Song, and Y Sembokuya. Marketing planning of small-sized farms by the fuzzy game theory. *Bulletin of the Faculty of Agriculture-Tottori University (Japan)*, 1996.

[35] H Katagiri, H Ishii, and T Itoh. Fuzzy random linear programming problem. In *Proc. Second European Workshop on Fuzzy Decision Analysis and Neural Networks for Management, Planning and Optimization*, pages 107–115, 1997.

[36] Hideki Katagiri and Masatoshi Sakawa. Interactive multiobjective fuzzy random programming through the level set-based probability model. *Information Sciences*, 181(9):1641–1650, 2011.

[37] Hideki Katagiri, Masatoshi Sakawa, and Hiroaki Ishii. Multiobjective fuzzy random linear programming using e-model and possibility measure. In *IFSA World Congress and 20th NAFIPS International Conference, 2001. Joint 9th*, volume 4, pages 2295–2300. IEEE, 2001.

[38] Hideki Katagiri, Masatoshi Sakawa, Kosuke Kato, and Ichiro Nishizaki. Interactive multiobjective fuzzy random linear programming: Maximization of possibility and probability. *European Journal of Operational Research*, 188(2):530–539, 2008.

[39] Hideki Katagiri, Masatoshi Sakawa, Kousuke Kato, and Syuuji Ohsaki. An interactive fuzzy satisficing method for fuzzy random multiobjective linear programming problems through the fractile optimization model using possibility and necessity measures. In *Asia Pacific Management Conference, 2003*, pages 795–802, 2003.

[40] Hideki Katagiri, Masatoshi Sakawa, and Syuuji Ohsaki. An interactive satisficing method through the variance minimization model for fuzzy random linear programming problems. In *Multi-objective Programming and Goal Programming: Theory and Applications*, pages 171–176. Springer Science & Business Media, 2003.

[41] Shinji Kataoka. A stochastic programming model. *Econometrica: Journal of the Econometric Society*, pages 181–196, 1963.

[42] Kosuke Kato and Masatoshi Sakawa. An interactive fuzzy satisficing method based on simple recourse model for multiobjective linear programming problems involving random variable coefficients. *International Journal of Innovative Computing, Information and Control*, 5(7):1997–2010, 2009.

[43] Kosuke Kato and Masatoshi Sakawa. An interactive fuzzy satisficing method based on variance minimization under expectation constraints for multiobjective stochastic linear programming problems. *Soft Computing*, 15(1):131–138, 2011.

[44] Ralph L Keeney and Howard Raiffa. *Decisions with multiple objectives: preferences and value trade-offs.* Cambridge university press, 1993.

[45] Pekka J Korhonen and Jukka Laakso. A visual interactive method for solving the multiple criteria problem. *European Journal of Operational Research*, 24(2):277–287, 1986.

[46] Huibert Kwakernaak. Fuzzy random variables-i. definitions and theorems. *Information Sciences*, 15(1):1–29, 1978.

[47] Young-Jou Lai. Hierarchical optimization: a satisfactory solution. *Fuzzy Sets and Systems*, 77(3):321–335, 1996.

[48] Young-Jou Lai and Ching-Lai Hwang. *Fuzzy mathematical programming.* Springer, 1992.

[49] E Stanley Lee and Hsu-Shih Shih. *Fuzzy and multi-level decision making: an interactive computational approach.* Springer Science & Business Media, 2012.

[50] Sang M Lee et al. *Goal programming for decision analysis.* Auerbach Philadelphia, 1972.

[51] Deng-Feng Li. A fuzzy multi-objective approach to solve fuzzy matrix games. *Journal of Fuzzy Mathematics*, 7:907–912, 1999.

[52] Deng-Feng Li. A fast approach to compute fuzzy values of matrix games with payoffs of triangular fuzzy numbers. *European Journal of Operational Research*, 223(2):421–429, 2012.

[53] MK Luhandjula and MM Gupta. On fuzzy stochastic optimization. *Fuzzy Sets and Systems*, 81(1):47–55, 1996.

[54] Takashi Maeda. On characterization of equilibrium strategy of two-person zero-sum games with fuzzy payoffs. *Fuzzy Sets and Systems*, 139(2):283–296, 2003.

[55] Imran Maqsood, Guohe Huang, Yuefei Huang, and Bing Chen. Itom: an interval parameter two-stage optimization model for stochastic planning of water resources systems. *Stochastic Environmental Research and Risk Assessment*, 19:125–133, 2005.

[56] Sakawa Masatoshi, Kosuke Kato, and Hideki Katagiri. An interactive fuzzy satisficing method through a variance minimization model for multiobjective linear programming problems involving random variables. *Knowledge-based Intelligent Information Engineering Systems & Allied Technologies KES2002*, 2:1222–1226, 2002.

[57] Sakawa Masatoshi, Ichiro Nishizaki, and Yoshio Uemura. Interactive fuzzy programming for multilevel linear programming problems. *Computers & Mathematics with Applications*, 36(2):71–86, 1998.

[58] Abu S Masud and Ching-Lai Hwang. Interactive sequential goal programming. *Journal of the Operational Research Society*, pages 391–400, 1981.

[59] Mansooreh Mollaghasemi and Julia Pet-Edwards. Making multi-objective decisions. *IEEE Computer Soc. Press*, 1997.

[60] Ichiro Nishizaki and Masatoshi Sakawa. Equilibrium solutions for multiobjective bimatrix games incorporating fuzzy goals. *Journal of Optimization Theory and Applications*, 86(2):433–457, 1995.

[61] Ichiro Nishizaki and Masatoshi Sakawa. Equilibrium solutions in multiobjective bimatrix games with fuzzy payoffs and fuzzy goals. *Fuzzy Sets and Systems*, 111(1):99–116, 2000.

[62] Ichiro Nishizaki and Masatoshi Sakawa. *Fuzzy and multiobjective games for conflict resolution*, volume 64. Physica-Verlag, 2001.

[63] Official Statistics of Japan. *Japan in Figures (URL : http://www.e-stat.go.jp/SG1/estat/)*.

[64] Don T Phillips and James J Solberg. *Operations research: principles and practice*. John Wiley & Sons, Inc., 1987.

[65] Madan L Puri and Dan A Ralescu. Fuzzy random variables. *Journal of Mathematical Analysis and Applications*, 114(2):409–422, 1986.

[66] Heinrich Rommelfanger. Fulpal-an interactive method for solving (multiobjective) fuzzy linear programming problems. In *Stochastic Versus Fuzzy Approaches to Multiobjective Mathematical Programming under Uncertainty*, pages 279–299. Kluwer Academic Publishers, 1990.

[67] Heinrich Rommelfanger. Fuzzy linear programming and applications. *European Journal of Operational Research*, 92(3):512–527, 1996.

[68] Thomas L Saaty. The analytic hierarchy process: planning, priority setting, resources allocation. *New York: McGraw*, 1980.

[69] Masatoshi Sakawa. Interactive multiobjective decision making by the sequential proxy optimization technique: Spot. *European Journal of Operational Research*, 9(4):386–396, 1982.

[70] Masatoshi Sakawa. Interactive computer programs for fuzzy linear programming with multiple objectives. *International Journal of Man-Machine Studies*, 18(5):489–503, 1983.

[71] Masatoshi Sakawa. *Fuzzy sets and interactive multiobjective optimization*. Springer Science & Business Media, 2013.

[72] Masatoshi Sakawa, Hideki Katagiri, and Kosuke Kato. An interactive fuzzy satisficing method for multiobjective stochastic linear programming problems using a fractile criterion model. In *Fuzzy Systems, 2001. The 10th IEEE International Conference on*, volume 2, pages 948–951. IEEE, 2001.

[73] Masatoshi Sakawa, Kosuke Kato, and Hideki Katagiri. An interactive fuzzy satisficing method for multiobjective linear programming problems with random variable coefficients through a probability maximization model. *Fuzzy Sets and Systems*, 146(2):205–220, 2004.

[74] Masatoshi Sakawa, Kosuke Kato, and Ichiro Nishizaki. An interactive fuzzy satisficing method for multiobjective stochastic linear programming problems through an expectation model. *European Journal of Operational Research*, 145(3):665–672, 2003.

[75] Masatoshi Sakawa and Takeshi Matsui. Interactive fuzzy multiobjective stochastic programming with simple recourse. *International Journal of Multicriteria Decision Making*, 4(1):31–46, 2014.

[76] Masatoshi Sakawa and Ichiro Nishizaki. Max-min solutions for fuzzy multiobjective matrix games. *Fuzzy Sets and Systems*, 67(1):53–69, 1994.

[77] Masatoshi Sakawa, Ichiro Nishizaki, and Hideki Katagiri. *Fuzzy stochastic multiobjective programming*, volume 159. Springer Science & Business Media, 2011.

[78] Masatoshi Sakawa, Ichiro Nishizaki, and Yoshio Uemura. Interactive fuzzy programming for multi-level linear programming problems with fuzzy parameters. *Fuzzy Sets and Systems*, 109(1):3–19, 2000.

[79] Masatoshi Sakawa, Ichiro Nishizaki, and Yoshio Uemura. Interactive fuzzy programming for two-level linear fractional programming problems with fuzzy parameters. *Fuzzy Sets and Systems*, 115(1):93–103, 2000.

[80] Masatoshi Sakawa and Fumiko Seo. Interactive multiobjective decision making in environmental systems using sequential proxy optimization techniques (spot). *Automatica*, 18(2):155–165, 1982.

[81] Masatoshi Sakawa and Hitoshi Yano. An interactive fuzzy satisficing method using augmented minimax problems and its application to environmental systems. *Systems, Man and Cybernetics, IEEE Transactions on*, (6):720–729, 1985.

[82] Masatoshi Sakawa and Hitoshi Yano. Interactive decision making for multiobjective nonlinear programming problems with fuzzy parameters. *Fuzzy Sets and Systems*, 29(3):315–326, 1989.

[83] Masatoshi Sakawa and Hitoshi Yano. An interactive fuzzy satisficing method for multiobjective nonlinear programming problems with fuzzy parameters. *Fuzzy Sets and Systems*, 30(3):221–238, 1989.

[84] Masatoshi Sakawa and Hitoshi Yano. Interactive decision making for multiobjective programming problems with fuzzy parameters. In *Stochastic Versus Fuzzy Approaches to Multiobjective Mathematical Programming under Uncertainty*, pages 191–228. Kluwer Academic Publishers, 1990.

[85] Masatoshi Sakawa and Hitoshi Yano. An interactive fuzzy satisficing method for generalized multiobjective linear programming problems with fuzzy parameters. *Fuzzy Sets and Systems*, 35(2):125–142, 1990.

[86] Masatoshi Sakawa and Hitoshi Yano. Trade-off rates in the hyperplane method for multiobjective optimization problems. *European Journal of Operational Research*, 44(1):105–118, 1990.

[87] Masatoshi Sakawa, Hitoshi Yano, and Ichiro Nishizaki. *Linear and multiobjective programming with fuzzy stochastic extensions*. Springer, 2013.

[88] Yoshikazu Sawaragi, Hirotaka Nakayama, and Tetsuzo Tanino. *Theory of multiobjective optimization*. Academic Press, 1985.

[89] Fumiko Seo and Masatoshi Sakawa. *Multiple criteria decision analysis in regional planning: concepts, methods and applications*, volume 10. Springer Science & Business Media, 2012.

[90] Lloyd S Shapley and Fred D Rigby. Equilibrium points in games with vector payoffs. *Naval Research Logistics Quarterly*, 6(1):57–61, 1959.

[91] Hsu-Shih Shih, Young-Jou Lai, and E Stanley Lee. Fuzzy approach for multi-level programming problems. *Computers & Operations Research*, 23(1):73–91, 1996.

[92] Wolfram Stadler. *Multicriteria Optimization in Engineering and in the Sciences*, volume 37. Springer Science & Business Media, 2013.

[93] IM Stancu-Minasian. *Stochastic programming with multiple objective functions*, volume 13. D. Reidel Publishing Company, 1984.

[94] IM Stancu-Minasian. Overview of different approaches for solving stochastic programming problems with multiple objective functions. In *Stochastic Versus Fuzzy Approaches to Multiobjective Mathematical Programming under Uncertainty*, pages 71–101. Kluwer Academic Publishers, 1990.

[95] Ralph E Steuer. *Multiple criteria optimization: theory, computation, and applications*. Wiley, 1986.

[96] Wei Sun, Chunjiang An, Gongchen Li, and Ying Lv. Applications of inexact programming methods to waste management under uncertainty: current status and future directions. *Environmental Systems Research*, 3(1):1–15, 2014.

[97] J Teghem, D Dufrane, M Thauvoye, and P Kunsch. Strange: an interactive method for multi-objective linear programming under uncertainty. *European Journal of Operational Research*, 26(1):65–82, 1986.

[98] Tasuku Toyonaga, Takesh Itoh, and Hiroaki Ishii. A crop planning problem with fuzzy random profit coefficients. *Fuzzy Optimization and Decision Making*, 4(1):51–69, 2005.

[99] V Vidyottama, S Chandra, and CR Bector. Bi-matrix games with fuzzy goals and fuzzy payoffs. *Fuzzy Optimization and Decision Making*, 3:327–344, 2004.

[100] V Vijay, S Chandra, and CR Bector. Matrix games with fuzzy goals and fuzzy payoffs. *Omega*, 33(5):425–429, 2005.

[101] David W Walkup and Roger J-B Wets. Stochastic programs with recourse. *SIAM Journal on Applied Mathematics*, 15(5):1299–1314, 1967.

[102] Guang-yuan Wang and Yue Zhang. The theory of fuzzy stochastic processes. *Fuzzy Sets and Systems*, 51(2):161–178, 1992.

[103] Guang-yuan Wang and Qiao Zhong. Linear programming with fuzzy random variable coefficients. *Fuzzy Sets and Systems*, 57(3):295–311, 1993.

[104] R Wang, Y Li, and Q Tan. A review of inexact optimization modeling and its application to integrated water resources management. *Frontiers of Earth Science*, 9(1):51–64, 2015.

[105] Ue-Pyng Wen and Shuh-Tzy Hsu. Linear bi-level programming problems–a review. *Journal of the Operational Research Society*, pages 125–133, 1991.

[106] Roger J-B Wets. Stochastic programs with fixed recourse: the equivalent deterministic program. *SIAM Review*, 16(3):309–339, 1974.

[107] Andrzej P Wierzbicki. The use of reference objectives in multiobjective optimization. In *Multiple Criteria Decision Making Theory and Application*, pages 468–486. Springer, 1980.

[108] Andrzej P Wierzbicki. A mathematical basis for satisficing decision making. *Mathematical Modelling*, 3(5):391–405, 1982.

[109] Hitoshi Yano. Interactive decision making for hierarchical multiobjective linear programming problems. In *Modeling Decisions for Artificial Intelligence*, pages 137–148. Springer, 2009.

[110] Hitoshi Yano. Interactive decision making for hierarchical multiobjective linear programming problems with random variable coefficients. In *Joint 5th International Conference on Soft Computing and Intelligent Systems and 11th International Symposium on Advanced Intelligent Systems*, pages 1032–1037, 2010.

[111] Hitoshi Yano. Fuzzy approaches for multiobjective stochastic linear programming problems considering both probability maximization and fractile optimization. In *Fuzzy Systems (FUZZ), 2011 IEEE International Conference on*, pages 1866–1873. IEEE, 2011.

[112] Hitoshi Yano. Interactive decision making based on fractile criterion optimization models for hierarchical multiobjective linear programming problems with random variable coefficients. *The Transaction of The Institute of Electronics, Information and Communication Engineers A (in Japanese)*, J94-A(8):568–576, 2011.

[113] Hitoshi Yano. Hierarchical multiobjective stochastic linear programming problems considering both probability maximization and fractile optimization. *IAENG International Journal of Applied Mathematics*, 42(2):91–98, 2012.

[114] Hitoshi Yano. Interactive decision making for fuzzy random multiobjective linear programming problems with variance-covariance matrices through probability maximization. In *Soft Computing and Intelligent Systems (SCIS) and 13th International Symposium on Advanced Intelligent Systems (ISIS), 2012 Joint 6th International Conference on*, pages 965–970. IEEE, 2012.

[115] Hitoshi Yano. Interactive fuzzy decision making for hierarchical multiobjective stochastic linear programming problems. In *Proceedings of the International MultiConference of Engineers and Computer Scientists*, volume 2, pages 1623–1628, 2012.

[116] Hitoshi Yano. Interactive fuzzy decision making for multiobjective stochastic linear programming problems with variance-covariance matrices. In *Systems, Man, and Cybernetics (SMC), 2012 IEEE International Conference on*, pages 97–102. IEEE, 2012.

[117] Hitoshi Yano. An interactive fuzzy satisficing method for multiobjective stochastic linear programming problems considering both probability maximization and fractile optimization. In *Intelligent Decision Technologies*, pages 119–128. Springer, 2012.

[118] Hitoshi Yano. Hierarchical multiobjective stochastic linear programming problems through a fractile optimization model using reference membership intervals. In *IAENG Transactions on Engineering Technologies*, pages 105–119. Springer, 2013.

[119] Hitoshi Yano. Fuzzy decision making for fuzzy random multiobjective linear programming problems with variance covariance matrices. *Information Sciences*, 272:111–125, 2014.

[120] Hitoshi Yano. Interactive decision making for fuzzy random multiobjective linear programming problems with variance covariance matrices through fractile optimization (selected papers from the 6th international conference on soft computing and intelligent systems and the 13th international symposium on advanced intelligent systems (scis&isis2012)). *Journal of Advanced Computational Intelligence and Intelligent Informatics*, 18(3):383–390, 2014.

[121] Hitoshi Yano. Fuzzy decision making for multiobjective stochastic programming problems. *Fuzzy Sets and Systems*, 2015.

[122] Hitoshi Yano. Multiobjective programming problems involving fuzzy coefficients, random variable coefficients and fuzzy random variable coefficients. *International Journal of Uncertainty, Fuzziness and Knowledge-Based Systems*, 23(04):483–504, 2015.

[123] Hitoshi Yano and Kota Matsui. Fuzzy approaches for multiobjective fuzzy random linear programming problems through a probability maximization model. In *Proceedings of The International MultiConference of Engineers and Computer Scientists*, pages 1349–1354, 2011.

[124] Hitoshi Yano and Kota Matsui. Two fuzzy approaches for multiobjective stochastic programming and multiobjective fuzzy random programming through a probability maximization model. *IAENG International Journal of Computer Science*, 38(3):234–241, 2011.

[125] Hitoshi Yano and Kota Matsui. Hierarchical multiobjective stochastic linear programming problems based on the fuzzy decision. In *Iaeng Transactions on Engineering Technologies Volume 7-Special Edition of the International Multiconference of Engineers and Computer Scientists 2011*, pages 1–14. World Scientific, 2012.

[126] Hitoshi Yano and Kota Matsui. Hierarchical multiobjective fuzzy random linear programming problems. *Procedia Computer Science*, 22:162–171, 2013.

[127] Hitoshi Yano and Kota Matsui. Multiobjective fuzzy random linear programming problems based on e-model and v-model. In *Transactions on Engineering Technologies*, pages 113–126. Springer, 2015.

[128] Hitoshi Yano, Kota Matsui, and Mikiya Furuhashi. Interactive decision making for multiobjective fuzzy random linear programming problems using expectations and coefficients of variation. In *Proceedings of the International MultiConference of Engineers and Computer Scientists*, volume 2, pages 1251–1256, 2014.

[129] Hitoshi Yano and Ichiro Nishizaki. Interactive fuzzy approaches for solving multiobjective two-person zero-sum games. *Applied Mathematics*, 7(05):387, 2016.

[130] Hitoshi Yano and Masatoshi Sakawa. Interactive fuzzy decision making for generalized multiobjective linear fractional programming problems with fuzzy parameters. *Fuzzy Sets and Systems*, 32(3):245–261, 1989.

[131] Hitoshi Yano and Masatoshi Sakawa. Interactive fuzzy decision making for multiobjective fuzzy random linear programming problems. In *4th International Joint Conference on Computational Intelligence*, pages 319–328, 2012.

[132] Hitoshi Yano and Masatoshi Sakawa. Interactive multiobjective fuzzy random linear programming through fractile criteria. *Advances in Fuzzy Systems*, 2012:21, 2012.

[133] Hitoshi Yano and Masatoshi Sakawa. Interactive fuzzy programming for multiobjective fuzzy random linear programming problems through possibility-based probability maximization. *Operational Research*, 14(1):51–69, 2014.

[134] Hitoshi Yano and Masatoshi Sakawa. Interactive fuzzy decision making for multiobjective fuzzy random linear programming problems and its application to a crop planning problem. In *Computational Intelligence*, pages 143–157. Springer, 2015.

[135] Hitoshi Yano and Rongrong Zhang. Interactive decision making for multiobjective fuzzy random simple recourse programming problems and its application to rainfed agriculture in philippines. In *Proceedings of the International MultiConference of Engineers and Computer Scientists*, volume 1, pages 912–917, 2016.

[136] Shigeki Yokoyama, SR Francisco, and Teruaki Nanseki. Optimum crop combination under risk: Second cropping of paddy fields in the philippines. *TECHNOLOGY AND DEVELOPMENT-TOKYO-ENGLISH EDITION-*, pages 65–74, 1999.

[137] Po-Lung Yu. *Multiple-criteria decision making: concepts, techniques, and extensions*, volume 30. Springer Science & Business Media, 2013.

[138] Milan Zeleny. Games with multiple payoffs. *International Journal of Game Theory*, 4(4):179–191, 1975.

[139] Hans-Jürgen Zimmermann. Fuzzy programming and linear programming with several objective functions. *Fuzzy Sets and Systems*, 1(1):45–55, 1978.

[140] Hans-Jürgen Zimmermann. *Fuzzy sets, decision making, and expert systems*, volume 10. Springer Science & Business Media, 2012.

[141] Stanley Zionts and Jyrki Wallenius. An interactive programming method for solving the multiple criteria problem. *Management Science*, 22(6):652–663, 1976.

[142] Constantin Zopounidis and Michael Doumpos. Multicriteria classification and sorting methods: A literature review. *European Journal of Operational Research*, 138(2):229–246, 2002.

Index

For Product Safety Concerns and Information please contact our EU
representative GPSR@taylorandfrancis.com
Taylor & Francis Verlag GmbH, Kaufingerstraße 24, 80331 München, Germany

www.ingramcontent.com/pod-product-compliance
Ingram Content Group UK Ltd.
Pitfield, Milton Keynes, MK11 3LW, UK
UKHW021505240926
475702UK00035B/446
* 9 7 8 1 4 9 8 7 6 3 5 4 7 *